Advances in Radio Science and Electromagnetics

Advances in Radio Science and Electromagnetics

Edited by **Claude McMillan**

New York

Published by Willford Press,
118-35 Queens Blvd., Suite 400,
Forest Hills, NY 11375, USA
www.willfordpress.com

Advances in Radio Science and Electromagnetics
Edited by Claude McMillan

International Standard Book Number: 978-1-68285-068-8 (Hardback)

Printed in the United States of America.

Contents

Preface

It is often said that books are a boon to mankind. They document every progress and pass on the knowledge from one generation to the other. They play a crucial role in our lives. Thus I was both excited and nervous while editing this book. I was pleased by the thought of being able to make a mark but I was also nervous to do it right because the future of students depends upon it. Hence, I took a few months to research further into the discipline, revise my knowledge and also explore some more aspects. Post this process, I began with the editing of this book.

The researches in radio science and electromagnetics have fuelled the advancement in sectors like defense, high speed electronics, etc. This book provides comprehensive knowledge into the field of radio science and electromagnetics. Most of the topics introduced in this book cover new techniques and the applications of radio science such as prediction, measurement, modeling and forecasting techniques along with concepts and theories pertaining to electromagnetics, such as electromagnetic induction, magnetic fields, etc. This book is apt for students and academicians pursuing radio science, electrical engineering, computer science, etc.

I thank my publisher with all my heart for considering me worthy of this unparalleled opportunity and for showing unwavering faith in my skills. I would also like to thank the editorial team who worked closely with me at every step and contributed immensely towards the successful completion of this book. Last but not the least, I wish to thank my friends and colleagues for their support.

Editor

1

Definition of a parameter for a typical specific absorption rate under real boundary conditions of cellular phones in a GSM network

D. Gerhardt

E-Plus Mobilfunk GmbH & Co. KG, E-Plus-Platz, D-40468 Düsseldorf, Germany

Abstract. Using cellular phones the specific absorption rate (SAR) as a physical value must observe established and internationally defined levels to guarantee human protection. To assess human protection it is necessary to guarantee safety under worst-case conditions (especially maximum transmitting power) using cellular phones.

To evaluate the exposure to electromagnetic fields under normal terms of use of cellular phones the limitations of the specific absorption rate must be pointed out. In a mobile radio network normal terms of use of cellular phones, i.e. in interconnection with a fixed radio transmitter of a mobile radio network, power control of the cellular phone as well as the antenna diagram regarding a head phantom are also significant for the real exposure.

Based on the specific absorption rate, the antenna diagram regarding a head phantom and taking into consideration the power control a new parameter, the typical absorption rate (SAR_{typ}), is defined in this contribution. This parameter indicates the specific absorption rate under average normal conditions of use. Constant radio link attenuation between a cellular phone and a fixed radio transmitter for all mobile models tested was assumed in order to achieve constant field strength at the receiving antenna of the fixed radio transmitter as a result of power control. The typical specific absorption rate is a characteristic physical value of every mobile model.

The typical absorption rate was calculated for 16 different mobile models and compared with the absorption rate at maximum transmitting power. The results confirm the relevance of the definition of this parameter (SAR_{typ}) as opposed to the specific absorption rate as a competent and applicable method to establish the real mean exposure from a cellular phone in a mobile radio network. The typical absorption rate provides a parameter to assess electromagnetic fields of a cellular phone that is more relevant to the consumer.

1 Introduction

Radio frequency electromagnetic fields are absorbed by the human body and lead to a temperature increase of biological tissue. This is characterised by the specific absorption rate SAR and is related to the electrical field strength E_{eff} in the biological tissue and its conductivity σ and density ρ:

$$SAR = \sigma \frac{E_{eff}^2}{\rho} \qquad (1)$$

The specific absorption rate can be determined according to equation 1, among other things, over the measurement of the electrical field strength in a human body (or head) phantom, which is filled with a tissue-simulating liquid (DASY, 1995; Kuster, 1997; ES 59005, 1998; EN 50360, 2001; EN 50361, 2001).

For radio systems in operation the specific absorption rate may not exceed a fixed exposure limit, so that personal safety is ensured (ICNIRP, 1998; EC Recommendation, 1999). The exposure limit of the specific absorption rate for cellular phones is at a value of 2 W/kg averaged over 10 g of tissue mass. For cellular phones the specific absorption rate SAR is proportional to the transmitting power P:

$$SAR = K\ P \quad \text{with} \quad K: \quad \text{proportionality constant} \qquad (2)$$

For the CE conformity tests the specific absorption rate is determined during maximum transmitting power (EN 50360, 2001; EN 50361, 2001). This does not correspond to a typical, average operating transmitting power during a phone call. The cellular phone works with the minimum technically necessary transmitting power that ensures an error free radio link to the fixed radio transmitter. The decisive parameter is the radio link attenuation, which essentially depends on the distance and on the angle between the antennas of the fixed radio transmitter and the cellular phone. The antenna gain of cellular phones in presence of a head phantom, measured after Schneider et al. (1998), indicates a pronounced variation depending on the azimuth angle φ and elevation angle ϑ (Fig. 1).

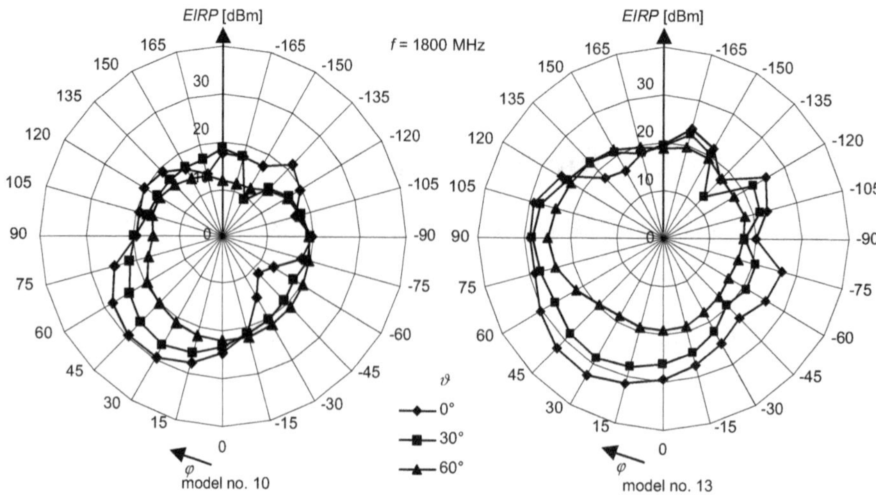

Fig. 1. Measured EIRP in dBm, examples for 2 mobile models.

If real exposure conditions are assessed with the typical use of a cellular phone, the specific absorption rate during maximum transmitting power of the cellular phone is not a meaningful value. This requires the definition of a so-called typical specific absorption rate (SAR_{typ}), which essentially considers both the antenna gain in presence of a head phantom and the power control.

2 Method

In order to derive the typical specific absorption rate the power of the cellular phone must be controlled, for each mobile model, such that with constant distance regarding radio link attenuation to the receiving antenna of the fixed radio transmitter the same field strength (or power density) is present. The distance r between cellular phone and fixed radio transmitter is large in relation to the wavelength of the electromagnetic field, so that far field conditions are present. The power density S is given by:

$$S = \frac{G(\varphi, \vartheta)P}{4\pi r^2} \tag{3}$$

The antenna gain G is angle dependent (φ: azimuth, ϑ: elevation) and takes into account the influence of a head phantom (see Figs. 1 and 2). A constant receiving power of the antenna of the fixed radio transmitter demands $4\pi r^2 S$ = const., so that the power control for the transmitting power P is dependent on the angle of the cellular phone according to Eq. (3):

$$P(\varphi, \vartheta) = \frac{4\pi r^2 S}{G(\varphi, \vartheta)} \tag{4}$$

The constant K can be determined from Eq. (2) with the measured specific absorption rate SAR_{max} during maximum transmitting power $P_{max} = 1\,\mathrm{W}$ (f = 1.8 GHz, GSM1800). The specific absorption rate from Eq. (4) and the equivalent isotropic radiating power $EIRP(\varphi, \vartheta) = G(\varphi, \vartheta)P_{max}$ is given by:

$$SAR(\varphi, \vartheta) = \frac{4\pi r^2 S}{EIRP(\varphi, \vartheta)} SAR_{max} \tag{5}$$

$EIRP$ can be determined by measurement and indicated as $EIRP_{\mathrm{dBm}}$. By using cellular phones for φ an evenly distributed probability is given, so that Eq. (5) is averaged over φ:

$$\overline{SAR}(\vartheta) = \frac{1}{N} \sum_{i=1}^{N} SAR(\varphi_i, \vartheta) \tag{6}$$

The typical power P_{typ} depends on the characteristics of the mobile radio network (Wiart et al., 2000; Vecchia et al., 2001) and from Eq. (3) the following condition results:

$$\frac{1}{N} \sum_{i=1}^{N} \frac{4\pi r^2 S}{G(\varphi_i, \vartheta)} \stackrel{!}{=} P_{typ} \tag{7}$$

From Eq. (7) the term $4\pi r^2 S$ can be determined as follows:

$$4\pi r^2 S = \overline{G}(\vartheta) P_{typ} \quad \text{with} \quad \overline{G}(\vartheta) := \frac{1}{\frac{1}{N} \sum_{i=1}^{N} \frac{1}{G(\varphi_i, \vartheta)}} \tag{8}$$

For the determination of $\overline{G}(\vartheta)$ a more realistic result can be obtained by averaging the function of $\overline{G}(\vartheta)$ also over a selection of mobile models. Since the antenna gain of the radio base station antenna depends on the elevation angle ϑ, Eq. (8) is evaluated explicitly for different values of $\vartheta = \{0°, 30°, 60°\}$ to make it possible to compare the mobile models under the same typical transmitting power. With this boundary condition averaging over ϑ is valid:

$$\overline{\overline{SAR}(\vartheta_j)} = \frac{1}{M} \sum_{j=1}^{M} \overline{SAR}(\vartheta_j) \tag{9}$$

Table 1. Measured and calculated radio properties for different mobile models

mobile model	SAR_{max} [W/kg]	SAR_{typ} [W/kg]	$EIRP$ [dBm]
no. 1	0.20	0.04	24.82
no. 2	0.14	0.03	22.84
no. 3	0.34	0.08	24.58
no. 4	0.41	0.07	23.95
no. 5	0.47	0.11	22.53
no. 6	0.32	0.10	20.83
no. 7	0.60	0.15	23.43
no. 8	0.57	0.16	22.97
no. 9	0.50	0.16	22.25
no. 10	0.39	0.18	21.02
no. 11	0.56	0.17	21.92
no. 12	0.75	0.18	23.38
no. 13	0.95	0.18	24.85
no. 14	0.74	0.22	23.30
no. 15	0.47	0.30	21.97
no.16	0.76	0.24	22.52

It is further noted that a call consist of talk time t_S and talk break t_{DTX} (DTX: discontinuous transmission), which has an influence on the power P_{typ}:

$$P_{typ} = P_S v + P_{DTX}(1-v) \quad \text{with} \quad v = \frac{t_S}{t_{DTX}} \tag{10}$$

During the talk time t_S the cellular phone works with the transmitting power P_S and during the talk break t_{DTX} with the transmitting power P_{DTX}. For $v > 0.5$ and $P_S \gg P_{DTX}$, to a good approximation P_{typ} is given by:

$$P_{typ} \cong P_S v \tag{11}$$

From the above equations the typical specific absorption rate can be calculated as follows:

$$SAR_{typ} = \frac{v P_S SAR_{max}}{MN} \frac{1}{1\text{ mW}} \sum_{j=1}^{M} \overline{G}(\vartheta_j)$$

$$\sum_{i=1}^{N} 10^{-\frac{EIRP_{\text{dBm}}(\varphi_i,\vartheta_j)}{10\,\text{dBm}}}$$

$$\text{with} \quad \overline{G}(\vartheta_j) = \frac{1}{\sum\limits_{i=1}^{N} \frac{1}{G(\varphi_i,\vartheta_j)}} \tag{12}$$

3 Results

The typical specific absorption rate (v = 0.7; P_S = 0.4 W; f = 1800 MHz) was calculated for 16 different mobile models from Eq. (12). Table 1 shows the results in the comparison to the specific absorption rate at maximum transmitting power and the averaged EIRP.

Equation 12 shows the relation between SAR_{typ} and SAR_{max}. In addition, a crucial parameter is the angle dependent EIRP, so that no pronounced correlation between

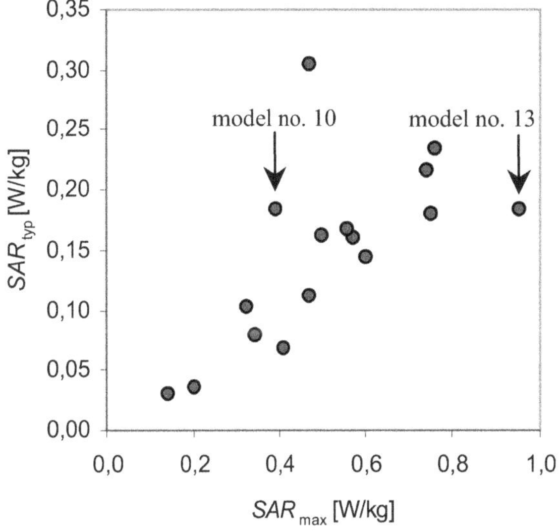

Fig. 2. SAR_{typ} and SAR_{max} of different mobile models from Table 1

SAR_{typ} and SAR_{max} exists (see Fig. 2). In Table 1 and Fig. 3 averaged over φ and ϑ $EIRP$ is shown and as can be seen, no pronounced correlation shows up between SAR_{typ} and $EIRP$. SAR_{typ} is thus an independent parameter for the characterisation of mobile models.

4 Discussions

A substantial result is to be recognised in Fig. 2. For some mobile models clear differences in the maximum specific absorption rate are visible, while the typical specific absorption rate changes by significantly less . There are mobile models, which show a far higher value for the maximum specific absorption rate but comparable (slightly lower) value for the

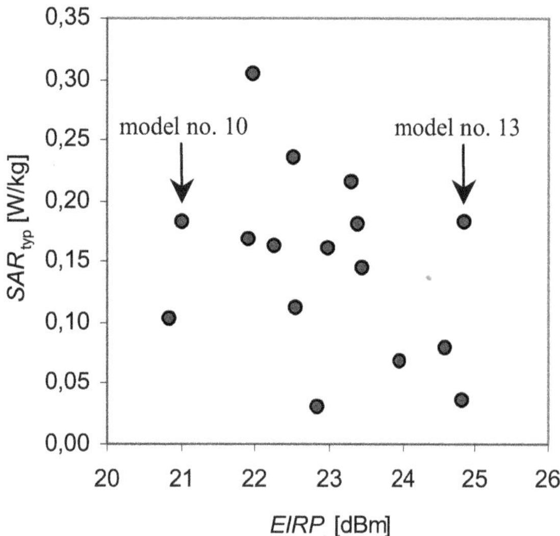

Fig. 3. SAR_{typ} and $EIRP$ of different mobile models from Table 1.

typical specific absorption rate (see Fig. 2 for mobile models 10 and 13 as an example).

An incorrect estimate of the real mean exposure would be produced if the mobile models were compared using only the specific absorption rate during maximum transmitting power. For the evaluation of real exposure conditions the typical specific absorption rate provides a suitable parameter for a reliable quantitative and qualitative comparison. The maximum specific absorption rate cannot supply this, since it was defined with a completely different objective in mind, namely safety aspects. It is also clear that the specific absorption rate for typical call conditions is lower, compared to the specific absorption rate during maximum transmitting power. The power control has a crucial influence in this regard.

Mobile models with a large EIRP value also exhibit good radio supply. This helps to provide a higher quality in the mobile radio network.

5 Conclusion

For the consumer it is important to have a criterion at hand by which the exposure conditions in a typical call situation can be judged. While the specific absorption rate supplies a statement about the safety of the cellular phone during maximum transmitting power, an estimate of real exposure conditions only becomes possible by the calculation of the typical specific absorption rate according to the presented model. For the estimate of good exposure and radio characteristics, this procedure offers clear advantages.

Acknowledgements. The author would like to thank Karsten Menzel of E-Plus Mobilfunk GmbH & Co. KG for the valuable discussions and suggestions.

References

DASY, Referenzliste des Herstellers, der Fa. Schmid & Partner Engineering AG, über installierte DASY-Systeme mit RX90 Robotern: Deutsche Telekom, Forschungs- und Technologiezentrum; Motorola Cellular – MRO; Motorola; Ericsson Mobile Communications AB; Nokia Mobile Phones LTD; IMST GmbH, 1995.

EN 50630, European Standard: Product standard to demonstrate the compliance of mobile phones with the basic restrictions related to human exposure to electromagnetic fields (300 MHz–3 GHz), CENELEC, Brussels, July 2001.

EN 50631, European Standard: Basic standard for the measurement of specific absorption rate related to human exposure to electromagnetic fields from mobile phones (300 MHz–3 GHz), CENELEC, Brussels, July 2001.

ES 59005, European Specification: Considerations for the evaluation of human exposure to electromagnetic fields (EMFs) from mobile telecommunication equipment (MTE) in the frequency range 30 MHz–6 GHz, 1998.

European Council Recommandation (1999/519/EC): Council recommandation of 12 July 1999 on the limitation of exposure of the general public to electromagnetic fields (0 Hz to 300 GHz), Official Journal of the European Communities L199/59, 1999.

International Commission on Non-Ionizing Radiation Protection (ICNIRP): Guidelines for limiting exposure to time-varying electric, magnetic, and electromagnetic fields (up to 300 GHz), In: Health Physics, Vol. 74 No. 4, pp. 494–522, 1998.

Kuster, N., Kästle, R., and Schmid, T.: Dosimetric evaluation of handheld mobile communications equipment with known precision, In: IEICE Trans. Commun., Vol. E80-B, No. 5, 645–652, 1997.

Schneider, M., Gehrt, M., Heberling, D., and Wicke, J.: Charakterisierung des Air-Interface von Handys fr das GSM 1800 Mobilfunknetz, ITG-Fachtagung Antennen, 21.–24.04.1998, München, In: ITG-Fachbericht 149, S. 323ff., 1998.

Vecchia, P., Ardoino, L., Bowman, J. D., Cardis, E., Mann, S., and Wiart, J.: Radiofrequency exposure of mobile phone users – the activity of the INTERPHONE study group, In: Proceedings of the 5th International Congress of the European BioElectromagnetic Association (EBEA), Helsinki, ISBN 951-802-440-5, 6.–8. September 2001

Wiart, J., Dale, Ch., Bosisio, A. V., and Le Cornee, A.: Analysis of the influence of the power control and discontinuous transmission on rf exposure with GSM mobile phones, IEEE Transactions on electromagnetic compatibility, Vol. 42, No. 4, November 2000.

2

Breakdown behavior of electronics at variable pulse repetition rates

S. Korte and H. Garbe

Institute of Electrical Engineering and Measurement Science, University of Hannover, Germany

Abstract. The breakdown behavior of electronics exposed to single transient electromagnetic pulses is subject of investigations for several years. State-of-the-art pulse generators additionally provide the possibility to generate pulse sequences with variable pulse repetition rate. In this article the influence of this repetition rate variation on the breakdown behavior of electronic systems is described. For this purpose microcontroller systems are examined during line-led exposure to pulses with repetition rates between 1 KHz and 100 KHz. Special attention is given to breakdown thresholds and breakdown probabilities of the electronic devices.

1 Introduction

As mentioned above several investigations concerning the breakdown and destruction behavior of electronics during single pulse impact have been performed. In this context failure rates for destruction and breakdown phenomena have been defined (Camp et al., 2002). As this investigation focuses on the breakdown effects without destructing the devices the Breakdown Failure Rate (BFR) is an interesting quantity to be measured. For single pulse measurement it is defined as:

$$\mathrm{BFR} = \frac{\text{number of breakdowns}}{\text{number of pulses}}. \tag{1}$$

This definition gives a quantity for the breakdown probability of the electronic devices during single pulse measurements.

Pulse sequences consist of a lot of pulses, so this fact will cause very small BFR. Therefore it makes more sense to relate the number of breakdowns to the number of tested devices. The definition of Eq. (1) been modified as shown in Eq. (2).

$$\mathrm{BFR} = \frac{\text{number of breakdowns}}{\text{total number of tested devices}} \tag{2}$$

Equation (2) is similar to the general definition of a probability. With this definition a qualitative analysis of the breakdown behavior can also be performed with smaller test series.

The characteristic of a BFR with increasing amplitude of the disturbance signal can be approximated with the probability mass function of a Weibull Distribution (Camp, 2004). In order to get a more specified description of the breakdown behavior two variables have been defined (Camp et al., 2002). The Breakdown Threshold (BT) is the amplitude where 5% of the electronic devices break down. The Breakdown Bandwidth (BB) is the difference between the amplitudes where the BFR reaches 95 and 5%.

Both variables provide the opportunity to get a quick comparison between different measurements.

In investigations with single pulses the following results where gained:

- BFR is increasing from 0 to 1 with increasing pulse amplitude (Camp, 2004)
- both BT and BB depend on the function of the investigated pin of the Equipment Under Test (EUT) (Camp, 2004)
- critical system states can be observed where the electronic system is highly vulnerable (Camp et al., 2004)

2 Measurement setup

In order to apply line-led disturbing signals to electronic devices a special measurement setup has been constructed in the past. Details of the microcontroller test circuit are shown in Fig. 1. It consists of a microcontroller PCB connected on both sides to a load which is constructed out of LEDs and resistors. With mounted chip this circuitry is able to run test programs which trigger the LEDs frequently to monitor the accurate function or a breakdown of the chips. Between

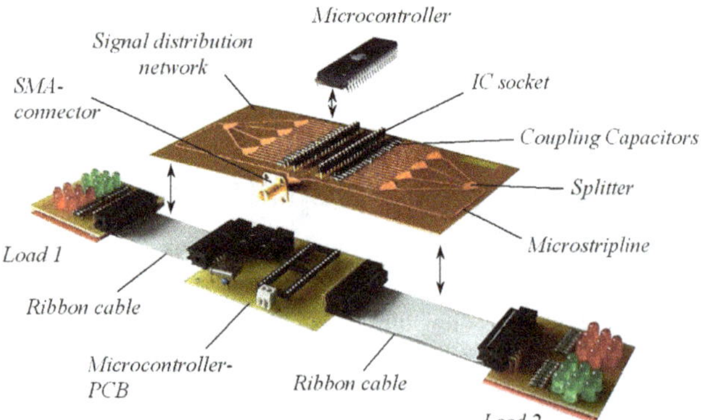

Fig. 1. Microcontroller Test Setup (Details).

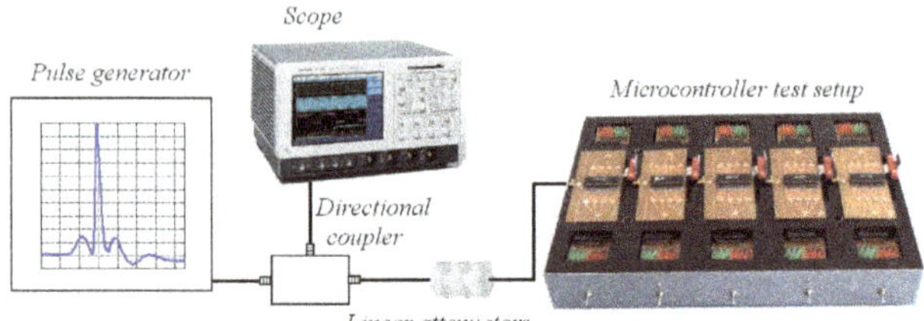

Fig. 2. Measurement setup (principle).

microcontroller and microcontroller PCB several signal distribution networks can be positioned. This networks have SMA connectors to connect arbitrary signal sources. Via microstriplines and coupling capacitors the signal is distributed to the considered pins. All other pins are terminated with 50 Ohm.

This circuitry is embedded in a microcontroller test setup which provides the possibility to apply disturbing signals line-led to several pins of the electronic devices in order to get information about the correlation between the function of the exposed pin and the breakdown behavior. The following pins are exposed:

- Supply pins
- Reset pin
- Clock pins
- I/O pins

As drafted in Fig. 2 a pulse generator is feeding via a directional coupler and several linear attenuators the test setup described above. Pulse shape and amplitude can be monitored with the scope connected to the directional coupler. The measurement setup is fed with pulse generators from Rheinmetall Waffe Munition GmbH. Rise time t_r, full width half max value t_{fwhm} and possible repetition rates f_{rep} of the pulse generators are shown in Table 1.

Generator 1 was used to generate pulses with repetition rates $f_{\mathrm{rep}} = 1$ kHz and $f_{\mathrm{rep}} = 10$ kHz while Generator 2 provided a fixed f_{rep} of 100 kHz. Both pulses show in spite of different pulse shapes similar behavior in the frequency domain as shown in Fig. 3. Both pulse generators amplitudes are adjustable. Additionally linear attenuators where used to get a coarse predivision.

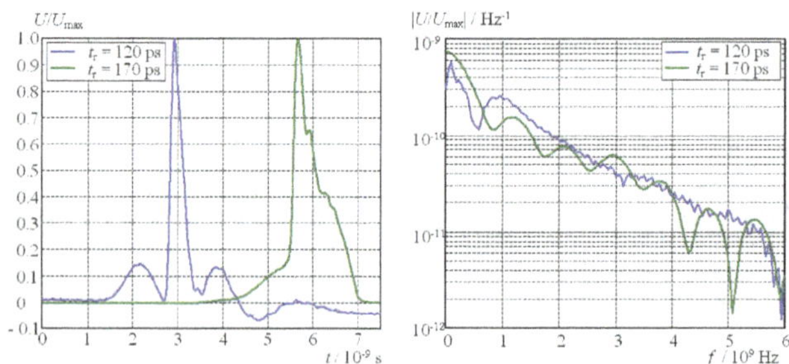

Fig. 3. Applied pulses in the time and frequency domains.

Table 1. Pulse generators - characteristics.

Generator	t_r	t_{fwhm}	f_{rep}
1	170 ps	500 ps	1 KHz, 10 KHz
2	120 ps	300 ps	100 KHz

All presented measurements have been carried out with microcontrollers of the type AT90S8515 from Atmel. This device is a 40-pin 8-bit RISC microcontroller with 512 Bytes SRAM and 512 Bytes EEPROM. It is equipped with 32 I/O pins which are organized in four 8-bit I/O ports.

3 Measurement results

Based on the possibilities of the measurement setup four measurements where carried out by exposing different pins of the microcontroller to pulses with three different repetition rates. Fig. 4 shows the breakdown behavior of the different pins with increasing pulse amplitude. All configurations show basically the same behavior with increasing amplitude. Noticeable is the difference between the thresholds of the several setups, a fact that has been observed also during single pulse investigations (Camp, 2004). The susceptibility of the electronic device is highly dependable on the function of the exposed pins. Furthermore this measurement shows a general decrease of the thresholds with increasing repetition rate.

The breakdown values of the tested microcontrollers for the different measurement configurations are presented in Fig. 5. The decrease of the breakdown value with increasing repetition rate of every single microcontroller is clearly observable. Furthermore the variation of the breakdown value between the several devices decreases with rising repetition rate. This variation is described with the Breakdown Band-

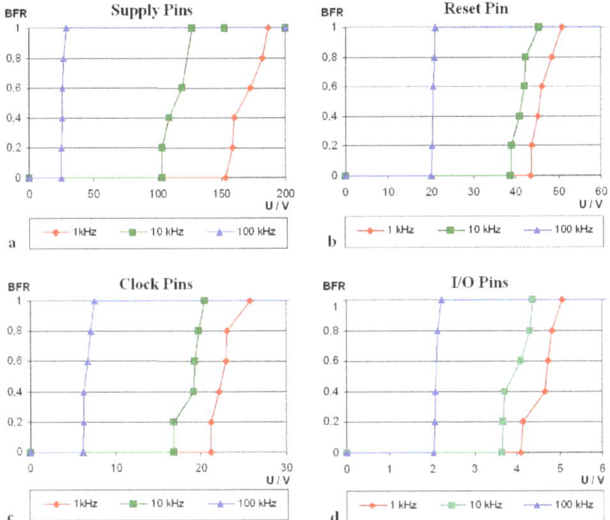

Fig. 4. BFR: **(a)** Supply Pins; **(b)** Reset Pin; **(c)** Clock Pins; **(d)** I/O Pins.

width (BB) as mentioned above. In order to demonstrate the described effects more clearly Fig. 6a shows the breakdown thresholds extracted from the appropriate Weibull Distribution while the breakdown bandwidths are presented in Fig. 6b. From both diagrams we can directly extract the information that BT as well as BB are decreasing dramatically with rising pulse repetition rate. Remarkable is that in the worst case (I/O pins) the value of the breakdown threshold at $f_{\text{rep}} = 100\,\text{kHz}$, $BT(\text{I/O}, 100\,\text{kHz}) = 2{,}01\,\text{V}$, is less than half the value of the supply voltage and the high level of the microcontroller ($U_{CC} = U_{\text{High}} = 5\,\text{V}$). With further increase of the repetition rate in the future a noteworthy continue of the decrease of the breakdown thresholds is unlikely. For 5 V logic devices the difference between lowest value for high-level and highest value for low-level $U_{H,\min} - U_{L,\max} = 3\,\text{V} - 1.5\,\text{V} = 1.5\,\text{V}$. The minimum voltage needed to switch a bit from one state to another is nearly reached.

Two characteristic values where defined in order to describe the variation of the breakdown behavior. For an arbitrary ratio of repetition rates A/B we define the value for Threshold Variation (TV) as follows:

$$TV(A/B) = \frac{BT_A}{BT_B}, \quad \text{with} \quad A > B. \tag{3}$$

The definition of the Breakdown Variation (BV) in Eq. (4) is similar.

$$BV(A/B) = \frac{BB_A}{BB_B}, \quad \text{with} \quad A > B \tag{4}$$

Table 2 shows both values at different ratios of the repetition rates. The TV decreases with increasing repetition rate. At $f_{\text{rep}} = 10\,\text{kHz}$ $TV(10/1) = 86\%$ of the pulse amplitude at $f_{\text{rep}} = 1\,\text{kHz}$ is needed for breakdown during I/O pins exposure. At ten times higher frequencies only $TV(100/10) = 53\%$ of the amplitude at $f_{\text{rep}} = 10\,\text{kHz}$ is necessary to gain breakdowns. If we compare minimum and maximum frequencies less than one half ($TV(100/1) = 46\%$) of the former amplitude is required. In the worst case (supply pins exposure) the $TV(100/1)$ value is only 16%.

The BV values also decrease with rising repetition rate. While during I/O pins exposure at $f_{\text{rep}} = 10\,\text{kHz}$ $BV(10/1) = 80\%$ is required for breakdown of all microcontrollers, only $BV(100/10) = 22\%$ is necessary at ten times higher frequencies. In comparison of minimum and maximum frequencies $BV(100/1)$ is only 18%. In the worst case (supply and reset exposure) $BV(100/1) = 12\%$. Thus the possibility to hit is increased by 88%.

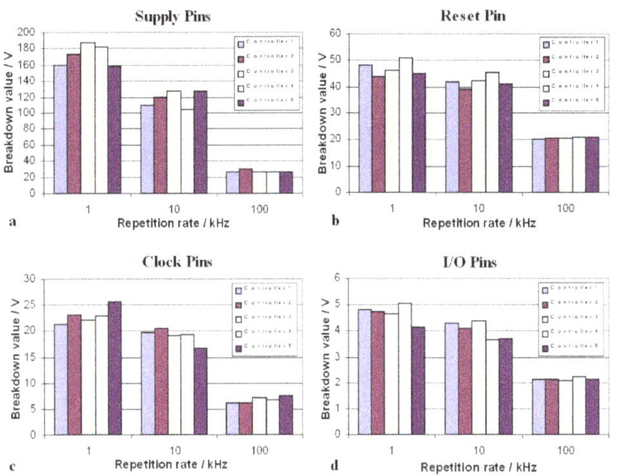

Fig. 5. Breakdown values: **(a)** Supply Pins; **(b)** Reset Pin; **(c)** Clock Pins; **(d)** I/O Pins.

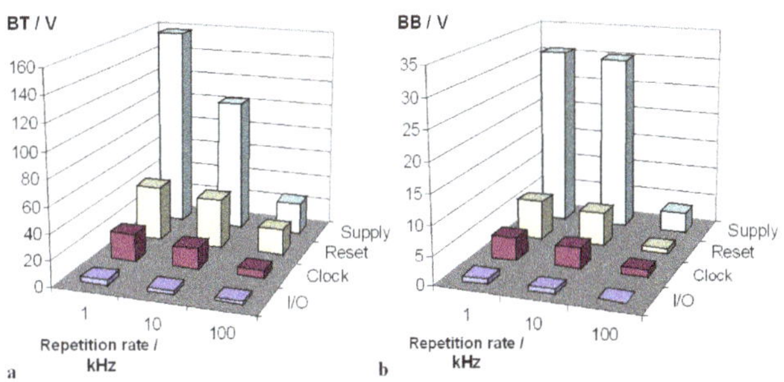

Fig. 6. Breakdown Thresholds (BT) and Breakdown Bandwidths (BB) of the different configurations with varying pulse repetition rate.

4 Summary

After several investigations (Camp et al., 2002; Camp, 2004; Camp et al., 2004) concerning the susceptibility of electronic devices to single transients this article shows a systematic

Table 2. Threshold Variation (TV) and Bandwidth Variation (BV) with increasing repetition rate.

Setup	TV (10 /1)	BV (10/1)	TV (100/10)	BV (100/10)	TV (100/1)	BV (100/1)
Supply	0,68	0,98	0,23	0,12	0,16	0,12
Reset	0,9	0,88	0,49	0,14	0,44	0,12
Clock	0,83	0,9	0,35	0,35	0,29	0,31
I/O	0,86	0,8	0,53	0,22	0,46	0,18

analysis of the breakdown behavior of microcontrollers exposed to repetitive pulses with varying repetition rate.

During this analysis it can be observed that the breakdown threshold is sinking with increasing repetition rate while the possibility to hit a certain microcontroller at a critical system state is higher. Therefore the electronic device is overall more susceptible to pulses if applied with higher repetition rates. Similar to the results of single pulse investigations it can be stated that all the observed effects are correlated with the exposed pins functions. The I/O pins of the investigated microcontrollers are the most susceptible pins while the supply pins are the least susceptible ones.

After definition of two characteristic values it is shown that in the worst case the breakdown threshold can be lowered to only 16% of the former amplitude by increasing the repetition rate from 1 KHz to 100 kHz. Remarkable is in this context the fact that the necessary breakdown threshold drops under less than one half of the supply voltage. A further noteworthy decrease under the value of 2 V is unlikely based on the electrical characteristics of the devices. The possibility to hit the device in a risky system state is increased due to the same repetition rate variation by up to 88%. Overall the increase of the repetition rate of applied pulses to an electronic device is a pretty good possibility to increase the ability to disturb the system and is only limited by the electrical characteristics of the electronics.

Acknowledgements. This investigation is part of the study "Protection of Electronic Systems against Electromagnetic Sources", commissioned by the Armed Forces Scientific Institute for Protection Technologies - NBC-Protection (Munster, Germany). The responsibility for the contents is with the authors.

The authors would like to thank Dr. M. Jung, Rheinmetall Waffe Munition GmbH for supporting these measurements.

References

Camp, M.: Empfindlichkeit elektronischer Schaltungen gegen transiente elektromagnetische Feldimpulse, Diss., Univ. of Hanover, Shaker Verlag, 2004.

Camp, M., Garbe, H., and Nitsch, D.: Influence of the Technology on the Destruction Effects of Semiconductors by Impact of EMP and UWB Pulses, IEEE Intern. Symposium on Electrom. Compatibility, USA, Minneapolis, 87–92, 2002.

Camp, M., Korte, S., and Garbe, H.: Classification of the Destruction Effects in CMOS-Devices after Impact of Fast Transient Electromagnetic Pulses, EUROEM, 14th High Power Electrom. Conf. (HPEM 14), 7th Ultra-Wide-Band Short-Pulse Electrom. Conf. (UWB SP7), 7th Unexploded Ordnance Detection and Range Remediation Conf. (UXO 7), Germany, 96, 2004.

3

Airborne field strength monitoring

J. Bredemeyer[1], T. Kleine-Ostmann[2], T. Schrader[2], K. Münter[2], and J. Ritter[3]

[1]FCS Flight Calibration Services GmbH, Braunschweig, Germany
[2]Physikalisch-Technische Bundesanstalt, Braunschweig, Germany
[3]EADS Deutschland GmbH, Military Aircraft, Bremen, Germany

Abstract. In civil and military aviation, ground based navigation aids (NAVAIDS) are still crucial for flight guidance even though the acceptance of satellite based systems (GNSS) increases. Part of the calibration process for NAVAIDS (ILS, DME, VOR) is to perform a flight inspection according to specified methods as stated in a document (DOC8071, 2000) by the International Civil Aviation Organization (ICAO). One major task is to determine the coverage, or, in other words, the true signal-in-space field strength of a ground transmitter. This has always been a challenge to flight inspection up to now, since, especially in the L-band (DME, 1GHz), the antenna installed performance was known with an uncertainty of 10 dB or even more. In order to meet ICAO's required accuracy of ±3 dB it is necessary to have a precise 3-D antenna factor of the receiving antenna operating on the airborne platform including all losses and impedance mismatching. Introducing precise, effective antenna factors to flight inspection to achieve the required accuracy is new and not published in relevant papers yet. The authors try to establish a new balanced procedure between simulation and validation by airborne and ground measurements. This involves the interpretation of measured scattering parameters gained both on the ground and airborne in comparison with numerical results obtained by the multilevel fast multipole algorithm (MLFMA) accelerated method of moments (MoM) using a complex geometric model of the aircraft. First results will be presented in this paper.

1 Introduction

Preceding work was done describing the antenna installed performance of a flight inspection aircraft (Bredemeyer et al., 2004). Those investigations mainly focussed on the modelling of the platform and the antennas to be installed as a prerequisite to apply the method of moments (Harrington, 1968). Additionally, measurements of selected aircraft antennas were done in a GTEM cell to obtain the non-installed antenna factors and to form a model which was then used to improve the on-platform far field simulation. The final validation of the model should include measurements in a well-defined environment to derive the effective performance since one cannot implicitly rely on simulations only. The results and critical aspects are discussed below.

First, we need to describe the electric far field antenna factor (AF) which is given by (Kraus, 1988; Balanis, 1997):

$$AF_{\text{electric}} = \frac{E_{\text{incident}}}{V_{\text{received}}} \tag{1}$$

In case of a loss-less antenna with a gain G and an impedance of 50 Ω the ideal AF then is given for the wavelength λ by

$$AF_{\text{ideal}} = \frac{9.73}{\lambda\sqrt{G}} \quad . \tag{2}$$

The radiation pattern of an antenna is mainly determined by the shape and material of the radiating element, while impedance matching networks in the input section determine the input reflection coefficient r of the antenna. DIN (DIN EN 45003, 1995) proposes the so called practical gain G_{pract} to be used in the definition of the antenna factor of antennas with a specific gain G_0 in case of non-ideally matched antennas:

$$G_{\text{pract}} = \left(1 - |r|^2\right) G_0. \tag{3}$$

In practice however, there are additional losses due to cable and connectors. An overall gain from the receiver's point of view must contain those values so that the effective gain can be written as

$$G_{\text{eff}} = (1 - |r|^2)\, \mathrm{e}^{-2\alpha l} \cdot \kappa \cdot G_0 \tag{4}$$

where l denotes the cable length, α is the attenuation constant according transmission line theory (Unger, 1996), and κ denotes the connector losses. Inserting G_{eff} as G in Eq. (2) then leads to the effective antenna factor:

$$AF_{\text{eff}} = \frac{9.73}{\lambda\sqrt{(1 - |r|^2) \cdot \kappa \cdot G_0}} \cdot \mathrm{e}^{\alpha l} \tag{5}$$

Fig. 1. Discretisation of the platform model with 15 680 triangles for analysis up to 118 MHz.

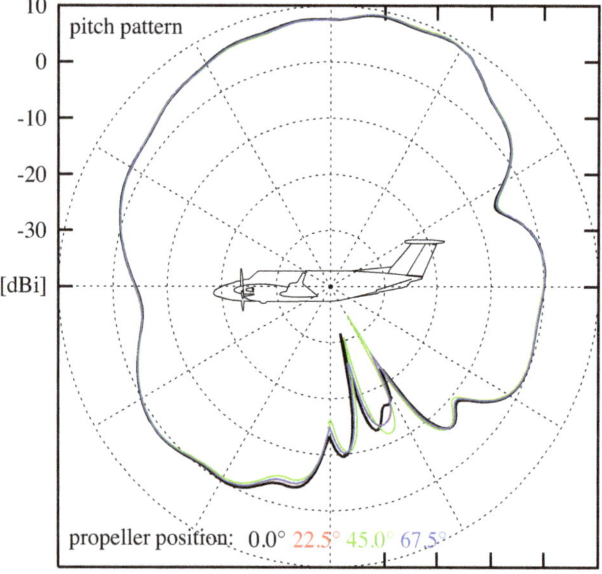

Fig. 2. Simulated pitch pattern of installed VHF top antenna at f=113 MHz.

2 Aircraft model and simulation

As stated above, a numerical simulation of the installed antenna performance using the method of moments was applied. In order to obtain the antenna factor, first the gain of the antenna is computed from the radiation pattern. For most of the relevant antennas, equivalent models can be created using analog radiation characteristics, i.e. gain G_0 of the original antenna. For complex antennas with unknown impedance matching networks, the reflection coefficient r can be determined from measurements.

Effectively, some relevant areas of the platform have geometrical variations within the order of the wavelength or

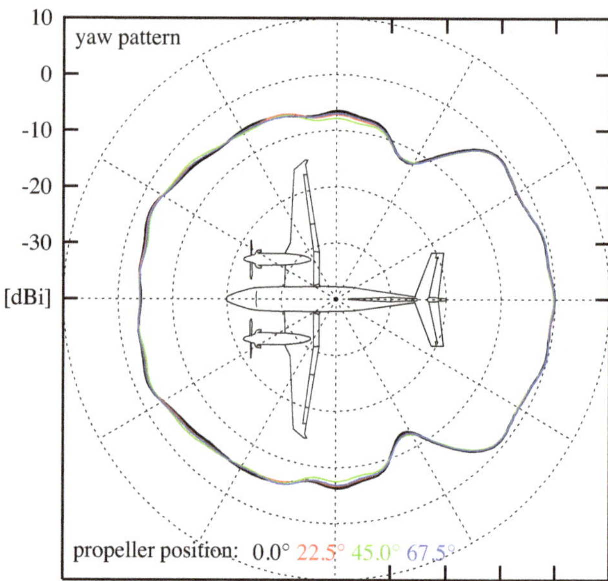

Fig. 3. Simulated yaw pattern of installed VHF top antenna at f=113 MHz.

less (propeller blades, leading edge curvatures, surface curvatures etc.), so a full-wave solution of Maxwell's equations is required as opposed to a (less computational intensive) asymptotic solution, assuming pure ray-optical behavior of the electromagnetic field. For this work a multilevel-fast-multipole accelerated method of moments (MLFMA) is used. The method of moments is a well known and established technique for antenna analysis (Harrington, 1968), while the MLFMA acceleration is a relative new addition allowing the analysis of much larger structures (Chew et al., 2000).

The method of moments requires the geometry to be meshed by elements of about a tenth of the wavelength in size. Here, triangular shaped elements are used. For the aircraft considered here, a Raytheon Beech Super King Air 350, a total of about 16 000 triangles is required to model the geometry at VHF NAV frequencies (108–118 MHz) and of about 600 000 at L-band frequencies around 1 GHz, respectively. An image of the VHF mesh is shown in Fig. 1.

Some 3-D-gain antenna patterns referencing an isotropic radiator (G_0=1) are exemplified in the following diagrams. The VHF antenna which is mounted on top of the Beech King Air gives a pitch pattern at 113 MHz shown in Fig. 2 and a corresponding yaw pattern in Fig. 3. In contrast, the concerned L-Band frequencies, e.g. for DME (Distance Measuring Equipment), are roughly 10 times higher than for VHF NAV so the resulting pattern has finer structures which can be seen in Figs. 4 and 5 at 1068 MHz. Since the L-band antenna is mounted underneath, the aircraft itself shields radiation in the upper direction which can be observed in the pitch pattern (Fig. 4). Relevant gain changes due to the

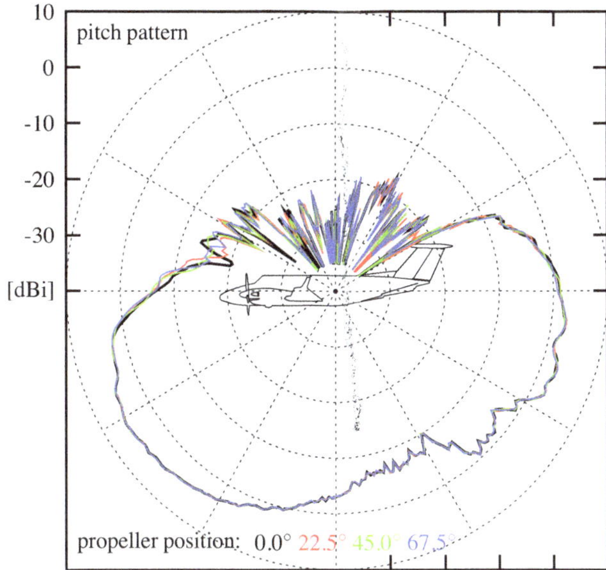

Fig. 4. Simulated pitch pattern of installed L-Band (DME) antenna at f=1068 MHz.

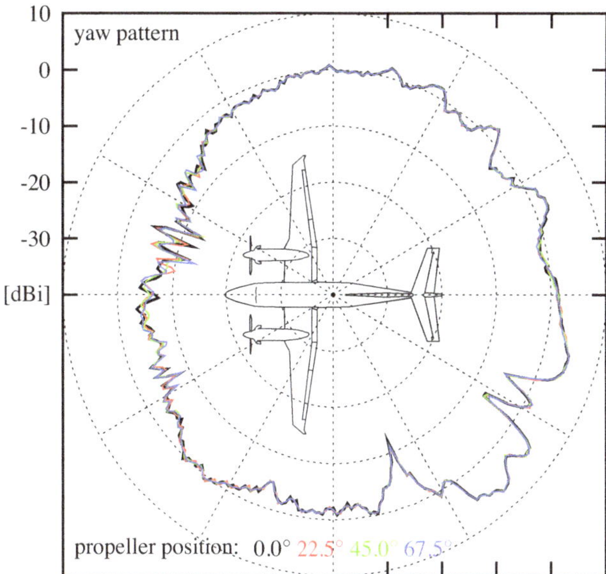

Fig. 5. Simulated yaw pattern of installed L-Band (DME) antenna at f=1068 MHz.

4-blade-propeller position can only be observed at directions which are not relevant for the purpose of the antenna, e.g. towards ground as shown in the VHF pitch pattern shown in Fig. 2.

3 Comparison to measurements

In order to verify the simulation results, near field measurements of three different antennas were performed by the Physikalisch-Technische Bundesanstalt (PTB) at the airport in Braunschweig. Two of the antennas were dipole antennas for the VHF (108–118 MHz) and UHF (329–335 MHz) band installed above the cockpit whereas the third was a L-band antenna installed underneath the hull of the aircraft.

For the measurements the aircraft was positioned on a 20 m·20 m ground plane made up of a conductive textile. The reference antenna was positioned on a circular arc with a radius of R=7.75 m facing the aircraft under an aspect angle α, as indicated in Fig. 6, between −60 and +60 degrees. The measurements for the L-band antenna were performed in a plane 90 cm above ground, whereas the measurements for the UHF and VHF antennas were performed in planes at three different heights 1.5, 2.5, and 3.5 m above ground. A photograph of the aircraft mounted on pillars above the ground plane and of the reference antenna mast is shown in Fig. 7. The ground plane acts as a co-radiator and therefore has surface currents which are visualized exemplarily at 118 MHz in Fig. 8.

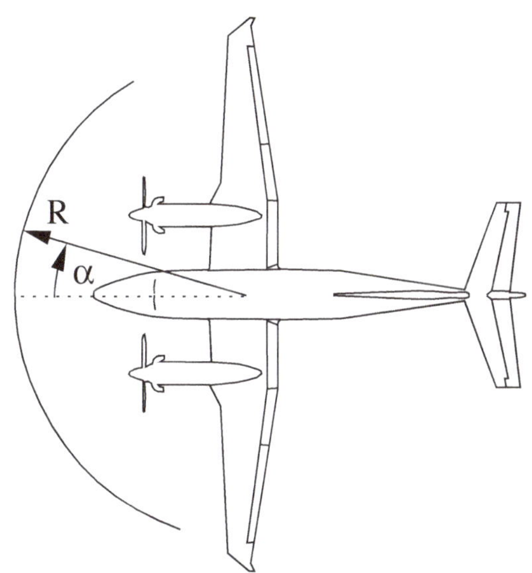

Fig. 6. Aspect angle.

3.1 Quantity for comparison

As quantity for the comparison we use the electrical field strength at the position of the reference antenna which has to be derived from the detected power P_d. This power is a quantity integrated over the effective aperture of the reference antenna. In order to be able to assign a detector field strength to a discrete point in space its value has to be sufficiently constant in an area comparable to the effective aperture.

If the range between transmission (aircraft) antenna and receiver (reference) antenna is considered as 2-port where

Fig. 7. Flight inspection aircraft and reference antenna mast positioned on a conductive textile ground plane.

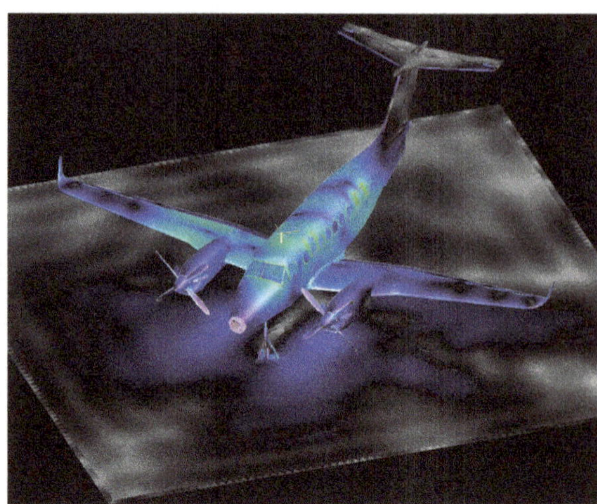

Fig. 8. Surface currents caused by the VHF NAV top antenna above a ground plane at f=118 MHz. The weakest currents are colored white.

port 1 is the transmitting antenna and port 2 is the receiving antenna, the scattering parameter S_{21} describes the relation between detected power and feed power of the transmitting antenna P_f as

$$P_d = |S_{21}|^2 P_f \ . \qquad (6)$$

Substituting the feed power P_f with the expression for the radiated power

$$P_t = (1 - |S_{11}|^2)\ \mathrm{e}^{-2\alpha l} P_f \ , \qquad (7)$$

where the first factor on the right side expresses the mismatch losses and the second factor the losses of the antenna cable with length l and attenuation coefficient α yields

$$P_d = \frac{|S_{21}|^2}{(1 - |S_{11}|^2)\ \mathrm{e}^{-2\alpha l}} P_t \ . \qquad (8)$$

The received power can be expressed as the product of radiation density in far field approximation S=$|E|^2/\eta_0$ (free-space impedance η_0) and the effective aperture of the receiver antenna A_r by

$$P_d = SA_r = \frac{|E|^2}{\eta_0} g_r \frac{\lambda^2}{4\pi} \qquad (9)$$

where g_r is the known reference antenna gain.

Using Eq. (8) the magnitude of the electric field strength $|E|$ at the location of the detector antenna can be calculated as

$$|E|^2 = \frac{|S_{21}|^2}{(1 - |S_{11}|^2)\ \mathrm{e}^{-2\gamma l}} \frac{4\pi \cdot \eta_0}{\lambda^2 g_r} P_t \qquad (10)$$

with known radiated power P_t and gain g_r of the detector antenna. The feed power is set to 1 Watt to be able to compare the measurements to the simulations which have been performed for this power level.

3.2 VHF antenna

The horizontally polarized V-shaped VHF dipole is located above the cockpit. According to its specifications it has a VSWR of 3:1 which corresponds to a reflection factor of $|S_{11}|$=0.5 (so mismatch loss is 1.25 dB). The RG-400 antenna cable has a length of 4.68 m and an attenuation of 0.73 dB in the frequency range from 108 to 118 MHz according to its specifications.

A fundamental reason for the discrepancy between simulation and measurement can be found in the inhomogeneity of the electromagnetic field at the location of the detector antenna. As already discussed in Sect. 3.1, a homogeneous field is the prerequisite for calculating the field strength level from the measured receiver power. Figure 9 shows the real part of the normal component of the radiation power density in comparison to the radiation power density obtained from the copolar electric field strength over the aspect angle at a height of 3.5 m and f=108 MHz. Obviously the assumption of a plane wave with $=|E|^2/\eta_0$ is no longer valid which means that a precise calculation of the field strength from the received power is no longer possible.

For an estimate of the resulting error we simulate the measurement situation with a 60 cm dipole at 108 MHz at a height of 3.5 m. The dipole length is comparable to the dimensions of the actual measurement antenna. From the calculated receiver power the field strength is then determined: The dipole has an input impedance of Z_D=(9.4−$\jmath$ 649.6) Ω and a gain of 1.84 dBi in free space. In the case of a matched load resistance of Z_L=Z_D^*=(9.4+$\jmath$ 649.6)Ω the field strength level determined in this way is only 0.13 dB below the directly simulated value without antenna. A mismatch in this impedance range has significant impact on the received power and hence on the field strength level. A load resistance

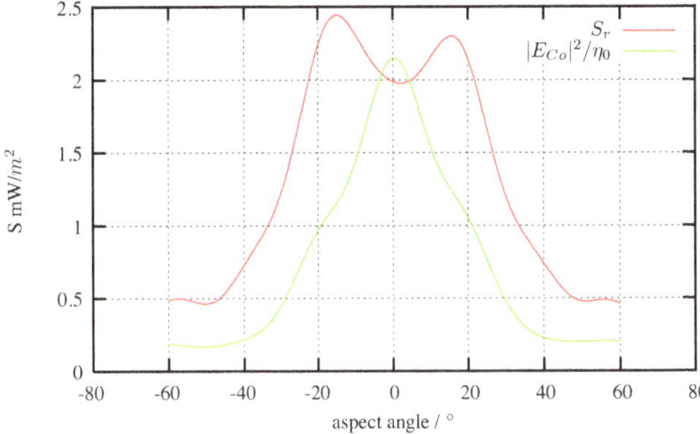

Fig. 9. Real part of the Poynting vector compared to the corresponding value derived from the copolar electric field strength.

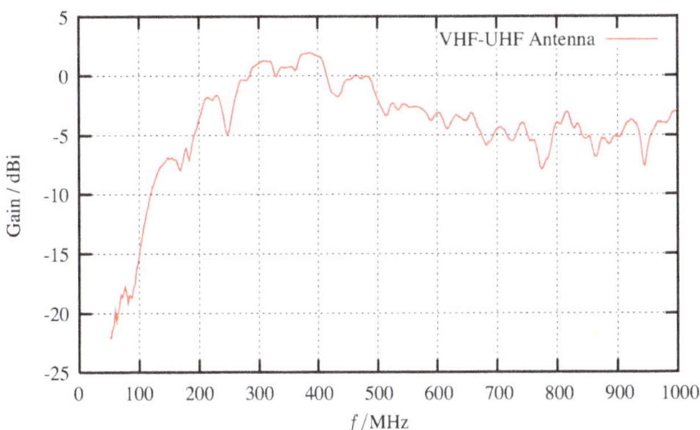

Fig. 10. Gain of the PTB reference antenna.

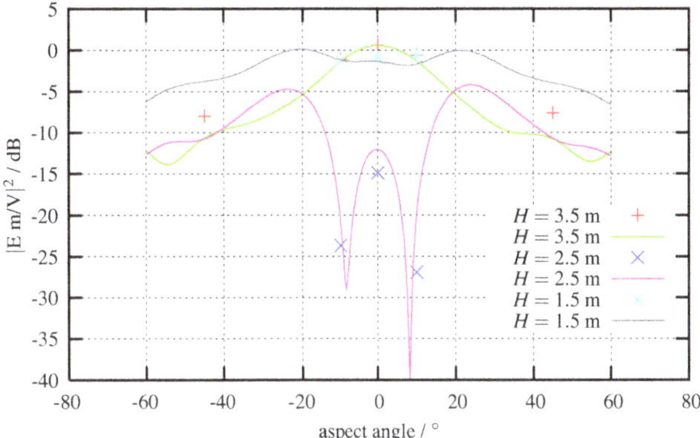

Fig. 11. Field strength level over aspect angle for different reference antenna heights at f=113 MHz (VHF NAV top antenna, lines: simulation).

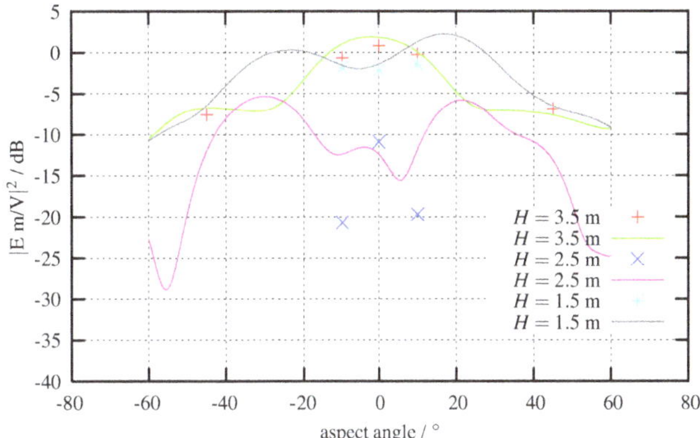

Fig. 12. Field strength level over aspect angle for different reference antenna heights at f=118 MHz (VHF NAV top antenna).

of $Z'_L=(9.82+j\ 682)\,\Omega$ with both real and imaginary values only 5% above the matching value causes the field strength level to decrease by 6 dB.

The reason for this sensitivity can be found in the unbalanced magnitude of real and imaginary part of the input impedance of the dipole. We assume that discrepancies between simulations and measurements can be attributed to such a situation. Figure 10 shows the measured gain of the receiver antenna over frequency. It decreases rapidly towards lower frequencies.

In order to be able to better match simulations and measurements, the measured values have been corrected by a constant offset of 2 dB. Figures 11 and 12 show the corrected measurements (dots) and the simulations (lines) for 113 and 118 MHz at different receiver antenna heights, respectively. For a height of 2.5 m a pronounced dip for a low aspect angle can be observed for both frequencies. We attribute this to the influence of the conducting ground plane. Although both simulations and measurements show this behavior, we observe the strongest discrepancy in this case, especially at 118 MHz. However, with respect to the deviations from the homogeneity of the field and the ideal conduction of the ground used in the calculations, the match between simulations and measurements is satisfactory.

3.3 UHF glideslope antenna

The horizontally polarized UHF dipole is located in front of the V-shaped VHF dipole. According to specifications it has a VSWR of 2:1, corresponding to a reflection factor of $|S_{11}|=1/3$ ($\hat{=}-9.5$ dB, mismatch loss 0.5 dB). The antenna is also fed via a RG-400 cable with a length of 5.12 m and a resulting attenuation of 1.5 dB at a frequency of 332 MHz (according to data sheet).

Figure 13 shows the comparison between measurements (dots) and simulations (lines) for the transmitting UHF antenna. The agreement is much better in this case because the receiver antenna operates under much better far field conditions. Expressed in wavelengths, it is located at approximately three times the distance from the transmitting antenna compared to the previous VHF measurement.

3.4 L-band antenna

The L-band antenna is a monopole. It is fixed underneath the hull of the aircraft between the wings and is vertically

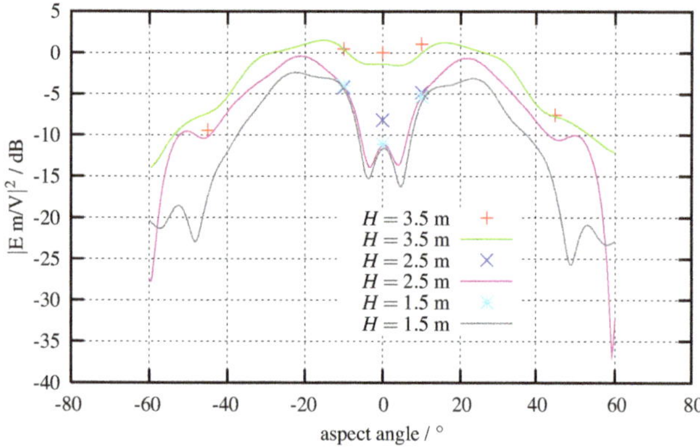

Fig. 13. Field strength level over aspect angle at different reference antenna heights at f=332 MHz (UHF Glideslope top antenna).

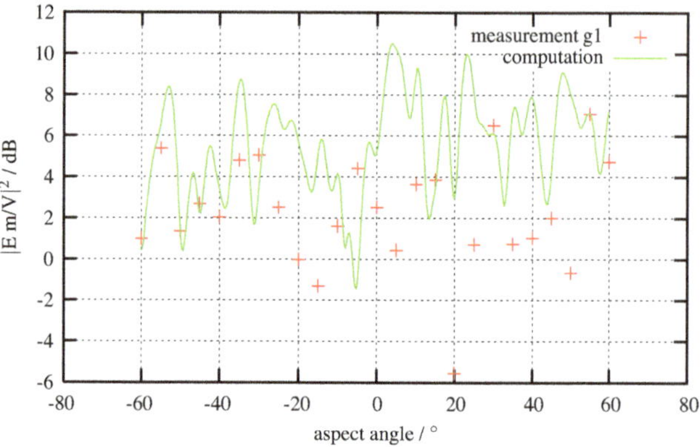

Fig. 14. Field strength level at 90 cm reference antenna height over aspect angle at f=1075 MHz.

polarized. According to specifications it has a VSWR of approximately 1.5:1 which corresponds to a reflection factor of $|S_{11}|$=0.2 ($\hat{=}-14$ dB, mismatch loss 0.18 dB). The attenuation of the feeding cable is yet unknown.

Figure 14 shows the calculated and measured field strength levels plotted over aspect angle at a frequency of 1075 MHz and at a height of H=90 cm over ground.

Over a vast range of aspect angles the field strength level difference between measurements and simulations reaches up to 10 dB. Due to the restricted number of measurement points the measurement characteristics lack from undersampling which makes it more difficult to interpret the curve characteristics. However, over small aspect angle ranges the measurements show a different trend than the simulations. A constant field strength level offset as expected from the contribution of the antenna cable attenuation which has not been considered here, would not give a satisfactory explanation.

In order to be able to evaluate the stability of the simulated model also in this case the modelling of the transition between wings and hull has been examined further. This area is close to the radiating aircraft antenna and partially non-conducting. Also the influence of the field inhomogeneity at the location of the reference antenna has been examined. This involved modelling the measurement situation where the receiver dipole, which has a size comparable to the aircraft antenna, was placed at the front aspect. The field strength level has then been calculated from the received power at the receiver antenna according to Eq. (10) and was compared to the directly determined field strength level, afterwards.

Although the field strength level calculated in this way varies with detector antenna height by 2.5 dB in this case, the field strength level calculated from the receiver power P_r and the directly determined value differ by less than 0.1 dB at H=90 cm over ground. Therefore the field inhomogeneity can be excluded as a possible reason for the difference between measurement and simulations.

A possible explanation for the difference between measurements and simulations could be found in the realization of the conducting ground plane. The reflection characteristics of the conducting textile remains to be determined in an adequate measurement setup for both polarization types. Also the local variation in flatness of the ground plane (as can be seen in Fig. 7) is not negligible.

3.5 Mutual L-band S-parameters

In addition to the ground plane measurements between an aircraft antenna and a remote detector antenna in-flight measurements between mutual aircraft L-band antennas underneath the hull were performed. In order to further verify the accuracy of the computer model, three pairs of antennas were measured. The amplitude of transmission factors $|S_{21}|$ were compared to simulated data obtained from the MLFMA model of the aircraft for the respective antennas.

The measurements have been performed during an 1 h test flight over Northern Germany using a HP8753D vector network analyzer. The measurements were performed at an altitude of 5000 ft. with a cloud ceiling far underneath and above the aircraft. The transponder was switched off to prevent unwanted interference with the measurements. After TOSM calibration of the vector network analyzer at the end of the two 2.5 m long measurement cables and after the consecutive measurements of the antennas under test the performance of the network analyzer was checked based on the E_n criterium (EAL, 1996). Measurements of calibrated PTB precision attenuators (10 dB, 20 dB, 30 dB and 50 dB) proved that the antenna measurements in between are within the expected measurement uncertainty of less than 0.25 dB.

Figure 15 shows the result of the measurements between the previously described L-band antenna centered between the wings used as transmission antenna (no. 23 in numeration of all aircraft antennas) and three other L-band antennas used as detector antennas (no. 26 underneath the right wing, no. 25 underneath the left wing and antenna 15 underneath

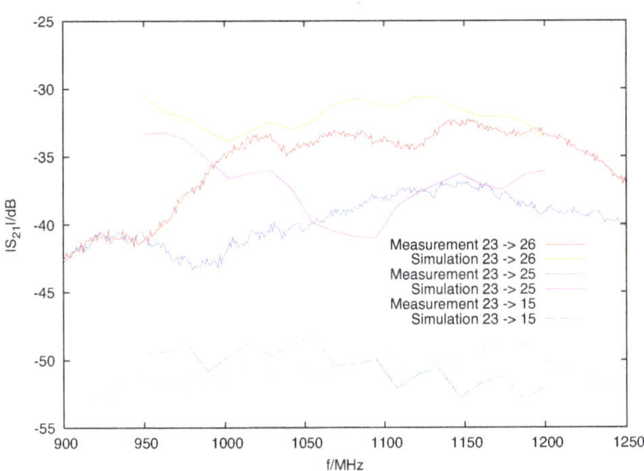

Fig. 15. Comparison of simulated and measured S-parameters of L-band antennas. The numbers 15, 23, 26 according to list of all aircraft antennas.

the rear hull). Measurements and calculations agree well in all cases but especially for the antenna pairs 23–25 and 23–15. The calculated attenuation has the tendency to be weaker than the measured values. We attribute this to the attenuation of several meters of RF cable which have been neglected so far. Further investigations to incorporate cable attenuation measurements into the calculation of effective antenna factors are underway.

4 Conclusions

We showed that a balanced combination of antenna measurements, platform modelling and MoM simulations is the key to obtain the effective antenna installed performance. However, some problems impede the comparison of measurement results and numerical solutions under certain circumstances: The VHF results were gained in a near field environment whereas the L-band antennas underneath the aircraft did not meet the ideal conditions above the ground plane as taken for granted in the model. In the VHF case one has to consider alternative measurement concepts. A solution could be to implement a near field scanner in context with adequate near to far field transformations. Further work is needed to implement such technologies and the authors will continue their work in upcoming projects.

Acknowledgements. The authors would like to thank M. Schwendener and J. Follop of Flight Calibration Services for their assistance during the measurement campaigns and J. Reiter of EADS for his contributions to the simulations.

References

Balanis, C. A.: Antenna Theory Analysis and Design, Wiley, New York, 1997.

Bredemeyer, J., Battermann, S., Garbe, H., and Ritter, J.: Antenna Installed Performance of a Flight Inspection Aircraft, Proceedings of International Symposium on Precision Approach and Automatic Landing (ISPA 2004), Munich, Germany, 2004.

Chew, W. C., Jin, J.-M., Michielssen, E., and Song, J. M.: Fast and efficient algorithms in computational electromagnetics, Artech House, Boston, London, 2000.

DIN EN 45003 (Akkreditierungssysteme für Kalibrier- und Prüflaboratorien; allgemeine Anforderungen für Betrieb und Anerkennung), 1995.

International Civil Aviation Organization: DOC 8071, Manual on Testing of Radio Navigation Aids, Volume I: Testing of Ground-based Radio Navigation Systems, Montreal, 2000.

European cooperation for Accreditation of Laboratories, Publication Reference EAL-P7, EAL Interlaboratory Comparisons, Edition 1, March, 1996.

Harrington, R. F.: Field Computation by Moment Methods, Cazenovia, N.Y., USA, 1968.

Kraus, J. D.: Antennas, Mc-Graw-Hill, Boston, 1988.

Unger, H.-G.: Elektromagnetische Wellen auf Leitungen, 4. Aufl., Heidelberg, 1996.

4

Investigation of the behavior of protection elements against field radiated line coupled UWB-pulses

R. Krzikalla and J. L. ter Haseborg

Institute of Measurement Technology and Electromagnetic Compatibility, Hamburg, Germany

Abstract. To protect electronic systems against electromagnetic interferences in general nonlinear protection circuits are used. These protection circuits are optimized mostly against special transient interferences such as lightning electromagnetic pulses (LEMP) or electromagnetic pulses caused by nuclear explosions (NEMP). Previous investigations have shown that these protection elements could be undermined by so-called ultra wideband (UWB) pulses. Thereby a direct charge of the UWB-pulse to the elements has been assumed. This assumption was a worst case approximation because in practice UWB-pulses only get into systems by coupling effects. In this investigation the behavior of typical nonlinear protection elements has been tested with field radiated line coupled UWB-pulses. For that line coupled UWB-pulses have been defined depending on the coupling behavior of typical electronic systems and a possibility of generation of this kind of pulses is presented. After it typical nonlinear protection elements such as spark gaps, varistors and protection diodes have been tested with the previously defined test pulses. Finally the measured behavior of the elements has been compared with the behavior by direct charged UWB-pulses and the protection effect of the elements against field radiated line coupled UWB-pulses is re-evaluated.

1 Introduction

The protection of electronic systems against external electromagnetic interferences is very important for an undisturbed functionality of the system. Especially in security relevant systems the requirements for a sufficient protection will be clear because in worst case a system breakdown would be very expensive or could threaten human life. There are sufficient protection circuits available which protect systems against natural interferences, such as lightning electromagnetic pulses (LEMP) or electrostatic discharges (ESD), or against man-made noise just like electromagnetic pulses caused by nuclear explosions (NEMP). In current investigations high amplitude signals with an extremely broad frequency spectrum became more and more important. These so-called ultra wideband (UWB) pulses can be described in time domain by a double exponential time characteristic with a rise time shorter than 150 ps and a pulse width of a few nanoseconds. Previous investigations have been demonstrated that first these kinds of extremely fast transients have destructive effects on electronic systems and second they could undermine traditional nonlinear protection elements (Krzikalla et al., 2003). In that case the behavior of traditional protection elements such as spark gaps, varistors and protection diodes have been determined by using directly the double exponential UWB-pulses. But this assumption is only a worst case approximation to get an idea of the behavior of the elements in a first approach. In practice the disturbance of a system is mostly caused by field radiated interferences which couples into the system (Sabath et al., 2002). Therefore in this investigation the behavior of protection elements has been described by impact with line coupled interferences caused by field radiated UWB-pulses. For this different test pulses have been defined which occur due to the coupling behavior of typical electronic systems. With these test pulses the response behavior of traditional protection elements has been determined which are often used in protection circuit against electromagnetic interferences such as LEMPs of NEMPs. But also new protection elements have been tested which can be used directly against double exponential UWB-pulses (Krzikalla et al., 2003). The results of the measured response behavior have been compared and evaluated with the behavior by impact of direct double exponential UWB-pulses.

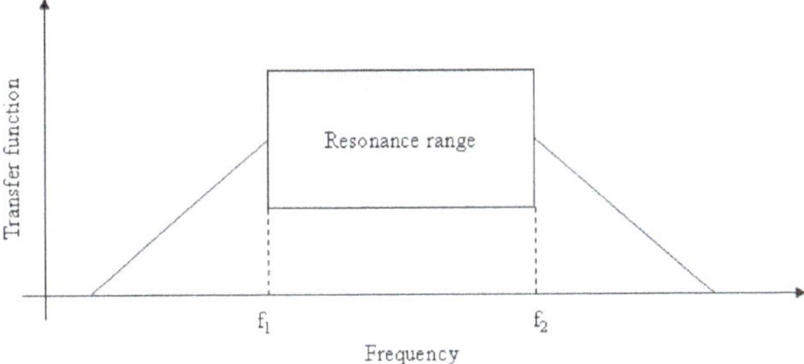

Fig. 1. General coupling behavior of a dipole structure.

2 Definition and generation of the test pulses

As mentioned before the disturbing interferences result from coupling effects of field radiated UWB-pulses into the system. To determine the time characteristic of the resulting interference analytically the transfer function of the system has to be known. As described by Nitsch (2005) the transfer function of a system with only a few different coupling paths can be very extensive. For large complex systems such as steering-, control- or computer-systems a transfer function only can be approximated. But this approximation can be done with the fact that the transfer function can be modeled by the transfer behavior of a dipole structure (Nitsch, 2005). The general transfer characteristic of a dipole structure is depicted in Fig. 1 (Baum, 1992). The characteristic corner frequencies f_1 and f_2 can be calculated with Eqs. (1) and (2) and they are related to the minimal and maximal coupling path lengths of the system $l_{\min}$ and $l_{\max}$:

$$f_1 = \left(\frac{3}{4}\right) \cdot \frac{c}{4 \cdot l_{\max} \cdot \sqrt{\varepsilon_r \mu_r}} \tag{1}$$

$$f_2 = \left(\frac{3}{2}\right) \cdot \frac{c}{2 \cdot l_{\min} \cdot \sqrt{\varepsilon_r \mu_r}}. \tag{2}$$

Below the lower corner frequency f_1 the system shows differential behavior and above the upper corner frequency f_2 the system has an integral behavior. The order depends mainly on the number of possible coupling paths into the system. Between both corner frequencies there is the resonance range of the system which has been assumed as constant over the frequency in the following.

For the generation of the test pulses it is necessary to model the previously described coupling behavior. To model the coupling behavior of different electronic systems the lower and upper corner frequency should be variable over a given range. The lower corner frequency is variable between 30 MHz and 1 GHz and the upper one can be changed between 300 MHz and 3 GHz. With Eqs. (1) and (2) these frequency ranges include systems with a maximal coupling path length of about 1.5 m(e.g. the length of a connection cable to a PC system) and a minimal coupling path length of about 10 cm (size of a typical handheld system). For that investigation an UWB-pulse-generator in combination with different

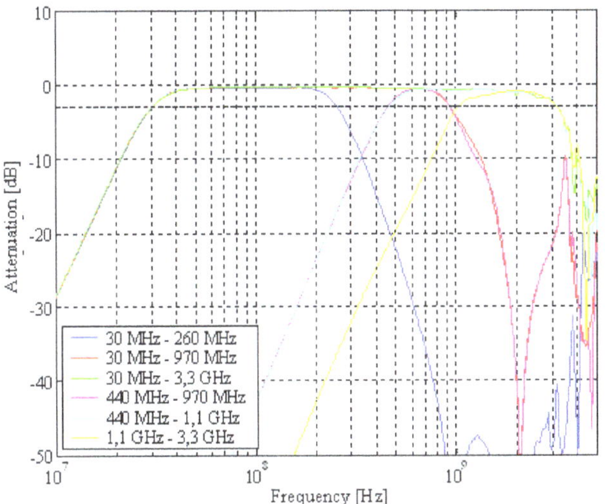

Fig. 2. Measured transfer functions in frequency domain of the realized linear filters to model the coupling behavior of complex electronic systems.

developed high- and lowpass filters has been used to generate the wanted test pulses. Figure 2 shows the transfer functions of the realized filter combinations in frequency domain. The behavior of the combination of high- and lowpass filters is very close to the described behavior of dipole structures. From the different combinations of high- and lowpass filters three representative combinations have been chosen to determine the response behavior of typical nonlinear protection elements by field radiated line coupled UWB-pulses. At first there is the whole frequency range from 30 MHz to 3 GHz, second a lower frequency range from 30 MHz to 300 MHz and at last an upper frequency range from 1 GHz to 3 GHz. According to Fig. 3 the test pulses have been generated with the chosen filter combinations in connection with an UWB-pulse-generator and a load. The generator produces double exponential pulses with a rise time of about 150 ps, a pulse width of 2.5 ns and a voltage amplitude of 12.5 kV. The combination of the UWB-pulse-generator and the high- and lowpasses generates the signal which occurs on the line due to coupling effects into system caused by field radiated UWB-pulses. To characterize the nonlinear behavior of the protection elements in time domain attenuators have been used in the signal path to provide voltage amplitudes of the test pulses above and below the breakdown voltage of the nonlinear elements. The load models the system which has to be protected. To avoid additional reflections on the line the load has been set to 50 Ω. In the presented measurement setup the test pulses have to be measured with a probe which provides a very broad frequency range. For that broadband current probes (e.g. F-2000 by FCC Fischer) or the new broadband high voltage probe picoTEM, developed by the Institute of Measurement Technology and Electromagnetic Compatibility at the Hamburg University of Technology (Weber, 2004)

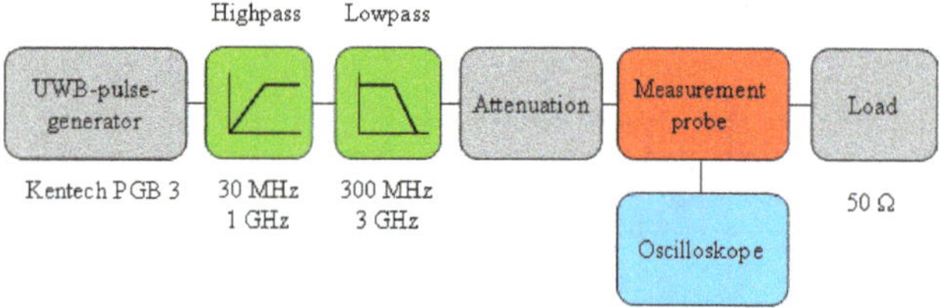

Fig. 3. Measurement setup to determine the time characteristic of the defined test pulses.

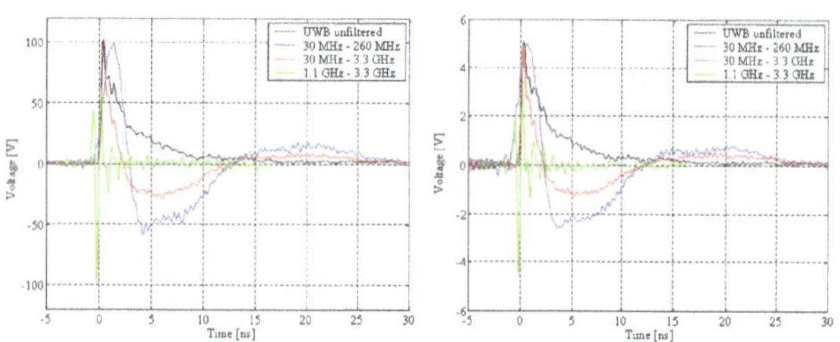

Fig. 4. Time characteristics of the generated test pulses for different coupling bandwidths and different voltage amplitude compared to a direct double exponential UWB-pulse.

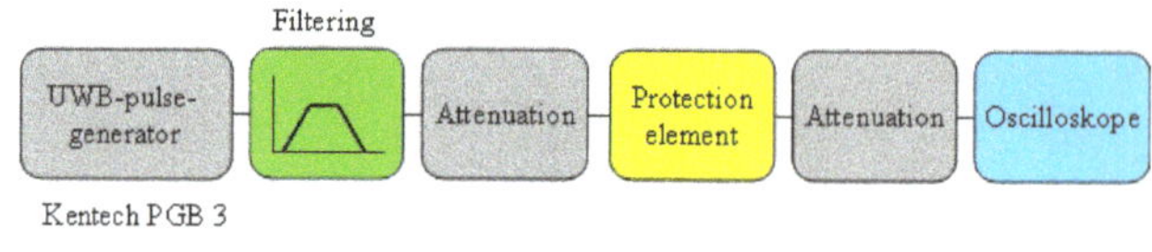

Fig. 5. Simplified test setup to determine the response behavior of nonlinear protection elements.

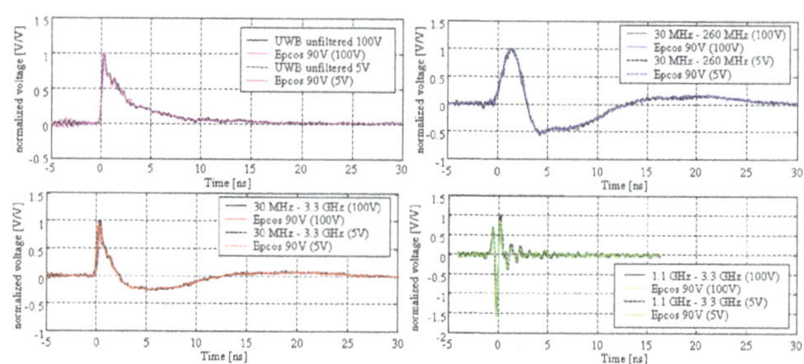

Fig. 6. Response behavior of the spark gap EPCOS90V normalized to the voltage amplitude of the respective interfering test pulse.

can be used. The measured test pulses in time domain are shown in Fig. 4 for different coupling bandwidths and different output voltages. For comparison the direct double exponential UWB-pulse is also shown in this figure.

3 Investigations of protection elements in time domain

In the following the response behavior of typical nonlinear protection elements will be determined by impact with the previously defined test pulses. In general typical protection circuits are combinations of spark gaps, varistors and protection diodes. For this investigation often used components have been chosen which have breakdown voltages between the amplitudes of the defined test pulses of 5 V and 100 V. Additionally a very fast protection diode with a very low parasitic capacitance has been used in the test. The following listed elements have been tested with respect to their response behavior:

- spark gap: EPCOS90V
- varistor: S14K30
- protection diode: BZX85C62
- fast protection diode: GBLC15C

A simplified test setup to measure the response behavior is shown in Fig. 5. The 50-Ω-load has been replaced directly by an oscilloscope. Additional attenuators have been used to protect the input port of the oscilloscope from over voltages. Now the obtained measurement results of the single protection elements will be discussed in detail. For a better comparison of the response behaviors of the elements at different voltage amplitudes at the same pulse shape the measured response behavior has been normalized to the voltage amplitude of the corresponding interfering pulse. If both normalized response behaviors have the same time characteristic the element shows at voltages above and below their breakdown voltage the same behavior. In that case the element has no nonlinear behavior and reacts only due to their parasitic linear elements. If the normalized response behaviors have different pulse shapes the element shows different behaviors at voltages above and below their breakdown voltage. This effect is typical for nonlinear behavior.

3.1 Spark gaps

The behavior of spark gaps is determined by the so-called breakdown characteristic which describes the relation between the breakdown voltage of the spark gap and the rise time of the interfering signal. Mostly spark gaps were used in lightning protection applications. Lightnings have rise times in the microsecond range and voltage amplitudes of some kV. The test pulses in this investigation have rise times in the subnanosecond range with maximal amplitudes up to 100 Investigation of the behavior of protection elementsV. Therefore it has not been expected that spark gaps show any reduction behavior on line coupled UWB-pulses. This has been verified by measurements which are shown in Fig. 6. Almost the normalized response behaviors fit exactly with the normalized interfering test pulses. So no protection effect is provided by spark gaps against field radiated line coupled UWB-pulses.

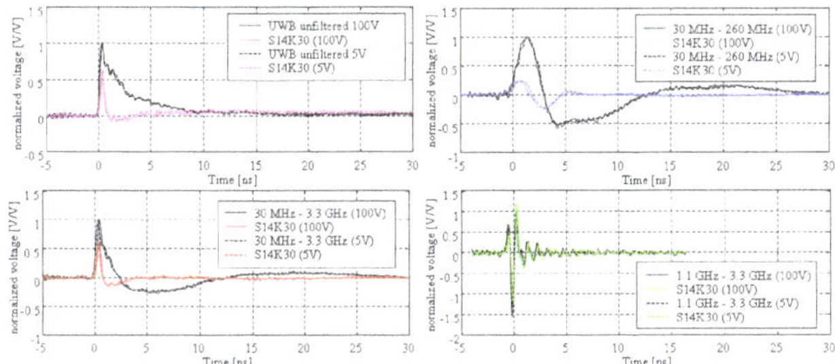

Fig. 7. Response behavior of the varistor S14K30 normalized to the voltage amplitude of the respective interfering test pulse.

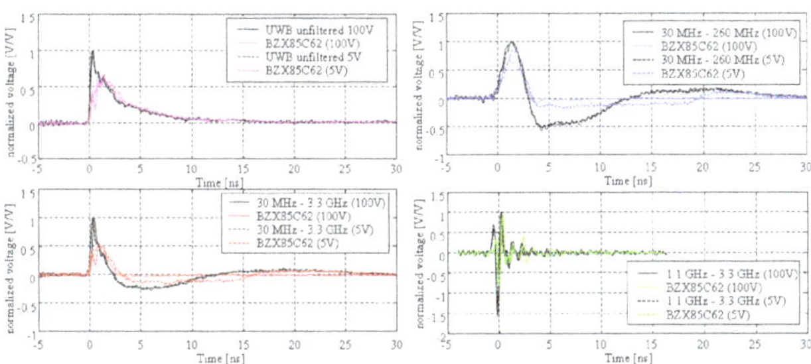

Fig. 8. Response behavior of the protection diode BZX85C62 normalized to the voltage amplitude of the respective interfering test pulse.

3.2 Varistors

Just as spark gaps varistors are used as protection elements due to their high energy absorption capability. But the high energy absorption capability is accompanied with a quite large parasitic capacitance of the varistor, whereby varistors only can be used as protection elements in low frequency applications. The measured and normalized response behavior of the chosen varistor is depicted in Fig. 7. Except to the upper frequency coupling range the varistor shows a clear response behavior to the interfering test pulses. However at both test pulses with different voltage amplitudes the normalized response behavior has exact the same pulse shape. As mentioned before in this case the varistor only reacts due to its parasitic linear elements. In fact there is a reduction of the voltage amplitude of the interfering test pulses but this reduction is not systematic and unintentional at both amplitudes of the defined test pulses. That also leads to an influence to a signal with amplitudes below the clamping voltage of the varistor.

3.3 Protection diodes

Protection diodes which have in contrast to spark gaps and varistors a much smaller energy absorption capability show again their typical nonlinear behavior at field radiated line coupled UWB-pulses (compare to Fig. 8). At the coupling range from 30 MHz to 300 MHz and from 30 MHz to 3 GHz

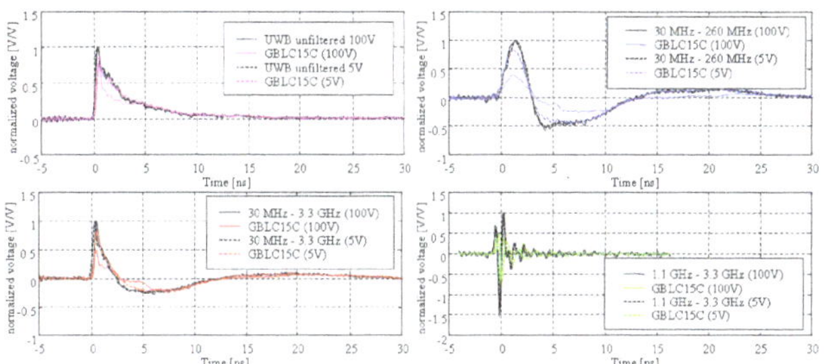

Fig. 9. Response behavior of the low capacitance protection diode GBLC15C normalized to the voltage amplitude of the respective interfering test pulse.

the normalized response behaviors of the protection diode show different pulse shapes which points out the typical nonlinear behavior of the diode. In the upper coupling range from 1 GHz to 3 GHz the tested diode does not show any voltage reduction effects. But at the other two test pulses a reduction of the voltage of the interfering pulses can be observed if the voltage amplitude of the pulses is higher than the breakdown voltage of the diode. Compared to the behavior of the diode at a direct double exponential UWB-pulse the diode shows its typical nonlinear behavior at field radiated line coupled UWB-pulses. However this behavior is highly related to the coupling bandwidth of the system which has to be protected.

3.4 Protection diodes with an extremely small parasitic capacitance

Because traditional protection diodes with a small parasitic capacitance have shown better behavior concerning field radiated line coupled UWB-pulses special protection diodes have been tested with an extremely small parasitic capacitance. The normalized measurements are shown in Fig. 9. In contrast to normal protection diodes this diode shows also a voltage limitation of the interfering test pulse at the highest coupling range from 1 GHz to 3 GHz and at the direct double exponential UWB-pulse. At each coupling range the diode shows the typical nonlinear behavior. At voltages above the breakdown voltage of the diode the voltage has been reduced and at voltages smaller that the breakdown voltage of the diode the signal has not been influenced significantly. With these kinds of fast protection diodes a sufficient protection element is available against field radiated line coupled UWB-pulses at different coupling bandwidths as well as against direct coupled double exponential UWB-pulses.

4 Conclusions

This investigation determines the response behavior of typical nonlinear protection elements concerning interfering signals which are caused by coupled field radiated UWB-pulses into complex electronic systems. For that a possibility was presented to generate field radiated line coupled UWB-pulses depending on the coupling bandwidth of the system which has to be protected. With different generated test pulses typical protection elements such as spark gaps, varistors and protection diodes have been tested. But also special protection diodes with extremely small parasitic capacitances have been used in this test. It could be observed that spark gaps did not show any effects at the defined test pulses. The tested varistor has shown a clear reduction of the interfering test pulses but the observed effect was not of nonlinear nature but only caused by the quite high parasitic linear elements of the varistor. Normal protection diodes have shown typical nonlinear behavior again in case of test pulses with lower frequency parts. At the test pulses with frequencies in the upper range the normal diodes also have been react mainly due to their parasitic linear elements. But special protection diodes with an extremely small parasitic capacitance have shown a very good response behavior at all kinds of test pulses so that they can be used against all kinds of UWB-interferences. Concluding it can be said that also field radiated line coupled UWB-pulses can undermine traditional protection circuits. This is mainly related to the coupled frequency parts of the field radiated UWB-pulse into the system. However with special low capacitance diodes the design of sufficient protection circuits should be possible.

References

Baum, C. E.: From the Electromagnetic Pulse to High-Power Electromagnetics, Proc. of the IEEE, USA, 80, 1992.

Krzikalla, R. and ter Haseborg, J. L.: HPEM protection on HF transmission lines, Kleinheubacher Tagung, Miltenberg, 2003.

Nitsch, D.: Die Wirkung eingekoppelter ultrabreitbandiger elektromagnetischer Impulse auf komplexe elektronische Systeme, Dissertation, Univ. of Technology Hamburg-Harburg, 2005.

Sabath, F.: UWB – Antenne für verbringbare Wirksysteme, Wehrtechnisches Symposium, Mannheim, 2002.

Weber, T.: Messverfahren und Schutzmaßnahmen für Elektromagnetische Pulse im UWB – Bereich, Dissertation, Univ. of Technology Hamburg-Harburg, 2004.

5

Considerations on radar localization in multi-target environments

H. Rabe[1], E. Denicke[1], G. Armbrecht[1], T. Musch[2], and I. Rolfes[1]

[1]Leibniz Universität Hannover, Institut für Hochfrequenztechnik und Funksysteme, Appelstr. 9a, 30167 Hannover, Germany
[2]Ruhr-Universität Bochum, Lehrstuhl für Elektronische Mess- und Schaltungstechnik, Universitätsstr. 150, 44801 Bochum, Germany

Abstract. In a multitude of applications like e.g. in automotive radar systems a localization of multiple passive targets in the observed area is necessary. This contribution presents a robust approach based on trilateration to detect point scatterers in a two-dimensional plane using the reflection and transmission information of only two antennas. The proposed algorithm can identify and remove ambiguities in target detection which unavoidably occur in certain target constellations in such a two-antenna configuration.

1 Introduction

In recent years, a number of applications for close-range imaging via millimeter wave radar sensors like automotive radar systems (Wenger, 1998) or through-wall imaging (Yang and Fathy, 2007) emerged for locating multiple targets in the environment. Current implementations often use either mechanical shifts of the antenna position to synthesize a virtually enlarged aperture thus exhibiting an increased lateral resolution or use electronically adjustable narrow beam antenna arrays in order to detect the position of passive radar targets in two or three spatial dimensions. However, these approaches suffer from the complexity in hardware equipment and signal processing (Yang and Fathy, 2007). Instead of using Synthetic Aperture Radar (SAR) or beam-steerable imaging, this contribution presents an approach related to the one described in Michael et al. (2000) which features a reduced system complexity by only using the information acquired from two fixed antennas for the detection of multiple scatterers in a plane. The wanted position information is reconstructed out of the complex reflection and transmission scattering parameters obtained over a certain bandwidth. The reconstruction is done by first extracting the dominant scattering centers out of all three data sets separately via Prony's method (Carriere and Moses, 1992) and combining them in a trilateration algorithm to allocate the scatterers according to their position in a plane. The advantage of this approach is the possibility to detect or even to remove ambiguities which inevitably appear as so called "ghost" targets (Helmbrecht and Biebl, 2005) in certain geometrical constellations of targets and antennas.

The paper is organized as follows. Section 2 describes the basic principle of point target reconstruction with the proposed trilateration approach. That section derives the accuracy in terms of the ambiguity area and presents a systematic approach for determining the occurrence of ghost targets. The Sect. 3 describes the implementation of the target reconstruction algorithm. Some characteristic simulation results are presented in Sect. 4 to confirm the expected properties of the algorithm using synthesized data sets. The paper ends with a conclusion.

2 Properties of point target reconstruction

The trilateration procedure for reconstructing point targets from the reflection and transmission scattering data of two antennas is done according to Fig. 1. The two transmit and receive antennas are situated on the x-axis of a local coordinate system at the positions $x_1=-20\lambda_c$ and $x_2=20\lambda_c$, where λ_c is the free-space wavelength at center frequency. The antennas ant_1 and ant_2 are assumed to radiate in the front half space which is valid for many applications so that the shaded area in the figure can be neglected. A point scatterer is situated at position $x=0\lambda_c$ and $y=20\lambda_c$ and is illuminated by antenna 1 and 2. The complex scattering parameters are extracted over the bandwidth B. The next step is to extract the Time Of Flight (TOF) information to the scatterer for both reflections and the transmission in order to calculate the distances r_1, r_2 and t_{21}. The propagation velocity is assumed to be the speed of light c_0. The TOF estimation can be done in

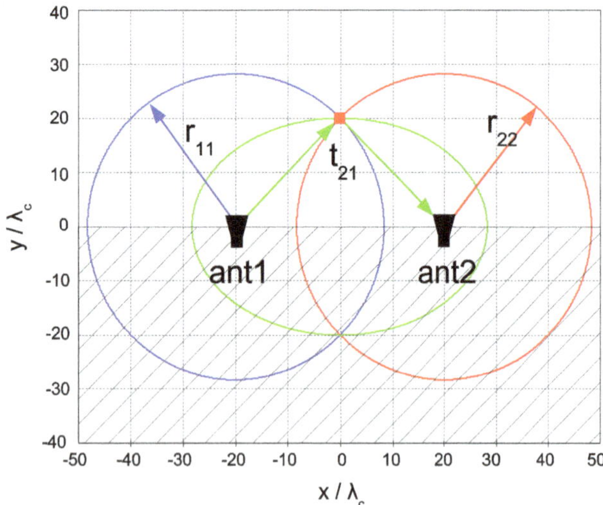

Fig. 1. Trilateration algorithm exploiting transmission and reflection information.

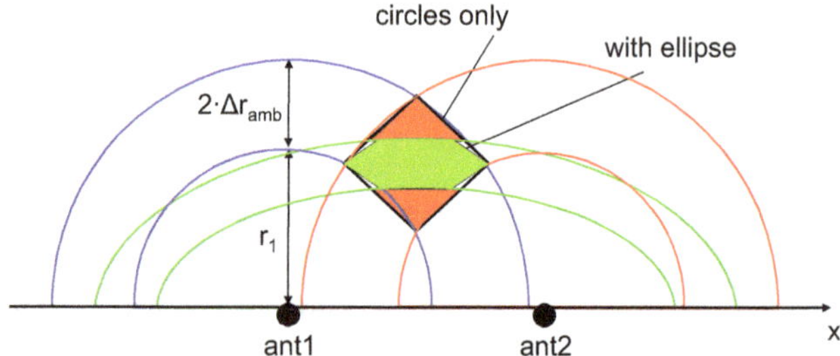

Fig. 2. Ambiguity area for intersection of circles only and in combination with transmission ellipses with radial uncertainty of $2\Delta r_{\mathrm{amb}}=0.1\lambda_c$.

several ways. We used Prony's method (Carriere and Moses, 1992). As depicted in Fig. 1 the point target can be reconstructed in the xy-plane by determining the intersection point between the two circles which determines the possible target position using the reflection data and the ellipse resulting from the TOF of the transmission signal[1].

In the following Sect. 2.1 the ambiguity area of a point target is derived in dependence of its position which is a direct measure for the precision of the reconstructed target location. Besides the ambiguity of the target position within a certain area other ambiguities arise in a scenario of multiple point scatterers. These ambiguities are called ghost scatterers. In Sects. 2.2 and 2.3 this kind of ambiguity will be examined. In some cases these ambiguities can be removed or at least be detected which will be explained in Sect. 3.

2.1 Ambiguity area of a single scatterer

The reconstruction accuracy of a point scatterer using the trilateration method depends on the SNR (Signal to Noise Ratio) of the collected data. The noise power causes an uncertainty in the estimation of the TOF values. Assuming a distance uncertainty of $2\Delta r_{\mathrm{amb}}$ applied to both reflection and transmission data the intersection point expands to an area wherein the real scatterer position is located. The ambiguity area for a scatterer directly placed in front of the antennas is shown in Fig. 2. The closest blue circle results from the real scatterer position and the second one is the result obtained with an uncertainty $2\Delta r_{\mathrm{amb}}$. By applying the same uncertainty on the right antenna four intersection points will occur between the two blue and the two red circles marking the edges of a rhombus which is approximately equivalent to the ambiguity area. The area consists of two red triangles and the hexagon in the middle of the area. Considering the transmission information with the same uncertainty of $2\Delta r_{\mathrm{amb}}$ the ambiguity area is shrinked to the green hexagon. As mentioned before, the area is only an approximation which gets worse in target positions near the x-axis due to the flat angles between the intersection of the red with the blue circles.

To get an impression on the expected precision for target reconstruction the ambiguity area has been examined in dependence on the scatterer position in the xy-plane. Assuming an uncertainty of $2\Delta r_{\mathrm{amb}}=0.1\lambda_c$, the values for the ambiguity area A related to λ_c^2 are given in Fig. 3. Due to the symmetry of the scenario only the results for positive values of x are displayed. The antennas are placed as before at $x_1=-20\lambda_c$ and $x_2=20\lambda_c$. The lines of same ambiguity area size are highlighted by the gradation. The lines of same area size exhibit shapes similar to ellipses. With increasing distance of the scatterer in x- and y-direction the ambiguity area enlarges. Especially for scatterers close to the x-axis the ambiguity area can reach big values. Therefore the proposed system works best in small distances from the antenna position and in noticeable distance from the x-axis. A further analysis has shown that a larger antenna distance leads to smaller ambiguity areas. This is obvious since a variation of the antenna distance can be regarded as a scaling of the scenery. At first sight it seems to be the best to separate the antennas in order to benefit from the shrinking ambiguity areas. However, with increasing antenna distance also the probability of ghost scatterers increases as shown in the following section.

2.2 Ambiguity in multi-target environment

Besides the ambiguity area of one scatterer another ambiguity arises by the presence of more than two point scatterers. The reconstruction results in a special constellation of three scatterers in a row along the y-axis is shown in Fig. 4. The figure shows some of the circles and ellipses that occur after processing the scattered signals. The real scatterers s_1, s_2 and s_3 are highlighted red and the ghost scatterers are marked

[1]The sum of the distances from ant_1 to the scatterer and back to ant_2 is constant when the scatterer is placed on an ellipse where the antennas correspond to its focal points.

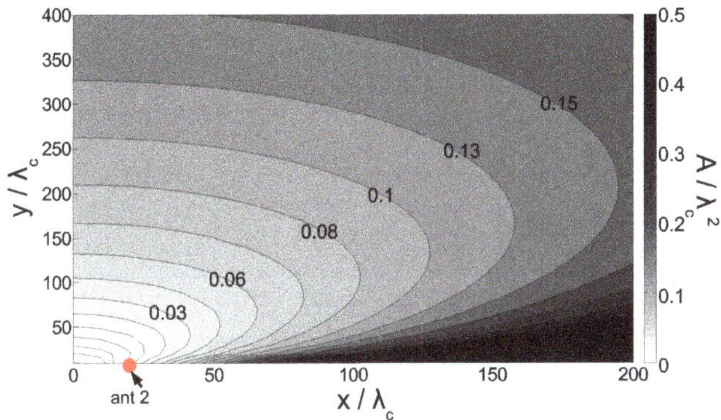

Fig. 3. Ambiguity area size in dependence of scatterer position with a radial uncertainty of $2\Delta r_{\text{amb}}=0.1\lambda_c$.

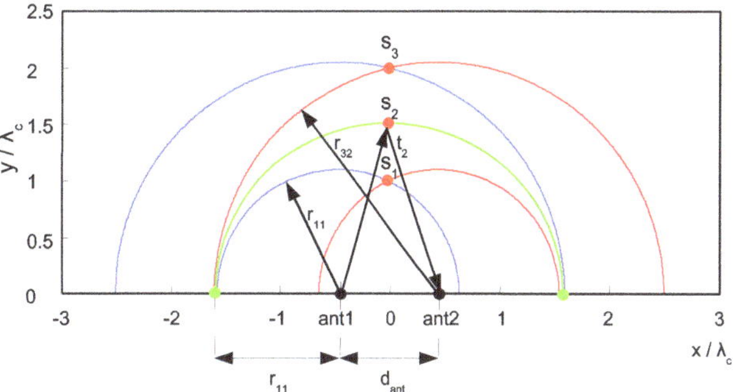

Fig. 4. Ghost scatterers resulting from intersections between two circles and one ellipse from three real scatterers.

green. The blue circle stands for the TOF between target s_1 and antenna 1 with the radius r_{11}. The first index denotes the target number and the second one is the antenna index. The circles r_{11}, r_{32} and the transmission t_2 (transmissions have a target index only due to channel reciprocity) generate two ghost targets on the x-axis. The occurrence of ambiguities depends on the scatterer distances as seen from the antennas. The condition in Eq. (1) for the radial distances of the scatterers from antenna 2 must be met to generate ambiguities on the x-axis.

$$
\begin{aligned}
r_{12} &= r_{11} \\
r_{32} &= r_{11} + d_{\text{ant}} \\
r_{22} &= \frac{t_2}{2} = r_{11} + \frac{d_{\text{ant}}}{2}
\end{aligned}
\qquad (1)
$$

The first scatterer distance can be arbitrarily chosen on the y-axis. Due to the symmetry of the problem the radius r_{12} is the same as r_{11}. If the radial distances of the following two scatterers increase in steps of $d_{\text{ant}}/2$ two ghost targets will occur symmetrically to the y-axis on the x-axis. Decreasing the radial distance below $d_{\text{ant}}/2$ causes the intersection point to move to positive values of y. Increasing the distance will otherwise cause no more intersections and therefore the generation of ghost targets will be avoided. This statement is also valid for other scatterer constellations generating a ghost target on the x-axis. These constellations can be achieved by moving the scatterers along the circles and the ellipse in Fig. 1. All these constellations offer at least one radial distance between the scatterers and the antennas that is closer than $d_{\text{ant}}/2$, so the minimal radial distance of the scatterers is $d_{\text{ant}}/2$ to avoid ambiguities and can be achieved by aligning the scatterers on the y-axis. All other constellations leading to ghost targets above the x-axis need a radial spacing of the scatterers of less than $d_{\text{ant}}/2$ so ghosts can be completely avoided by scatterer spacing larger than $d_{\text{ant}}/2$. If the transmission and reflection signals allow a reconstruction in range cells with the size of $d_{\text{ant}}/2$ or larger there would not be an ambiguity. Utilizing for example a Fourier transform to get the band-limited impulse response will result in the well known resolution cells that have a minimum size depending on the bandwidth B and the propagation velocity c_0 according to Eq. (2).

$$
\Delta r_{\text{res}} = \frac{c_0}{2B} \qquad (2)
$$

To avoid the occurrence of ambiguities the value of Δr_{res} has to be larger than $d_{\text{ant}}/2$ according to the preceding explanations. This relationship is shown in Eq. (3) and can be regarded as a design criterion for the antenna distance for a given signal bandwidth to suppress ghosts completely.

$$
d_{\text{ant}} \leq \frac{c_0}{B} \qquad (3)
$$

The expression can vary with different window functions in the frequency domain before applying the Fourier transform or with the beamwidth of the used antennas. The window function causes an increase of the size of a resolution cell. By reducing the beamwidth to less than in the omnidirectional case the ambiguities occurring in the area which is not covered by the antenna's beam can be removed by plausibility. Equation (3) can therefore be regarded as a worst case without windowing and applying an omnidirectional beam.

2.3 Higher order ambiguities

It was shown how ghost targets arise and how they can be avoided. However, in some cases the choice of a small antenna distance to avoid such ambiguities probably does not fulfill the requirements of reconstruction accuracy, as a small antenna distance leads to bigger ambiguity areas as explained in the preceding Sect. 2.1. If the number of scatterers is known in a certain application, an upper bound for the number of ghost scatterers can be determined that can assist the reconstruction algorithm to limit the detected scatterers to a physically plausible value as explained in Sect. 3. By adding one more target to the scene in Fig. 4 on the y-axis in a distance less or equal to $d_{\text{ant}}/2$ from target s_3 relative to the antenna positions, further two ambiguities will be generated symmetrically to the y-axis. The situation gets more complicated if the number of scatterers approaches five or more

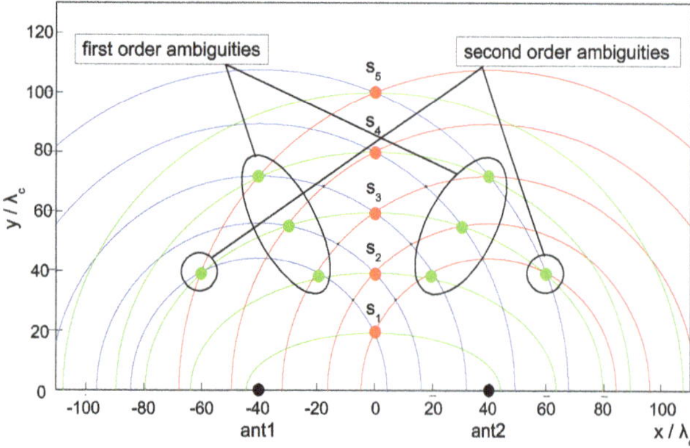

Fig. 5. Occurrence of higher order ghost scatterers in a scenario of five real scatterers.

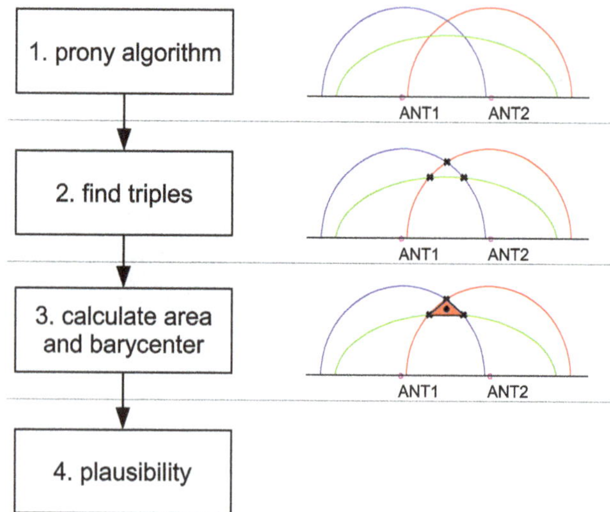

Fig. 6. Algorithm steps for point target reconstruction in a plane.

and the distance between the first and the last scatterer stays $d_{\text{ant}}/2$. In the case of five scatterers not only the neighboring scatterers cause ambiguities as shown in Fig. 5. Besides the expected ambiguities formed by the adjacent scatterers that now move to positive y-values due to the smaller scatterer distances there will occur two more ambiguities named "second order" ambiguities. These new ghosts are formed by the information r_{11}, t_3 and r_{52} which now satisfy the condition of a distance smaller than $d_{\text{ant}}/2$ and therefore lead to ambiguities as in the three scatterer scenario above occurring again on the x-axis. It can be shown easily that seven scatterers lead to "third order" scatterers, nine scatterers lead to "fourth order" scatterers and so on if the radial distances between the outer scatterers stay below $d_{\text{ant}}/2$. Assuming no restriction in the system's resolution an upper limit for the ghost scatterers N_g can be derived in dependence of the number of the real scatterers N_s and the maximal order O of the scenario according to Eq. (4).

$$N_g = 2 \cdot \sum_{n=3}^{2O+1} (N_s + 1 - n) \quad \text{for} \quad n = 3, 5, ..., 2O + 1$$
$$O = \frac{d_{\text{ant}}}{2x_s} \tag{4}$$

The maximal order O is the largest uneven value in $d_{\text{ant}}/2x_s$ where x_s is the radial distance between neighboring scatterers. This formula is valid for equally spaced scatterers aligned on the y-axis with a minimal distance of $d_{\text{ant}}/2$ between the nearest and the farthest scatterer and is an upper limit for the occurring ambiguities. The order can be reduced by lowering the antenna distance or increasing the scatterer distance. In Sect. 4 N_g is displayed for maximal order in Fig. 8 up to N_s=15. The number of ghost scatterers is increasing exponentially with higher values of N_s. In Sect. 4 the benefit of knowing the upper limit for ambiguities is discussed in order to reduce the number of detected targets.

3 Algorithm

The results from the previous section have been integrated as a part of the presented algorithm for point target reconstruction. The single steps of the algorithm are shown in Fig. 6. As an example, the reconstruction is explained for one point scatterer. At first the complex scattering parameters in the frequency domain of the two reflections and the transmission are inserted into a Prony algorithm according to Carriere and Moses (1992). This parametric approach directly extracts the scatterer distances for the three parameter sets. The parameters are obtained by approximating the signal in the frequency domain as a sum of weighted exponential functions corresponding each to a single scattering center. The approximation is done in a least squares sense. The Prony's method offers good results at high SNR. At low SNR (20 dB and lower) however, other algorithms like peak detectors applied on the time domain signal obtained by a Fourier transform have offered more robustness. In the given example the reflection and the transmission information will each contain one value corresponding to the only scatterer in the scenario. These values are inserted into an analytical expression calculating the intersection points between the blue circles from antenna 1 and the transmission ellipse. In contrast to the reconstruction using perfect data the intersection points of all three geometries do not hit a single point but will expand to three points accordingly as displayed in the figure when using noisy data. Hence, the corresponding triples belonging to one point scatterer have to be found in step 2 using a nearest-neighbor method. Furthermore, a barycenter is calculated out of the spanned triangle as an approximation for the real target position in step 3. Due to the imperfections of point scatterer reconstruction it is sometimes not possible to find the correct triple of one scatterer using the nearest-neighbor

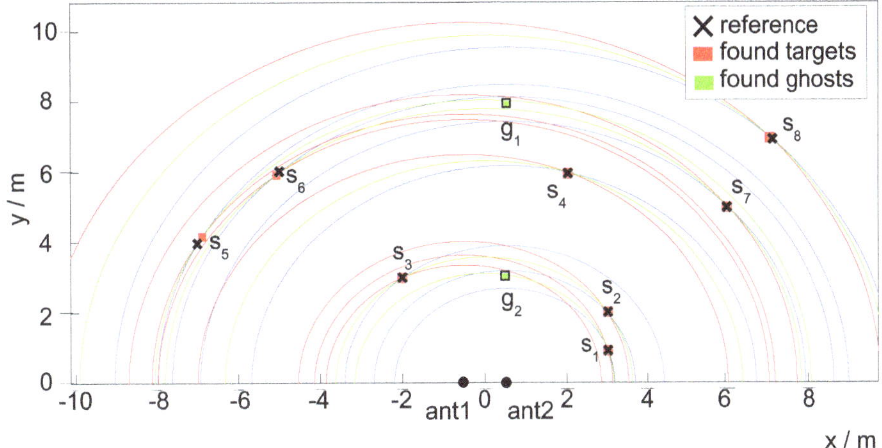

Fig. 7. Example of the reconstruction of 8 scatterers using the trilateration method with plausibility check.

method. Likewise the probability to detect ghost scatterers increases as an exact intersection of both circles and the ellipse is no more necessary for a scatterer detection. Therefore a plausibility check is performed at the end of the algorithm in step 4.

If the number of the found triangles is larger than the sum of the upper limit for the number of ghost scatterers (see Eq. 4) plus the number of real scatterers (the number of real scatterers is given by the number of targets detected by the Prony algorithm) the algorithm has detected more scatterers than physically possible for the current scenario. The algorithm therefore removes the biggest triangles until the upper limit is reached. A further reduction can be achieved by removing triangles with ambiguous information. If a triangle is constructed by either a circle or an ellipse only crossing a single scatterer then the point belonging to the spanned triangle is unambiguous. An example is depicted in Fig. 4 where the ghost targets on the x-axis consist of the circles and the ellipse of the real targets. The real targets all have at least one circle or ellipse that is uniquely used for their scatterer reconstruction and can therefore be marked as "real" target. This information allows to identify real scatterers in the whole pool of possible scatterers. This principle does not work with all configurations as the position of a ghost target can for example coincide with a real scatterer which is then marked as ghost target. The next section shows some simulation results showing the detection capability of the algorithm.

4 Simulation results

To verify the capability of target reconstruction using the trilateration algorithm a scenario with 8 scatterers is chosen which are placed in a rectangular field of the size 20 m×10 m. In Fig. 7 the 8 scatterers are placed at the positions marked as black crosses. The antennas are placed in a distance of 1 m symmetrical to the origin. The scattering parameters have been determined over a bandwidth of 2 GHz. To obtain realistic values a white noise signal was added to the scattering data. The result for an SNR of 35 dB is shown by the red and green rectangles which mark the reconstructed scatterer positions. The red rectangles have at least one unique information so they can be regarded as real scatterers. The green rectangles mark the scatterers which only use information that is already used for other scatterers. The scatterer g_2 (g stands for "ghost") for example is generated by the ellipse from s_1, the blue circle from s_2 and the red circle from s_3. As explained in Sect. 2.2 the ambiguities originate from a radial scatterer distance of less than $d_{\text{ant}}/2$ which is true for the scatterers s_1, s_2 and s_3. The same situation leads to the ghost g_1 which is formed by the scatterers s_5, s_6 and s_7. Although these scatterers have a large distance among each other, they are separated very closely as seen from the antennas. In contrast the scatterers s_4 and s_8 are separated with a higher radial distance so they do not generate ambiguities. The figure also gives a hint regarding the accuracy of the reconstruction. As shown in Sect. 2.1 the accuracy degrades with higher radial distance to the antennas as can be seen for scatterer s_8 where the rectangle is a slightly shifted out of the reference position. Another degradation can be seen at scatterer s_5 at approximately the same radial distance but a position closer to the x-axis. These conditions lead to a stronger deviation of the reconstructed point to the reference position due to the larger ambiguity area.

In the example of Fig. 7 the number of ghost scatterers is much less than the theoretical limit of 24 as is shown in Fig. 8. The black curve shows the relation between the number of scatterers N_s and the maximum number of ghosts N_g for this case. The blue and the red curve are the results for the found ghosts by placing the scatterers randomly in the scenery for two different adjustments in the algorithm. By regarding the deviation of both curves from the black curve it is obvious that the theoretical limit of ghost scatterers is

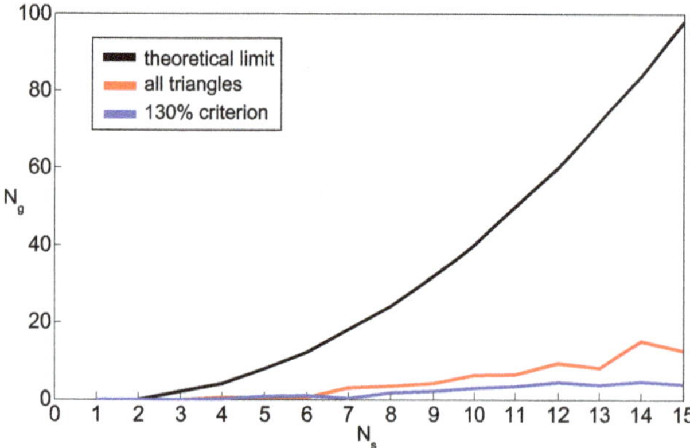

Fig. 8. Number of ghost targets in dependence of real scatterers.

seldom reached in scenarios with randomly spaced scatterers. For this reason the theoretical limit is not a reasonable choice for reducing the number of triangles found by the algorithm as expected in Sect. 2.3. The red curve shows the result of the algorithm by considering all triangles found in the nearest-neighbor search and is significantly lower than the theoretical limit. In order to reduce the ambiguities due to the number of triangles found by the nearest-neighbor algorithm beside the real scatterers and the ghosts, a limit for the number of triangles has been introduced to 130% of the scatterers resolved by the Prony algorithm. Only the smallest triangles are considered which reduces the number of ghost scatterers. The value of 130% has been found to be a good compromise between the number of reconstructed ghost targets and the certainty of reconstructing all physically available scatterers.

The simulations showed that the algorithm detects ghost targets as expected. Due to the generation of further ambiguities by the search for triangles it is useful to reduce the number of detected scatterers. The proposed algorithm showed good results with a fixed limit of the resolved triangles.

5 Conclusions

A trilateration algorithm was introduced that reduces the problem of ghost scatterers in a two-antenna setup for imaging purposes in the plane. The basis of this approach is the systematic description of the generation of ghost targets in dependence of the scatterer placing and the choice of the antenna distance. It can be concluded that a two-antenna setup provides imaging in a tradeoff between reconstruction precision and the occurrence of ghost targets. It was shown that by reducing the system's resolution and under-run a certain antenna distance on the one hand leads to a complete elimination of ghosts. On the other hand a possibility of detecting ghost scatterers was presented that provides a save detection of at least some of the real targets in the scenery without limitation in resolution or antenna distance allowing the enhancement of the reconstruction accuracy by choosing high antenna distances.

References

Carriere, R. and Moses, R.: High resolution radar target modeling using a modified Prony estimator, IEEE T. Antenn. Propag., 40, 13–18, 1992.

Helmbrecht, E. and Biebl, E.: Radar imaging using noncoherent sensors, Proc. IEEE MTT-S International Microwave Symposium Digest, 4 pp., 2005.

Michael, B., Menzel, W., and Gronau, A.: A real-time close-range imaging system with fixed antennas, IEEE T. Microw. Theory, 48, 2736–2741, 2000.

Wenger, J.: Future Trends in Automotive Radar/Imaging Radar, Proc. 28th European Microwave Conference, 1, 636–664, 1998.

Yang, Y. and Fathy, A.: Design and Implementation of a Low-Cost Real-Time Ultra-Wide Band See-Through-Wall Imaging Radar System, Proc. IEEE/MTT-S International Microwave Symposium, 1467–1470, 2007.

6

Influence of GSM900 electromagnetic fields on the metabolic rate in rodents

A. El Ouardi[1], J. Streckert[1], A. Lerchl[2], K. Schwarzpaul[2], and V. Hansen[1]

[1]Chair of Electromagnetic Theory, University of Wuppertal, 42097 Wuppertal, Germany
[2]School of Engineering and Science, Jacobs University Bremen gGmbH, 28759 Bremen, Germany

Abstract. The development of exposure devices for investigating possible effects of mobile communication systems to non-restrained animals aims at a homogenous field distribution in the area the animals occupy. In the presented 900 MHz exposure device a quite good field homogeneity of 5% (including the standing wave contribution due to internal reflections) is reached in the cage region mainly by flattening the transverse field. For the standard waveguide (WR1150) without dielectric sheets this value reads 14%. The desired maximal whole body specific absorption rate (SAR) of 4 W/kg in the Djungarian hamster model is achieved at an input power of only 3.7 W.

1 Introduction

Several experiments were performed in the last few years in order to investigate possible biological effects of radio frequency (rf) signals in the non-thermal range, i.e. with a maximum rf-induced temperature increase fairly below 1°C. Some published results in the frequency region of mobile communication signals indicating body weight alterations might support the hypothesis that rf signals can influence the metabolic rate of rodents. Therefore, an experiment was set up by the Federal Office for Radiation Protection within the framework of the German Mobile Telecommunication Research Programme (BfS) to confirm or disapprove this thesis. This paper presents details of the engineering part of the project.

Four plastic cages (dimensions (width × height × length) in mm: 147×117×355), each housing up to three hamsters for one week, are exposed to a GSM test signal at 900 MHz. Air of defined speed (30–40 l/h) flows through the cage. In the supplied air the temperature and the concentration of oxygen and carbon dioxide is measured. The analysis of the data gives information about the metabolism of the animals. Due to the rf field whole body SAR values of 4 W/kg, 0.4 W/kg, and 0.08 W/kg are applied added by sham exposure phases (0 W/kg). The choice of the SAR is done randomly and in a blinded manner.

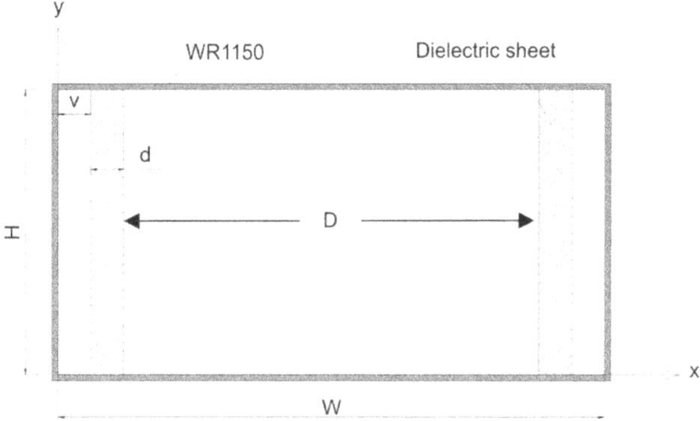

Fig. 1. Cross section of waveguide WR1150 with inserted dielectric sheets.

2 Exposure device

The field distribution in the volume of a cage should be homogeneous in order to obtain uniform exposure conditions for the animals regardless of their location. This can be achieved by inserting each cage into a rectangular waveguide (type WR1150: internal dimensions W=292.1 mm, H=146.05 mm) equipped with additional inner dielectric sheets (Fig. 1).

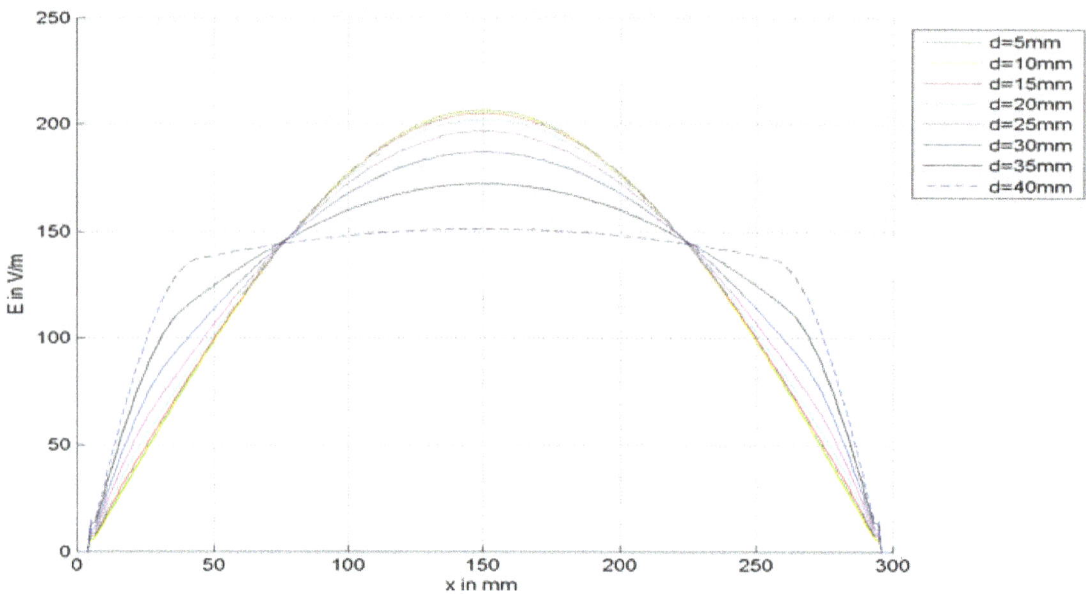

Fig. 2. Field distribution in transverse direction for v=0, ε_r=5, and different sheet widths d (P=1 W).

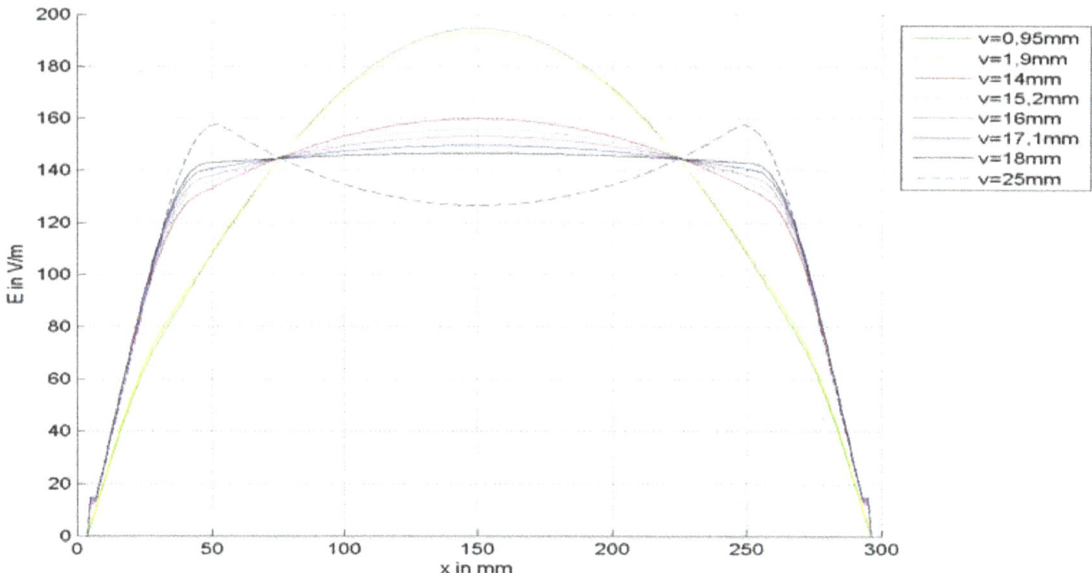

Fig. 3. Field distribution in transverse direction for d=25.4mm, ε_r=5, and different distances v (P=1 W).

2.1 Concept

In a rectangular waveguide with standard dimensions for the carrier frequency of 900 MHz only the fundamental H_{10}-wave can propagate. However, it has a sinusoidal transverse field. In order to flatten the field, two dielectric sheets of thickness d, distance D and height H are positioned symmetrically to the waveguide axis. In principal, by use of matters in waveguides the propagation of higher-order wave types can be provoked, because the dielectric material reduces their cut-off frequencies. This is also the case in the structure shown in Fig. 1. To overcome this drawback, the fundamental mode is excited selectively by using a transition from the empty metallic standard waveguide to the optimized structure with dielectric sheets. Both measures – flattening of the field and mode-selective excitation – are considered in the following.

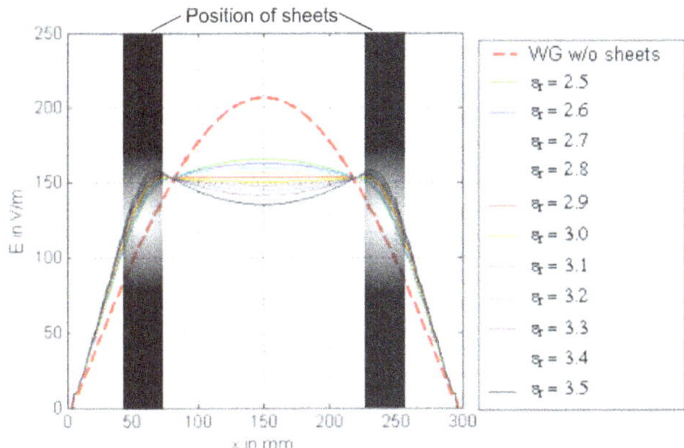

Fig. 4. Field distribution in transverse direction for d=30 mm, v=38.5 mm, and variable permittivity ε_r (P=1 W).

2.1.1 Improved field homogeneity

Due to the width of the cage including its cover, the distance D must not be smaller than 155 mm. Material characteristics, thickness and position of the sheets are modified to achieve a good field homogeneity between those walls. For optimising the dielectric sheets different points are discussed.

First, with the permittivity ε_r=5 and the distance v=0 mm (i.e. the dielectric sheets are in contact with the interior side of the metal walls of the waveguide) the thickness d is varied in order to find the optimum thickness for a homogeneous field distribution in the area between both dielectric sheets. The calculation of the field distribution is performed with help of the software package MICROWAVE STUDIO™ (CST, 2006). This software has a graphical user interface based on Windows and the calculation uses the FI method (Finite Integration technique (Weiland, 1990)). For an increasing thickness d the electric field distribution of the fundamental mode becomes more and more homogeneous in transverse direction reaching a rather constant field for d=40 mm (Fig. 2), but for d>20 mm also higher-order waves become propagable.

Another possibility is to keep the thickness d constant and to increase the distance v (v>0). In Fig. 3 the field distribution in transversal direction is plotted for different distances v, for d=1″=25.4 mm and for the permittivity ε_r=5 as before. This solution seems better than the first, because the field is quite homogeneous at v_{opt}=18 mm with less material usage.

The last point concerns the reduction of the permittivity of the dielectric sheets with the aim to use standard material available on the market at moderate costs. For this investigation the thickness d and the position v of the dielectric sheets are fixed, while ε_r is varied. From the transverse electrical field distribution shown in Fig. 4 for an input power of 1 W, d=30 mm, and v=38.5 mm the permittivity ε_r=2.9 is found to be the best solution for a homogeneous field distribution. With these parameters a quasi TEM waveguide can be designed.

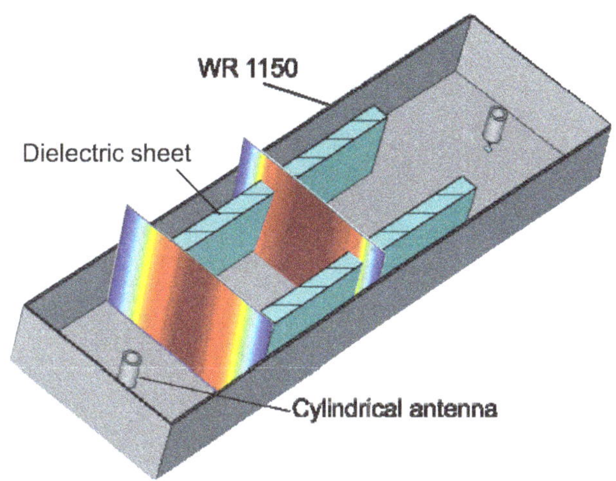

Fig. 5. Field distribution in two cross-sections of the total configuration.

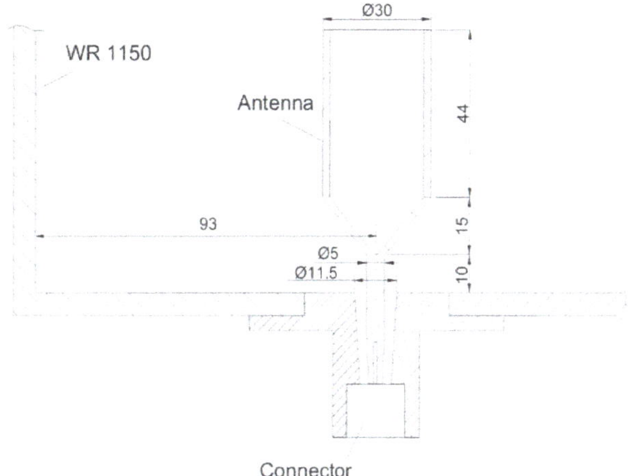

Fig. 6. Longitudinal cut through the coaxial-to waveguide adapter for excitation of the fundamental mode, optimized for 900 MHz.

2.1.2 Selective wave excitation

For the opimization given above, the first-order higher mode (H_{20}) is able to propagate in addition to the fundamental mode, but only the propagation of the fundamental mode is desired. This can be achieved by using a section of standard waveguide with an integrated coaxial-to-waveguide adapter in front of the waveguide section with the dielectric sheets (Fig. 5). From cross-sectional cuts of the calculated field distribution, like the two plotted in Fig. 5, it turns out that

1. in a distance of 160 mm from the exciting cylindrical antenna, whose dimensions are given in Fig. 6, the standard waveguide propagates solely the H_{10}-mode (with the sinusoidal transverse field) yielding a good excitation efficiency for the fundamental mode at the transition to the waveguide section with dielectric sheets,

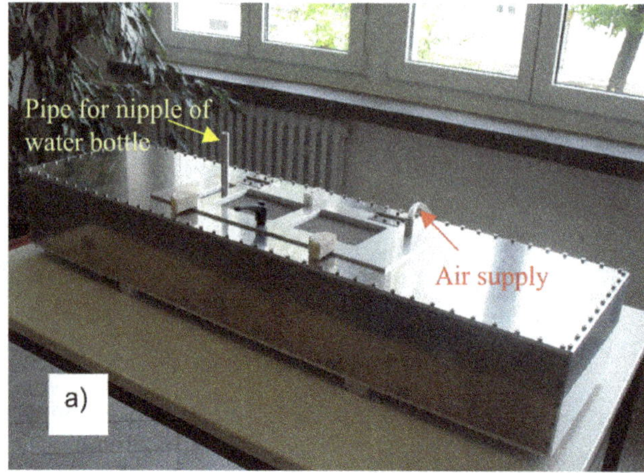

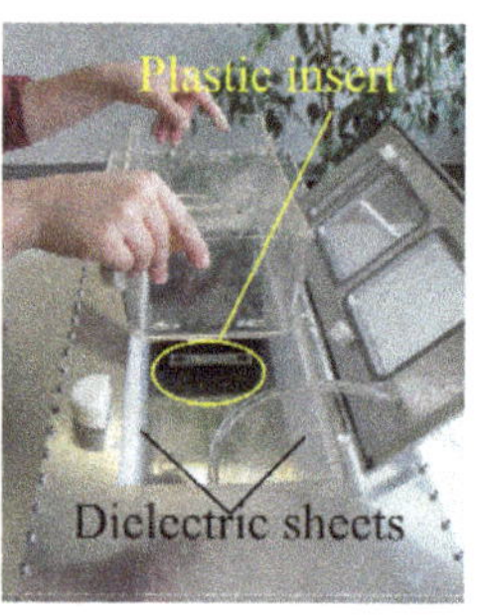

b)

Fig. 7. Built-up rectangular waveguide **(a)** with dielectric sheets **(b)**.

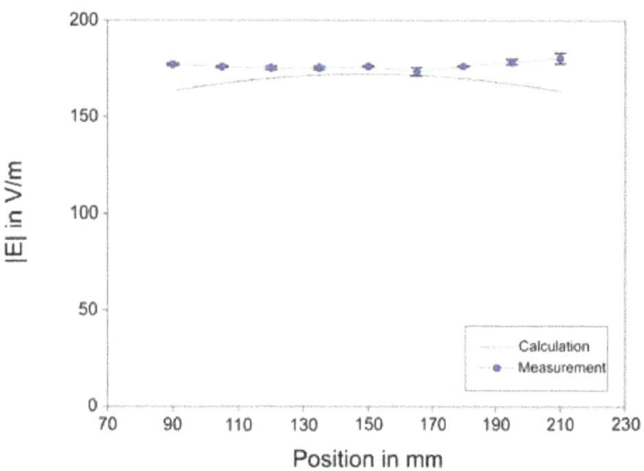

Fig. 8. Comparison of measured and calculated transverse field distribution between the dielectric sheets for d=30 mm, v=38.5 mm, and ε_r=2.5 (P_{in}=1 W).

while the overlap integral between H_{10}- and H_{20}-mode gives zero, and

2. in a distance of 72.5 mm from the beginning of the dielectric sheets a very good homogeneity of the field between both sheets (area of the cage) is obtained.

It should be mentioned that for reasons of comparison explained below, the field distributions in Fig. 5 are not presented for sheets with the optimum permittivity of 2.9 (leading to a perfectly homogeneous transverse field between the sheets) but for ε_r=2.5.

The waveguide is designed symmetrically, i.e. a second coaxial-to-waveguide adapter is placed behind the section with the dielectric sheets.

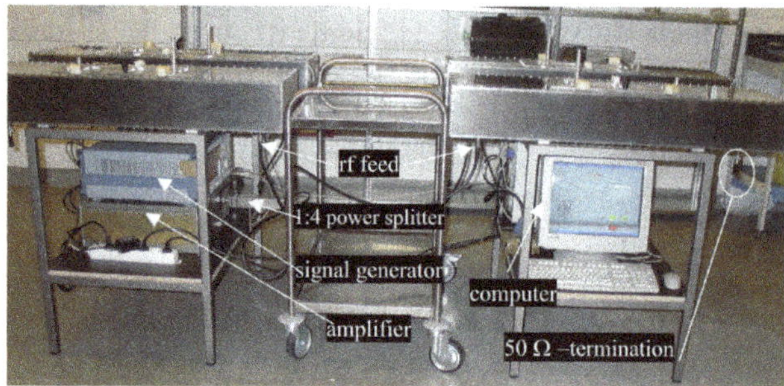

Fig. 9. Photograph of the entire installed rf exposure set-up.

2.2 Results

Figure 7a shows a photograph of the built-up rectangular waveguide. The total length is 1.1 m, the width 30 cm and the height about 18 cm. The top plate is closed by a hinged lid with a fine metal grid for reasons of lighting and simultaneous electromagnetic shielding. The pipe on the left is used for inserting the nipple of the water bottle, the second pipe on the right is needed as a feed-through for the flexible tube of the air supply. These pipes are dimensioned as circular waveguides-beyond-cut-off, thus being open or shut from an mechanical or respectively an electromagnetical point of view. Figure 7b shows the waveguide with the lid opened. The dielectric sheets and one of the plastic inserts for positioning the cage at a defined position are visible.

Using a simulation model of the waveguide including the coax-to-waveguide adapters, the dielectric sheets, and a 50 Ω termination at the output port, a total reflexion coefficient of approx. −20 dB is calculated at the input port. This value is confirmed by S-parameter measurements with the help of a network analyzer.

Since the permittivity measurement of the actual dielectric sheets yielded a value of 2.5, deviating from the optimum value of 2.9, the former was used for the field computations. Moreover, measurements of the field within the waveguide were performed with the help of a monopole probe inserted through bores in an adapter plate covering the opening in the top side of the waveguide. Figure 8 shows the comparison between the measured and calculated transverse field distribution for an input power of 1 W. The agreement within 9% is rather satisfying, particularly since an uncertainty of the permittivity measurement of 5% must be considered.

Figure 9 shows the entire exposure system installed at the Jacobs University Bremen. It consists of four rectangular waveguides each housing one cage, of a signal generator/modulator/amplifier-combination and of a 1:4-power splitter in order to distribute the output power to the four waveguides. 50 Ω terminations are mounted to the output connectors of the waveguides. The computer is used to select the power level (and thereby the applied SAR) and to record a control signal from a rf detector-diode. This signal is permanently analysed with the view to check the functionality of the equipment.

The determination of the field- and SAR-distribution in case of the waveguides filled with animals is only possible by numerical computations. A FDTD (Finite Difference Time Domain) in-house code (Bitz, 2004) was used involving a dielectric computer model of a hamster developed from ten MRI cut views. 21 different tissues have been considered.

From the analysis of the computations for different configurations and positions of the hamsters within their cages the necessary input power for the desired SAR-values was found.

In order to achieve the maximum whole body SAR of 4 W/kg the rf amplifier has to provide a total average power of 15,6 W to feed the four waveguides. Comparing this value with the waveguides' total input power of 4×3.7 W=14.8 W indicates that only 0.8 W are lost by absorption and reflection in the splitter, the cables and connectors. The uniformness of input powers amongst the four waveguides is within 6%.

3 Conclusions

A 900 MHz exposure set-up for the investigation of the metabolic rate of rodents was developed at the Bergische Universität Wuppertal and installed at the Jacobs University Bremen. The set-up consists of four rectangular waveguides designed for quasi TEM operation with a nearly homogeneous transverse field distribution due to specially dimensioned dielectrical sheets. The input powers needed for generating the demanded SAR values within the animals were found by means of computer simulations involving a model of the entire waveguide with one cage and up to three hamsters in different arrangements and postures. The detailed dosimetrical results will be presented elsewhere. The modulation signal was chosen according to the generic GSM signal defined in Schüller et al. (2000).

References

BfS: German Mobile Telecommunication Research Programme, http://www.emf-forschungsprogramm.de, March 2008.

Bitz, A.: Numerische Feldberechnung im biologischen Gewebe: Exposition von Personen, Tieren und isolierten biologischen Systemen in elektromagnetischen Feldern, Dissertation, Bergische Universität Wuppertal, 11–20, 2004.

CST GmbH: MICROWAVE STUDIO™ Version 2006B, Darmstadt, 2006.

Schüller, M., Streckert, J., Bitz, A., Menzel, K., and Eicher, B.: Proposal for generic GSM test signal, Proc. 22st BEMS Annual Meeting, Munich, Germany, 122–123, 2000.

Weiland, T.: Maxwell's grid equations, Frequenz, 44, 9–16, 1990.

7

Identification of the complex relative dielectric constant of porous polymers at different degrees of humidity

K. Haake and J. L. ter Haseborg

Hamburg University of Technology, Institute for Metrology and EMC, Germany

Abstract. In order to investigate the complex relative dielectric constant of different materials over a frequency range of 50 MHz to 3 GHz a setup has been built that allows rf-measurements in a climate chamber. Within this chamber the temperature and humidity can be changed and be measured by accurate sensors. The construction of the rf-part of the setup has been made with an outer cylindrical shell that allows climate interactions through its large orifices. Methods for calculating the complex relative dielectric constant from the measured scattering parameters are presented and the changes of the relative constants due to different humidity conditions are shown.

1 Introduction

Constructions in rf-metrology often use wood as non-conducting material. Wood combines the advantages of easy workability, low weight, high mechanical stability, low cost and a low relative permittivity compared to a lot of other non-conducting materials. The last point results on the fact that wood is a more or less porous material. This is the reason for the assumed disadvantage of wood, making it acting like a sponge for humidity. This might alter its electric behavior causing different diffraction and especially raises the dielectric losses of the material. To know the influence of the different materials under different environmental conditions is highly interesting for rf-metrology. Hence the investigation of this effect and especially the influence of humidity towards the dielectric behavior is the goal of this paper. Due to the fact that the relative dielectric constant is strongly dependent on frequency, a setup has been built that permits measurements up to 3 GHz. Further the humidity of the environment can be altered. The result of the measurement is a determination of the complex relative dielectric constant:

$$\underline{\varepsilon_r}(f) = \varepsilon_r'(f) - j \cdot \varepsilon_r''(f) \qquad (1)$$

2 Measurement setup

The setup, depicted in Fig. 1, consists of a coaxial transmission line (Unger, 1995) that is shown in Fig. 2. It is tapered at both sides in order to enlarge the diameter in the middle segment without drastically change the wave impedance, keeping it in the vicinity of 50 Ω. The dielectric in this section is the specimen to study. Due to the fact that the influence of humidity towards the dielectric is to be investigated, the middle section's cylindrical shell is been simulated by using wires symmetrically positioned around the circumference as shown in Fig. 1. Such an outer wire conductor keeps up the coaxial structure and electrical behavior of the transmission line as shown by Rojas-Coto and ter Haseborg (2006) and further permits an interchange of the environmental climate. The coaxial setup is placed inside a sealed box, which is only been penetrated by cables and a small tube for the water supply. Inside are also an evaporator and a fan (with heating) for homogenizing the climate conditions within the box. Two identical sensors are placed inside the box, which are able to measure the relative humidity, temperature and dew point. Further parts of the setup are a vector network analyzer and an interface to digitalize the temperature and humidity signals digitally for further processing. A PC controls the data acquisition and stores the measurements.

3 Theory

The complex relative dielectric constant will be calculated using the scattering parameters (Zamow, 2005). These have been measured beforehand by the vector network analyzer

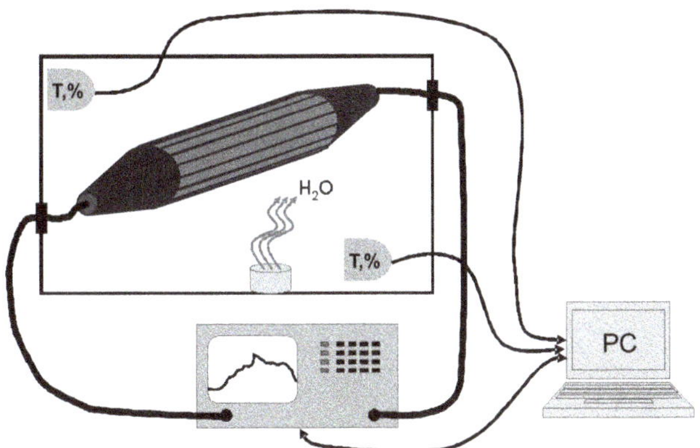

Fig. 1. Measurement setup with climate chamber.

Fig. 2. Coaxial setup with outer weir conductor in the middle section.

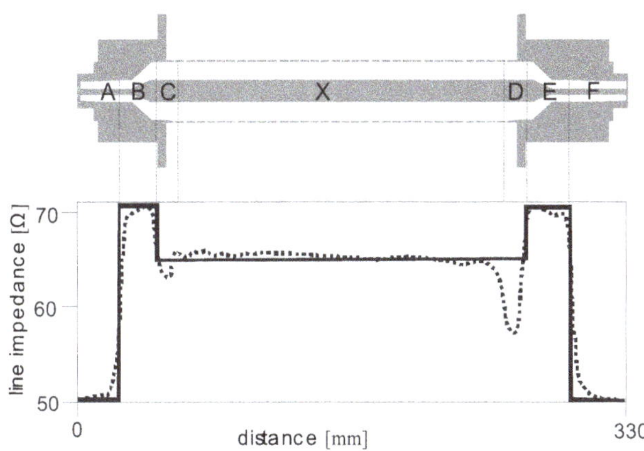

Fig. 3. Coaxial measurement section (top) with TDR measurement. TDR plot has been stretched piecewise making the x-axis linearly scalable.

and are put into the 2×2-scattering-matrix **T**. The coaxial setup consists of different sections (*A-B-C-X-D-E-F*) with different wave impedances, but only the middle part *X* is filled with the specimen. Measuring the coaxial setup with a TDR, it can be divided into sections as depicted in Fig. 3.

Section *X* denotes the specimen section. Hence **T** has to be transformed to chain matrix notation and Eq. (2) can be applied as shown in Schiek (1984).

$$\mathbf{T} = \mathbf{A} \cdot \mathbf{B} \cdot \mathbf{C} \cdot \mathbf{X} \cdot \mathbf{D} \cdot \mathbf{E} \cdot \mathbf{F} \quad (2)$$

In order to calculate the middle section X, Eq. (3) is used and a back transformation to scattering parameters has to be done afterwards.

$$\mathbf{X} = \mathbf{C}^{-1} \cdot \mathbf{B}^{-1} \cdot \mathbf{A}^{-1} \cdot \mathbf{T} \cdot \mathbf{F}^{-1} \cdot \mathbf{E}^{-1} \cdot \mathbf{D}^{-1} \quad (3)$$

This is necessary, because the Nicholson-Ross method (Eq. 4) as used by Zamow (2005) uses the scattering parameters of X to calculate the complex relative dielectric constant of this section.

$$\varepsilon_{r,\mathrm{eff}} = \sqrt{\frac{-\left[\frac{c_0}{\omega l_x}\ln\left(\frac{S_{X11}+S_{X21}-R_{12}}{1-(S_{X11}+S_{X21})R_{12}}\right)\right]}{\left(\frac{1+R_{12}}{1-R_{12}}\right)^2}} \quad (4)$$

The terms for W and R_{12} are given in Eqs. (5) and (6); the absolute value of Eq. (5) is supposed to be equal or smaller than 1 in order to get physically sensible results.

$$W = \frac{1-\left(S_{X21}^2 - S_{X11}^2\right)}{2S_{X11}} \pm \sqrt{\left(\frac{1-\left(S_{X21}^2 - S_{X11}^2\right)}{2S_{X11}}\right)^2 - 1} \quad (5)$$

$$R_{12}(\omega) = W \pm \sqrt{W^2 - 1} \quad (6)$$

The Nicholson-Ross method tends to provide not very smooth curves (Zamow, 2005). To achieve this, the Debye function (Eq. 7) is applied. This function is supposed to express the general run of the relative dielectric constant of a material over the frequency.

$$\begin{aligned}\varepsilon_r(\omega) &= \varepsilon_r' - j\varepsilon_r'' \\ &= \left(\varepsilon_\infty + \frac{\varepsilon_s - \varepsilon_\infty}{1+\omega^2\tau_e^2}\right) - j\left(\frac{\sigma_e}{\omega\varepsilon_0} + \frac{(\varepsilon_s - \varepsilon_\infty)\,\omega\tau_e}{1+\omega^2\tau_e^2}\right)\end{aligned} \quad (7)$$

The parameters to determine in this function approximating the resulting run of the curve gotten by the Nicholson-Ross method are: Static permittivity ε_s, specific conductivity σ_e and relaxation time τ_e.

The relative permittivity at very high frequencies ε_∞ approximates 1 due to the fact that at very high frequencies no polarization is possible (Zamow, 2005).

3.1 Humidity

The relative humidity is defined as the percentage of the absolute mass of water that can be solved in a gas at a certain temperature without condensing (Hilbrunner, 2005). This depends of course on the used gas. Hence the absolute humidity is the mass of water that is solved in a gas. Figure 4 depicts the absolute humidity for air over the temperature. The different curves show different saturation degrees. Nevertheless in Central Europe the absolute humidity hardly exceeds 20 g/m^3.

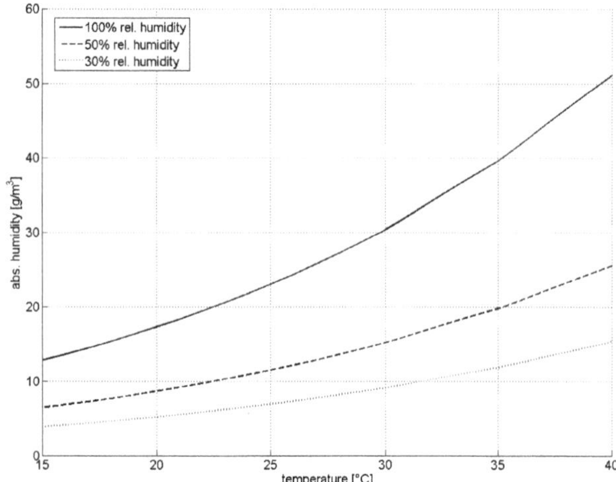

Fig. 4. Absolute humidity of air for three different saturation conditions (Hilbrunner, 2005).

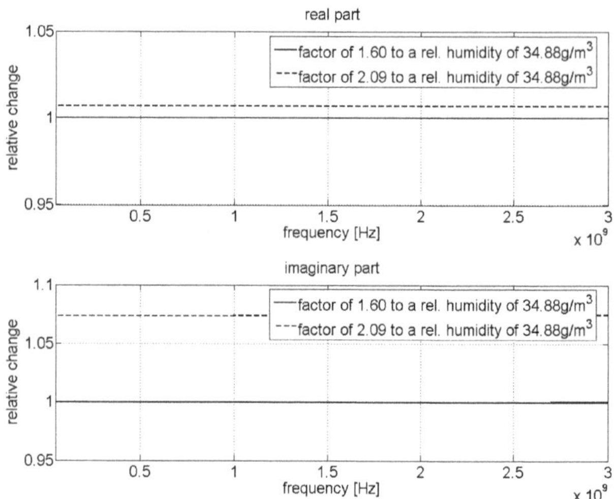

Fig. 5. The relative change of the complex relative dielectric constant of balsa shown as real part (top) and negative imaginary part (botton).

4 Results

Two different materials have been tested that are mostly used in rf-constructions in the institute. These are balsa and beech. The electrical characteristics of different types of wood are described in Vidémé Bossou (2007). This investigation deals only with the relative change of the electrical characteristics. The results have been obtained by measuring the scattering parameters at a steady state of the wood's humidity. This has been accomplished by recording over a long time simultaneously the data of the outer environmental humidity and the scattering parameters. When the change goes to zero the data has been taken.

Figures 5 and 6 show the relative change of the complex relative dielectric constant of balsa and beech over frequency. Normalization has been done to the lowest value of measured humidity. For reasons of depiction, the real and the negative imaginary part are plotted separately. It can be observed, that despite of drastic change in humidity, the change in the relative dielectric constant is not significantly. Due to the use of Eq. (7) the resulting curves are smoother than the actual measured data.

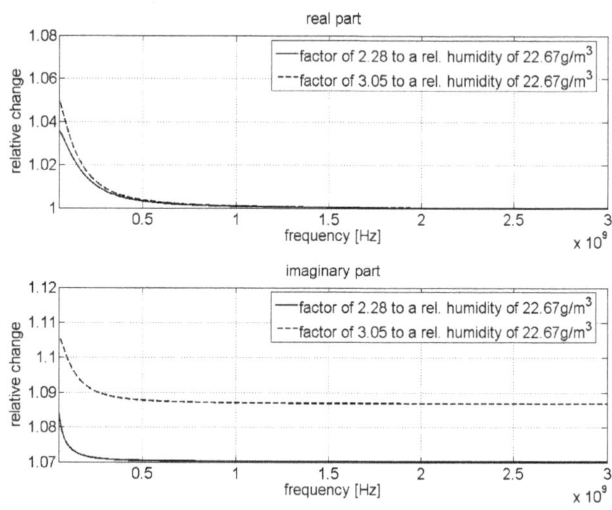

Fig. 6. The relative change of the complex relative dielectric constant of beech shown as real part (top) and negative imaginary part (botton).

5 Conclusion

A setup has been built that allows measuring the change of the complex relative dielectric constant over a frequency range of 50 MHz to 3 GHz. The measurements have been taken out under different humidity conditions. The specialty of the setup is that the construction of the quasi coaxial conductor lets the climate have a direct effect on the specimen while still allowing measurements up to high frequencies. The results show that the effect of humidity towards the dielectric behavior and the loss of the materials is rather small.

References

Hilbrunner, F.: Ein Beitrag zur Feuchtekompensation von Präzisionsmeßgeräten, PhD. Thesis, Technische Universität Ilmenau, Ilmenau, 2005.

Rojas-Coto, A. and ter Haseborg, J. L.: Bestimmung der Induktivität komplexer großer metallischer Strukturen mittels Reusenleiteraufbau für die EMV-Analyse, Int. Fachmesse und Kongress für EMV, VDE-Verlag GmbH, Offenbach, 2006.

Schiek, B.: Meßsysteme der Hochfrequenztechnik, Hüthig, Heidelberg, 1984.

Unger, H.-G.: Elektromagnetische Wellen auf Leitungen Hüthig, Heidelberg, 1995.

Vidémé Bossou, O.: A simple & accessible front-end satellite receiver for communication and e-learning, PhD. Thesis, Ecole Polytechnique Fédérale de Lausanne, Lausanne, 2007.

Zamow, D.: Untersuchung und Simulation des Absorptionsverhaltens in TEM-Wellenleitern, diploma thesis, Universität Hannover, 2006.

8

Time domain reflectrometry measurements using a movable obstacle for the determination of dielectric profiles

B. Will, M. Gerding, S. Schultz, and B. Schiek

Ruhr-University Bochum, Institute of High Frequency Engineering, Universitätsstr. 150, 44801 Bochum, Germany

Abstract. Microwave techniques for the measurement of the permittivity of soils including the water content of soils and other materials, especially TDR (time domain reflectometry), have become accepted as routine measurement techniques. This summary deals with an advanced use of the TDR principle for the determination of the water content of soil along a probe. The basis of the advanced TDR technique is a waveguide, which is inserted into the soil for obtaining measurements of the effective soil permittivity, from which the water content is estimated, and an obstacle, which can mechanically be moved along the probe and which acts as a reference reflection for the TDR system with an exactly known position. Based on the known mechanical position of the reference reflection, the measured electrical position can be used as a measure for the effective dielectric constant of the environment. Thus, it is possible to determine the effective dielectric constant with a spatial resolution given by the step size of the obstacle displacement.

A conventional industrial TDR-system, operating in the baseband, is used for the signal generation and for the evaluation of the pulse delay time of the obstacle reflection. Thus, a cost effective method for the acquisition of the dielectric measurement data is available.

1 Introduction

The aim of this investigation is to obtain representative profile measurement results from a material under test. Time domain reflectometry (TDR) is known as a method to obtain a reliable estimation of the soil water content from a measurement of the real part ε' as well as the imaginary part ε'' of the relative permittivity ε_r (Kupfer, 2005; Hoekstra, 1975). In many cases TDR is used in order to get just one integral value for the relative permittivity of the material (Robinson, 2000). In this case the reflection from the end of the probe is

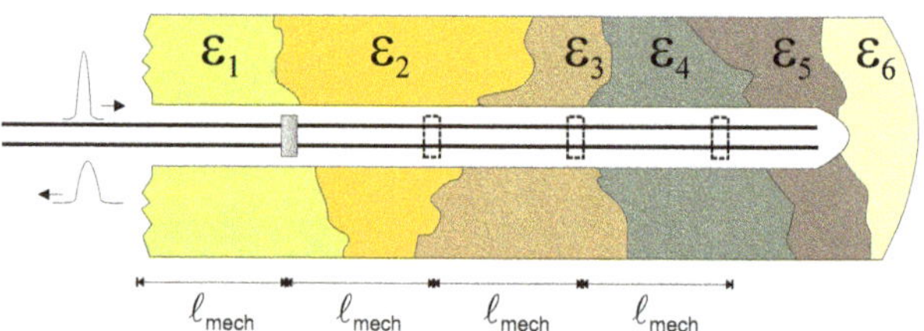

Fig. 1. Functional principle.

measured and converted into a value for the relative permittivity. This summary deals with a method that uses different reflections. Thus, it is possible to measure a dielectric profile along the probe. The different reflections are caused by an obstacle which is moved mechanically along the probe. The step size of the motion of this obstacle is one important parameter for the possible spatial resolution of the system.

2 Functional principle

The propagation speed of electromagnetic waves is directly related to the effective permittivity of the penetrated medium. In closed waveguide structures like in circular waveguides or coaxial waveguides with a relative permittivity ε_r the propagation velocity is given by:

$$c = \frac{c_0}{\sqrt{\varepsilon_r}} \ , \qquad (1)$$

while the propagation velocity along the probe of the test setup shown in Fig. 1 is determined by the effective permittivity $\varepsilon_0 \cdot \varepsilon_{\text{eff}}$. The effective permittivity describes the permittivity of the dielectric mixture of the penetrated medium. The volume of e.g. soil for which the estimation is valid, primarily depends on the design of the probe (Heimovaara, 1993). By moving a reflecting obstacle stepwise along the probe, a permittivity profile can be determined by using the relation between the free space pulse delay time Δt_{ref} corresponding to the mechanical length l_{mech} of the obstacle displace-

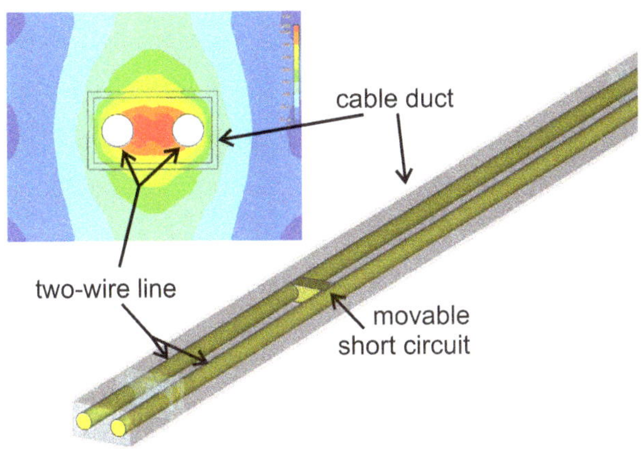

Fig. 2. Probe design and field distribution.

ment and the measured delay time Δt_{meas} corresponding to the electrical length and the evaluation of the TDR signal. The measured pulse delay time is proportional to the relative effective permittivity ε_{eff} of the penetrated medium around the analyzed probe length l_{mech}:

$$\Delta t_{meas} = \frac{2 l_{mech}}{c_0} \sqrt{\varepsilon_{eff}} \ . \tag{2}$$

By comparison with the theoretical pulse delay time Δt_{ref}:

$$\Delta t_{ref} = \frac{2 l_{mech}}{c_0} \tag{3}$$

the effective permittivity along the analyzed probe length l_{mech} can be determined:

$$\varepsilon_{eff} = \left(\frac{\Delta t_{meas}}{\Delta t_{ref}} \right)^2 \ . \tag{4}$$

Obviously the spatial resolution of the system directly depends on the length interval l_{mech} and on the pulse delay time Δt_{ref} corresponding to the obstacle displacement, respectively. The higher the spatial resolution the shorter the pulse delay time. If the pulse delay time is in the order of the inherent time jitter Δt_{jitter} of the TDR system, the upper resolution limit is reached and can be determined by:

$$\Delta l_{min} = \frac{1}{2} c_0 \Delta t_{jitter} \ . \tag{5}$$

The minimum resolution length is a multiple of Δl_{min} for accurate measurement results. To keep the estimated error below 1%, the practical minimum resolution length will be approximately $100 \Delta l_{min}$. In relation to this, the effective time jitter of ±500 fs without an averaging of the TDR prototype system data would lead to a minimum resolution length of 30 mm.

The basic functional principle of the measuring system is illustrated in Fig. 1.

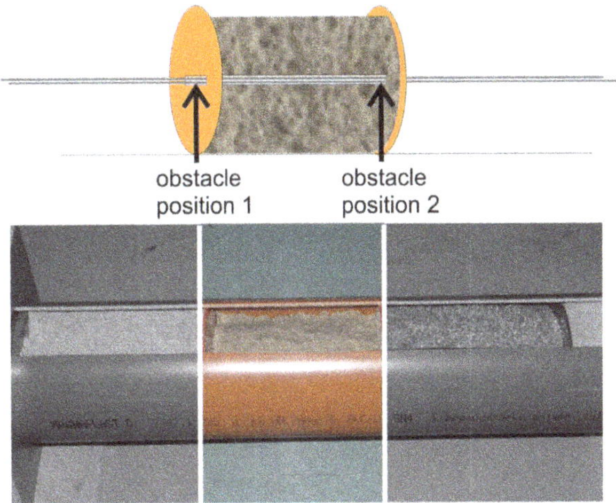

Fig. 3. Measurement setup for a fixed obstacle position.

In this case the movable obstacle is realized as a movable short circuit. This short is moved stepwise by l_{mech} and in each position the pulse delay time is measured. With this pulse delay time and the pulse delay time measured without a dielectric surrounding one gets the effective dielectric constant for each step. Thus, one can perform a measurement for the characterization of dielectric profiles with a spatial resolution given by the step size with which the obstacle is moved.

3 Probe Design

In order to obtain good measurement results it is very important to design a probe, which allows measurements in an environment with high permittivity values on the one hand and high resolution measurements on the other hand (Knight, 1992; Zegelin, 1989). Thus it is necessary to have a good interaction between the electromagnetic field and the environment. On the other hand it is necessary to have a certain gap between the conductors and the environment, otherwise high dielectric constants in the environment would cause a strong reflection. Due to these requirements the probe shown in Fig. 2 was designed.

This two-wire probe with a short as a movable obstacle is a good compromise between maximum interaction of the electromagnetic field and the environment on the one hand and the detection of high permittivities on the other hand.

4 Measurement results

For first measurements the obstacle was placed consecutively in two positions as shown in Fig. 3. In a first step the delay time in position 1 was measured as a reference. Then the obstacle was moved into the second position and the delay time

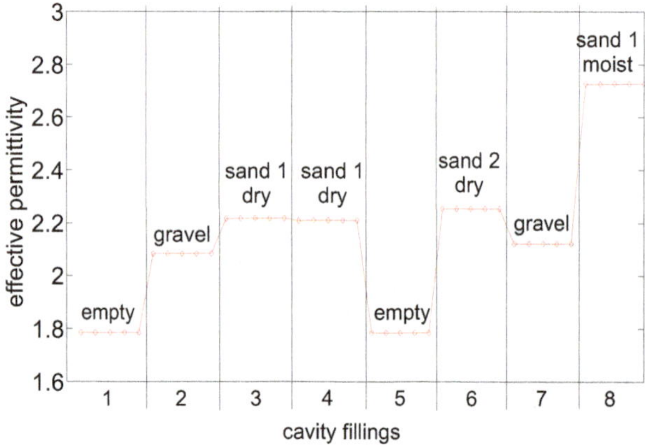

Fig. 4. Measurement results for a fixed obstacle position and individual cavity fillings.

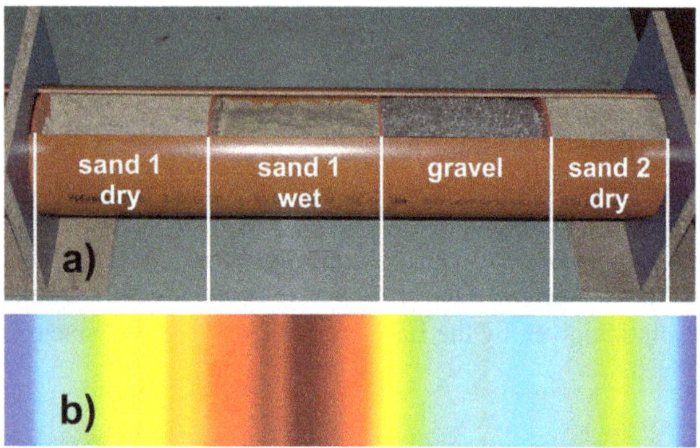

Fig. 5. Measurement setup with a movable obstacle **(a)** and measurement results as a color bar **(b).**

with air as environment was measured. Then the cavity between position 1 and position 2 was consecutively filled with different materials. The delay time was measured for each material and compared with the delay time without filling. For these measurements the obstacle was fixed in position 2. Thus measurement errors caused by a displacement of the obstacle could not occur.

Figure 4 shows the measurement results for a fixed obstacle position. The cavity between position 1 and position 2 was filled with five different materials. For each material the delay time was measured five times. The effective permittivity of each material was calculated by using Eq. 4.

Thus, these first measurement results show that the comparison of delay time intervalls is a very useful method for the characterization of effective permittivities. In a next step a measurement with a movable obstacle, in this case a short, was performed. Figure 5 a) shows the measurement setup.

A tube with a diameter of 20 cm was divided into four cavities, which are filled with different types of gravel and sand. Fig. 5 b) shows the measurement results as a color bar. The

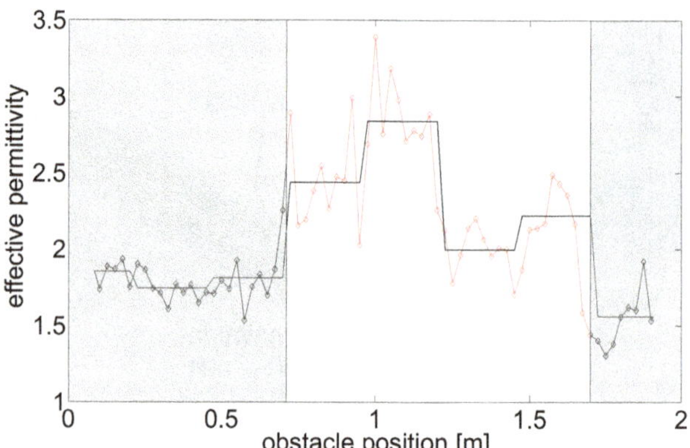

Fig. 6. Measurement results with a movable obstacle.

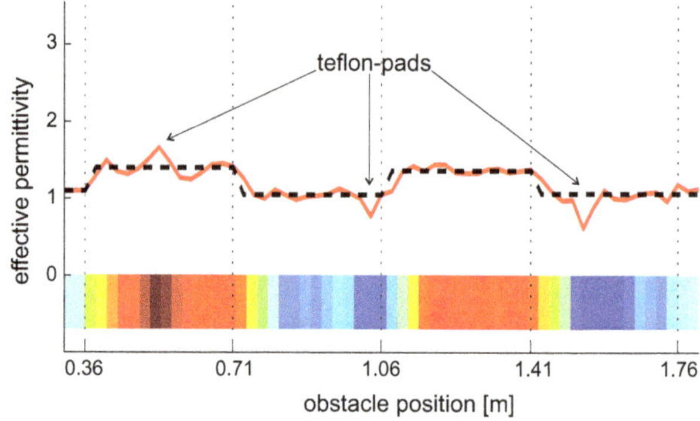

Fig. 7. Automated measurement with sand in the different cavities.

two-wire probe is located in the radial center of the tube. The dielectric profile of the environment was measured with a step size of 2.5 cm and the length of each cavity was 25 cm. Figure 6 shows the measurement results.

We observe some deviations concerning the effective permittivity, which can partially be explained by the manual moving of the obstacle. Nevertheless the different materials are distinguishable from each other. In order to reduce these uncertainties, the obstacle has to be optimized. For certain measurement results it is very important to know the position of the obstacle exactly. If the obstacle is realized as a short circuit, one should ensure, that the obstacle does not tilt between the two transmission lines. Thus, the obstacle which is moved should be very thin and it should provide a good contact between the transmission lines. Furthermore the obstacle should be moved automatically to reduce uncertainties during the positioning.

Figure 7 shows first results for an automated measurement with a proper short circuit. In this case every other cavity was filled with dry sand. This figure shows that there are only small uncertainties and the different cavities are well defined. The transmission lines are fixed by three pairs of teflon pads, which cause small deviations. In a next step the separators

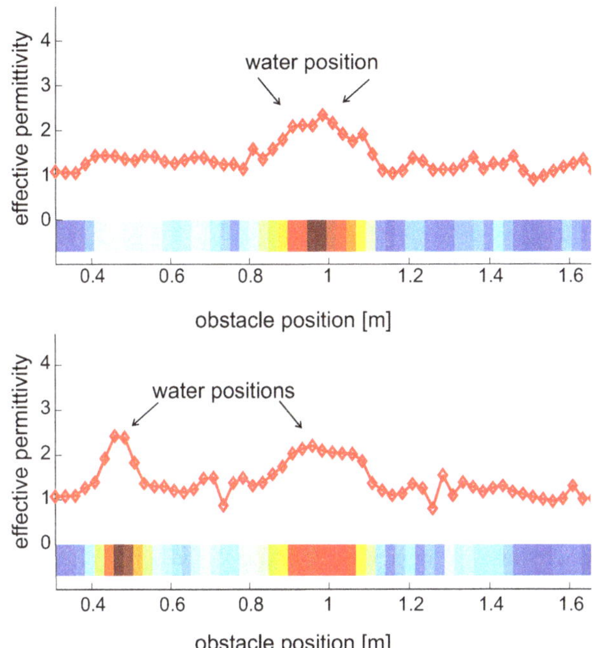

Fig. 8. Measurement results with different moist positions.

between the different cavities are removed. The entire tube was filled with dry sand and in several positions water was locally added. Figure 8 shows the measurement results with one and with two moist areas. Thus, this measurement setup is a very proper method for the measurement of dielectric profiles in the surrounding of the probe.

5 Conclusions

The results in this summary allow an estimation of the capability of the presented measurement concept for the determination of dielectric profiles and profiles of the water content of soils, respectively.

In addition to well known TDR moisture sensing systems, which only determine the integral value of the moisture content along the sensor, the presented concept is capable of measuring moisture profiles by use of a conventional industrial TDR-system or a vector network analyzer, respectively, for the signal generation and the signal evaluation. To achieve the additional information for reconstructing a dielectric profile from the measured data, a special sensor is necessary, which includes a movable reflecting target, a so-called obstacle, which can manually or automatically be moved along the probe, in order to act as a reference. In this way a mapping of the effective permittivities is possible.

References

Heimovaara, T. J.: Design of triple wire domain reflectometry probes in practice and theory, Soil Sci. Soc. Am., 57, 1410–1417, 1993.

Hoekstra, P. and Delanay, A.: Dielectric Properties of soils at UHF and microwave frequencies, J. Geophys. Res., 79, 1699–1708, 1975.

Knight, J. H.: Sensitivity of time domain reflectometry measurements to lateral variations in soil water content, Water Resour. Res., 28, 9, 2345–2352, 1992.

Kupfer, K.: Electromagnetic Aquametry: Electromagnetic Wave Interaction with Water and Moist Substances, Springer Verlag, Berlin, 2005.

Robinson, D. A. and Friedman, S. P.: Parallel Plates Compared with Conventional Rods as TDR Waveguides for Sensing Soil Moisture, Subsurface Sensing Technologies and Applications Journal, 1, 497–511, 2000.

Zegelin, S. J., White, I., and Jenkins, D. R.: Improved field probes for soil water content and electrical conductivity measurement using Time Domain Reflectometry, Water Resour. Res., 25, 2367–2376, 1989.

9

Comparison of methods for broadband electromagnetic characterization of Molded Interconnect Device materials

C. Orlob, D. Kornek, S. Preihs, and I. Rolfes

Leibniz Universität Hannover, Institut für Hochfrequenztechnik und Funksysteme, Appelstr. 9A, 30167 Hannover, Germany

Abstract. Combining the Molded Interconnect Device technology with the Laser Direct Structuring technology exhibits the potential of designing electrical and mechanical components on three-dimensional surfaces to increase functionality, level of integration and to reduce costs. When taking advantage of this technology especially in the design of RF devices, a precise knowledge of the electromagnetic parameters of the MID material is required, as the complex permeability and permittivity strongly influence the device performance. At present time, these materials are not electromagnetically characterized in the RF frequency range. In this paper different methods are therefore presented and compared with respect to their potentials for broadband electromagnetic characterization of Molded Interconnect Device materials.

1 Introduction

The Molded Interconnect Device technology (MID) in combination with Laser Direct Structuring (LDS) offers the possibility of designing electrical and mechanical components on three-dimensional surfaces to increase functionality, level of integration and to reduce costs. For example, transmission lines, antennas, switches, and connectors can be integrated on carriers like the covers of cellular phones or the cases of exterior mirrors of a car. However, applying this novel technology to RF circuits implies the precise knowledge of the electromagnetic properties of the MID material. Especially for MID based development of antennas dedicated for different applications like IEEE 802.11 WLAN, IEEE 802.15.1 Bluetooth or IEEE 802.16 WiMAX it is necessary to determine the complex permittivity $\epsilon_r = \epsilon_r' - j\epsilon_r''$ and complex permeability $\mu_r = \mu_r' - j\mu_r''$ over a wide frequency range.

Since these parameters are not quantified for RF frequencies at the present time, this paper presents a first characterization of an exemplarily chosen MID material Pocan DP T7140 LDS (Lanxess AG) demonstrating the features of the measurement methods, which were chosen with respect to the constraint of available samples geometries. Measurement principle, capability, bandwidth, uncertainties and sample requirements of each method are pointed out.

In Sect. 2 the known properties of the considered MID material are indicated and the MID LDS processing is shortly described. In the subsequent section the measurement methods are presented in detail followed by the measurement results shown and discussed in Sect. 4. The paper ends with a conclusion.

In the following, the investigated materials are generally categorized with the terms low permittivity, medium loss and low loss. The term low permittivity refers to materials where $\epsilon_r' \leq 5$ and the terms medium loss and low loss correspond to $3 \times 10^{-4} \leq \tan\delta \leq 3 \times 10^{-2}$ and $\tan\delta < 3 \times 10^{-4}$, respectively.

2 MID LDS materials and processing

This section gives information on relevant properties of the investigated material Pocan DP T7140 LDS and how it is applied in the MID LDS process. According to the data sheet, Pocan DP T7140 LDS is a solid polyethylene terephthalate/polybutylene terephthalate (PET/PBT) polymer with low permittivity and medium loss. In detail, a dielectric constant $\epsilon_r' = 4.1$ and loss factor $\tan\delta = \epsilon_r''/\epsilon_r' = 1.38 \times 10^{-2}$ are stated at $f = 1$ MHz.

Like any other MID LDS material it is provided with an additive in form of an organic metal complex (Schlueter et al., 2002). For usage in the MID LDS process, first its surface is activated by a laser beam as shown in Fig. 1. This separates the metal atoms from the organic ligands and roughens the surface enabling copper coating with a strong grip.

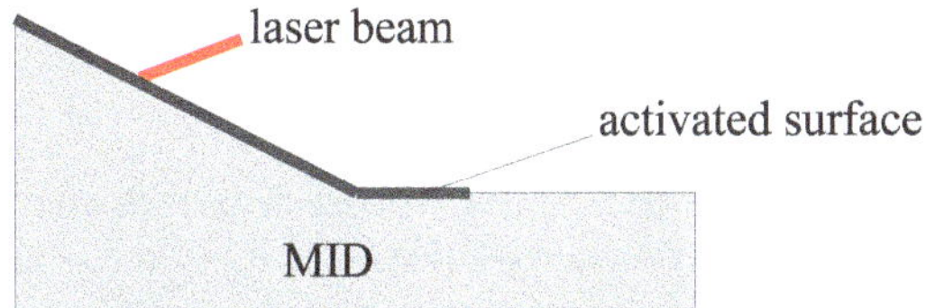

Fig. 1. MID LDS activation step.

Subsequently, the cleaning step follows, in which the debris from laser structuring is removed. Finally, the metallization is build up on the activated area by use of current-free copper baths. Depending on application, the copper layer can additionally be plated with gold or other metals by means of electroforming baths.

3 Measurement methods

In this section, the analyzed measurement methods are shortly presented in terms of functionality, measurement capability, frequency range of operation, measurement uncertainty and sample requirements. Further information about each method can be found in Clarke et al. (2002).

The first method considered is a commercially available admittance cell Agilent 16453A connected with the impedance analyzer Agilent E4991A. As shown in Fig. 2a, the dielectric is sandwiched by two electrodes forming a capacitor. The complex permittivity ϵ_r is determined as the ratio of the admittances for the material filled electrodes and for the air filled electrodes. This method can be used in the range from a few MHz to 1 GHz and requires planar sheets with a thickness $0.3\,\text{mm} \leq h \leq 3\,\text{mm}$ and a diameter $d \geq 15\,\text{mm}$. In accordance to the data specification of the impedance analyzer, a minimal uncertainty of approximately $\Delta\epsilon_r'/\epsilon_r' = 7\%$ can at best be achieved for measurement of low permittivity dielectrics.

The second technique refers to the coaxial probe HP 85070A, which is pressed against a flat surface of the dielectric, as shown in Fig. 2b. Assuming an isotropic, homogeneous and non-magnetic sample, the complex permittivity ϵ_r is derived from the reflection coefficient based on an quasi static model of the probe termination. Corresponding to the manufacturer's data, the theoretical operating range is $200\,\text{MHz} \leq f \leq 20\,\text{GHz}$, which is practically shortened by the effect of falling sensitivity with decreasing frequency. A minimum uncertainty of $\Delta\epsilon_r'/\epsilon_r' = 5\%$ in the upper part of the frequency range can be accomplished assuming appropriate contacting and calibration. This method requires samples with at least one flat surface and dimensions $h \geq 20/\sqrt{|\epsilon_r|}\,\text{mm}$ and $d > 20\,\text{mm}$.

The third method relies on on-wafer scattering-parameters measurement for coplanar waveguides (CPW) of different lengths (Arz and Leinhos, 2008). A CPW is formed from a signal conductor and a pair of groundplanes, all arranged

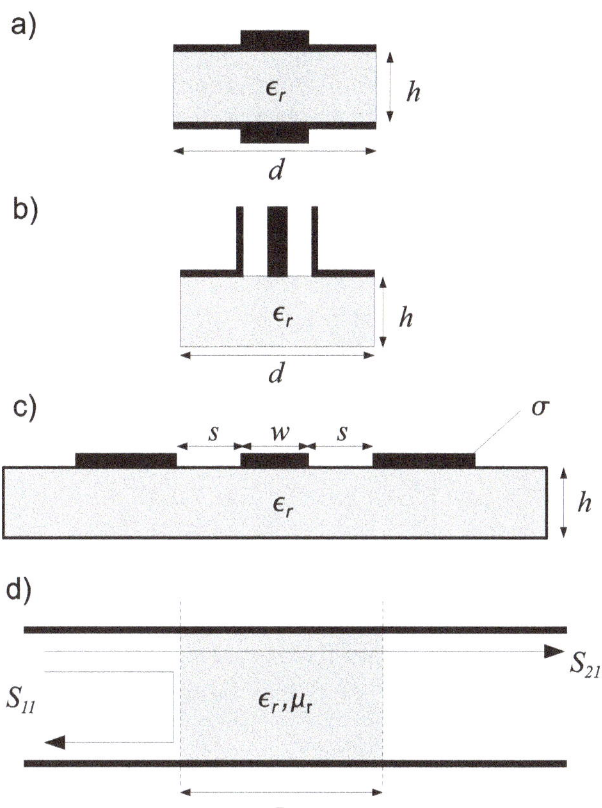

Fig. 2. Cross-sections of measurement setups: **(a)** admittance cell, **(b)** coaxial probe, **(c)** coplanar waveguide and **(d)** rectangular waveguide.

on top of the dielectric, as shown in Fig. 2c. Concerning the measurement procedure, first of all an effective propagation constant γ_{eff} is determined from the measured raw data on the basis of a multiline-TRL algorithm (Marks, 1991). In a second step this result is used in a quasi-TEM model for coplanar lines (Heinrich, 1993), which assumes an isotropic, homogeneous and non-magnetic substrate material, to finnaly extract the complex permittivity ϵ_r of the substrate material. The chosen CPW dimensions should hold the condition $2(w+2s) < h$ in order to approximately meet the assumption of an infinite thick substrate as demanded in the quasi-TEM model (Heinrich, 1993). For accurate loss factor extraction the model requires the value of the metallization's conductivity σ. This method covers a broad frequency range, whose lower limit depends on the maximal realizable length differences between the shortest and longest line and the upper limit is given by the measurement equipment like the vector network analyzer or the contact probes. A definite statement about the measurement uncertainty can not be issued. However, as shown in Arz and Leinhos (2008) this method exhibits a measurement accuracy comparable to the accuracy, which can be achieved with a split-cylinder resonator.

The last method investigated is based on the rectangular waveguide technique. The sample, filling a section of waveguide of length L as demonstrated in Fig. 2d, is electromagnetically characterized by the scattering-parameters

$$S_{11} = \frac{(1 - z^2)\Gamma}{1 - (\Gamma z)^2} \tag{1}$$

$$S_{21} = \frac{(1 - \Gamma^2)z}{1 - (\Gamma z)^2} \tag{2}$$

with the transmission coefficient

$$z = e^{-\gamma L} \tag{3}$$

and the reflection coefficient

$$\Gamma = \frac{\frac{\mu}{\gamma} - \frac{\mu_0}{\gamma_0}}{\frac{\mu}{\gamma} + \frac{\mu_0}{\gamma_0}} . \tag{4}$$

The terms γ_0 and $\gamma = f(\epsilon_r, \mu_r)$ stand for the waveguide propagation constants of the fundamental mode TE_{10} in air and material, respectively. In comparison with the other three discussed techniques this method features the capability to measure the relative permittivity as well as the relative permeability by solving the system Eqs. (1)–(2) via the Nicolson-Ross solution (Nicolson and Ross, 1970). In the case of measuring long samples this approach leads to the drawback of instability at frequencies corresponding to integer multiples of one-half wavelength in the material. Alternatively, when a non-magnetic material can be assumed the permittivity can be determined in a stable way by solving Eq. (2) numerically according to the procedure of Baker-Jarvis et al. (1990). In both approaches this method covers a frequency range of a single waveguide band, which is in this paper $8.4\,\text{GHz} \leq f \leq 12\,\text{GHz}$, and requires samples, which should be precisely matched to the inside dimensions of the waveguide to avoid air gaps. In this paper for the numeric solution a differential uncertainty analysis accounting for uncertainties in scattering-parameters and sample length L is accomplished. In accordance with Baker-Jarvis et al. (1990), the total uncertainty is defined as

$$\Delta\epsilon_r' = \sqrt{\left(\frac{\partial\epsilon_r'}{\partial|S_{21}|}\Delta|S_{21}|\right)^2 + \left(\frac{\partial\epsilon_r'}{\partial\Theta_{21}}\Delta\Theta_{21}\right)^2 + \left(\frac{\partial\epsilon_r'}{\partial L}\Delta L\right)^2} \tag{5}$$

where $\Delta|S_{21}|$ is the uncertainty in magnitude of S_{21}, $\Delta\Theta_{21}$ is the uncertainty in phase of S_{21} and ΔL is the uncertainty in sample length L. The total uncertainty for the loss factor is defined respectively. For low permittivity samples with medium or low loss, uncertainties below $\Delta\epsilon_r'/\epsilon_r' = 1\%$ can be achieved for lengths L longer than half the wavelength in the material. For this class of materials it can be expected by trend that the larger the sample length L is, the lower the total uncertainty hence the higher the measurement reliability will be (Baker-Jarvis et al., 1990).

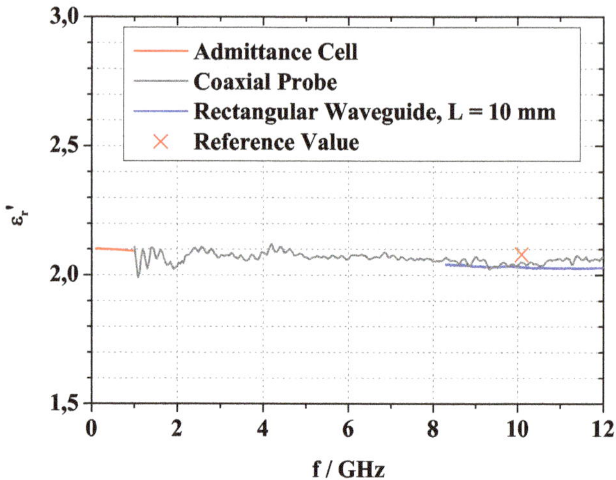

Fig. 3. Real part of the relative permittivity ϵ_r' of Teflon.

4 Measurements and comparison

Except for the measurements utilizing the admittance cell, all measurements were carried out on the basis of a vector network analyzer (VNA). Before measuring, the admittance, coaxial probe and the rectangular waveguide setup were calibrated.

For the CPW measurements five lines with lengths between 2 mm and 55 mm were structered on a sheet of Pocan DP T7140 LDS with a thickness $h = 2\,\text{mm}$ by the use of the MID LDS technology. The CPW dimensions were chosen to $w \approx s \approx 300\,\text{um}$, which hold the required condition $2(w + 2s) < h$ for the assumption of an infinite thick substrate. The lines were contacted via probes and a separate DC measurement for the determination of the conductivity σ was performed.

For the analysis of the measurement results achieved with the rectangular waveguide setup, the appearance of higher order modes, wall losses and air gaps between sample and waveguide walls were neglected. The uncertainties were calculated according to Eq. (5) for the given scattering-parameters uncertainties of the VNA and an estimated uncertainty in specimen length of $\Delta L = 50\,\mu\text{m}$.

Furthermore, the admittance cell and coaxial probe methods were only used for the determination of the real part of the permittivity ϵ_r', because of their low accuracy in loss measurements for an expected medium loss.

Before characterizing a MID material with almost unknown electromagnetic properties the performances of the admittance cell, the coaxial probe and the rectangular waveguide were tested by measuring ϵ_r' of the reference material Teflon (PTFE). The results and a reference value from Krupka et al. (1998) are shown in Fig. 3. The corresponding uncertainties for the admittance cell, coaxial probe and rectangular waveguide are $\Delta\epsilon_r'/\epsilon_r' = 14\%$, $\Delta\epsilon_r'/\epsilon_r' = 5\%$ for $f \geq 9\,\text{GHz}$ and $\Delta\epsilon_r'/\epsilon_r' = 1\%$ respectively. The measurement

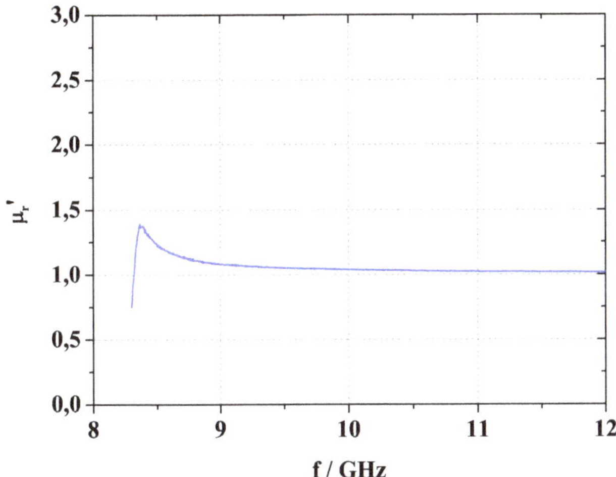

Fig. 4. Real part of the relative permeability μ_r' as determined by the rectangular waveguide method using Nicolson-Ross solution.

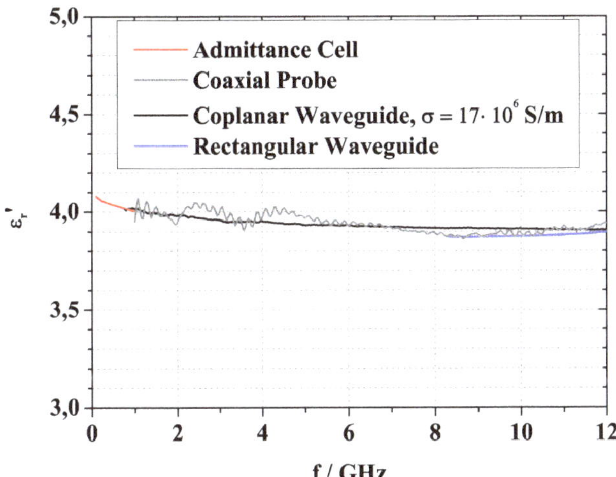

Fig. 5. Measured real part of the relative permittivity ϵ_r' of Pocan DP T7140 LDS.

curve achieved with the coaxial probe shows in comparison to the other curves significant ripples in this frequency region due to low accuracy at low frequencies. However, all measurement results lie in the typical range for teflon $2 \leq \epsilon_r' \leq 2.1$ and differ less than 5% from the reference value implying a reasonable performance of each method.

Based on this, the MID material Pocan DP T7140 LDS, which was assumed to be isotropic and homogeneous, was measured with all four methods. Since this thermoplastic includes an organic metal complex for LDS processing, it was first checked if the added complex caused a significant permeability. The result calculated with the Nicolson-Ross solution for a sample with a maximal available length of $L=10\,\text{mm}$ is displayed in Fig. 4. Except for lower frequencies near $f=8.5\,\text{GHz}$, where $\lambda/2$-resonance occurs and the Nicolson-Ross solution fails, the condition $\mu_r \approx 1$ holds.

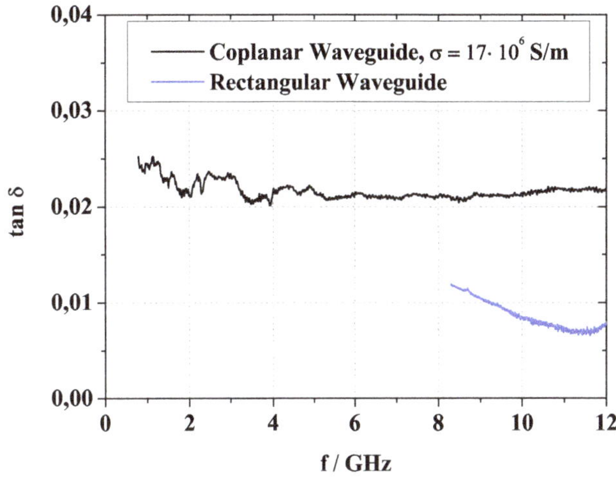

Fig. 6. Dielectric loss tangent $\tan\delta$ of Pocan DP T7140 LDS measured with both waveguide methods.

Consequently, the further analysis could focus on the determination of the complex permittivity only. The ϵ_r' results determined with all four methods are presented in Fig. 5. In the case of the rectangular waveguide method the result was calculated with the numeric solution of Eq. (2) as described in the previous section. The yielded uncertainties for the admittance cell, coaxial probe and rectangular waveguide are $\Delta\epsilon_r'/\epsilon_r'=13\%$, $\Delta\epsilon_r'/\epsilon_r'=5\%$ for $f \geq 7\,\text{GHz}$ and $\Delta\epsilon_r'/\epsilon_r'=1\%$, respectively. In particular, the results achieved with the admittance cell, CPW and rectangular waveguide method show a considerable mutual consistency indicating a permittivity ϵ_r', which continuously falls from approximately 4 at $f=1\,\text{GHz}$ to approximately 3.9 at $f=12\,\text{GHz}$. By contrast the coaxial probe measurement curve shows the strongest variations for frequencies below $f=7\,\text{GHz}$. This could be a result of low accuracy as well as imperfect contact between the probe and the hard sample of Pocan DP T7140 LDS.

In addition, the loss factor could be determined with both waveguide methods as shown in Fig. 6. Which of the both curves is closer to the true value can not be stated so far. On the one hand the rectangular waveguide method exhibits a mean uncertainty within the X-Band of $\Delta\tan\delta/\tan\delta \approx 50\%$ and on the other hand the CPW loss measurement could be affected by the roughness of the laser structured surface, which is not implicated in the used quasi-TEM model. A final statement could be issued by measuring a longer sample with the rectangular waveguide method or an additional resonator method. Unfortunately both possibilities could not be realized so far, due to lack of adequate samples. Nevertheless, a loss factor $\tan\delta$ in the dimension of 10^{-2} can definitely be assumed.

5 Conclusions

In this paper four measurement methods for the electromagnetic characterization of MID materials have been presented and compared with each other. Each method has been investigated in terms of measurement setup, functionality, bandwidth, uncertainties, sample requirements and finally measurement performance.

The admittance cell enables measurement of ϵ_r' with highest uncertainty. Since it requires only little sample and setup preparation, it is a suitable method for a first, less time consuming permittivity characterization. The coaxial probe is also a simple technique, but with a capability of higher measurement bandwidth. It makes little demands on the sample, but larger on the user having to regard to appropriate contacting and calibration, which were the reasons for noticeable ripples in the measurement curves. It is also a convenient method for a first broadband valuation of ϵ_r'. The CPW measurements also exhibit an adequate possibility of broadband permittivity characterization, but they invoke high expenses in sample preparation and measurement complexity. Especially for MID LDS structured CPWs a surface roughness probably has to be taken into account for accurate loss determination. Finally, the rectangular waveguide method represents the most powerful of the four discussed methods concerning measurement capability. With the potential of determination of both the complex permittivity and the complex permeability it is well suited for a basic classification of a material with unknown electromagnetic properties. Furthermore for low permittivity materials with medium or low losses it is well suited for the determination of ϵ_r', but not necessarily for loss characterization. For accurate measurements long samples with respect to the wavelength are required, which are not always available.

Next to the comparison of the measurement methods, the MID material Pocan DP T7140 LDS was characterized within a frequency range of f=0.1 GHz to f=12 GHz for the first time. Measurement results indicate a permeability $\mu_r \approx 1$, a dielectric constant falling from $\epsilon_r' \approx 4$ to $\epsilon_r' \approx 3.9$ and a dielectric loss factor $\tan\delta$ in the dimension of 10^{-2}. Finally, by considering the facts that different samples of this material were measured with four methods, each having a different electromagnetic field configuration, and that the results in ϵ_r' show a mutual consistence, the assumption of a homogeneous and isotropic material could be confirmed.

Acknowledgements. The authors wish to thank LPKF Laser & Electronics AG, Garbsen for structuring the CPW substrate. We also thank Rosenberger Hochfrequenztechnik GmbH & Co. KG for sponsoring the GSG Z-probes.

References

Arz, U. and Leinhos, J.: Broadband Permittivity Extraction from On-Wafer Scattering-Parameter Measurements, 12th IEEE Workshop on Signal Propagation on Interconnects, Avignon, France, 1–4, 2008.

Baker-Jarvis, J., Vanzura, E. J., and Kissick, W. A.: Improved Technique for Determining Complex Permittivity with the Transmission/Reflection Method, IEEE T. Microw. Theory, 38, 1096–1103, 1990.

Clarke, B., Gregory, A., Cannell, D., et al.: A Guide to the characterisation of dielectric materials at RF and microwave frequencies, National Physical Laboratory, Institute of Measurement and Control, London, 2003.

Heinrich, W.: Quasi-TEM Description of MMIC Coplanar Lines Including Conductor-Loss Effects, IEEE T. Microw. Theory, 41, 45–52, 1993.

Krupka, J., Derzakowski K., Riddle B., and Baker-Jarvis, J.: A dielectric resonator for measurements of complex permittivity of low loss dielectric materials as a function of temperature, Meas. Sci. Technol., 9, 1751–1756, 1998.

Marks, R. B.: A Multiline Method of Network Analyzer Calibration, IEEE T. Microw. Theory, 39, 1205–1215, 1991.

Nicolson, A. M. and Ross, G. F.: Measurement of the Intrinsic Properties of Materials by Time-Domain Techniques, IEEE T. Instrum. Meas., 19, 377–382, 1970.

Schlueter, R., Roesener, B., Kickelhain, J., and Naundorf, G.: Completely Additive Laser-Based Process for the Production of 3D MIDs, The LPKF LDS Process, 5th International Congress Molded Interconnect Devices, Erlangen, Germany, 2002.

10

Determination of the permittivity of soils by use of double transmission measurements

B. Will and M. Gerding

Ruhr-Universität Bochum, Institute of High Frequency Engineering, Universitätsstr. 150, 44801 Bochum, Germany

Abstract. Delay time measurements, e.g. time domain reflectometry (TDR), are a well-established method for the measurement of permittivity of various materials, especially soils. However, common measurement systems only provide one average value of the dielectric constant along the length of the TDR probe. This contribution deals with an advanced application of the TDR principle, the so-called double transmission method, for the determination of the water content of soil along a probe. To apply the advanced TDR technique, a probe, realized by a combination of a transmission line and a dielectric obstacle, which can mechanically be moved along the probe, is used. The probe is inserted into the soil to measure the effective soil permittivity. Thus, the water content along the probe can be estimated by means of the effective permittivity. Based on the known mechanical position of the reflection at the end of the probe and the position of the obstacle, the measured delay time can be used as a measure for the effective dielectric constant of the environment surrounding the obstacle. Thus, it is possible to determine the effective dielectric constant with a spatial resolution given by the step size of the obstacle displacement.

1 Introduction

Several well-established techniques have been developed for the determination of the permittivity in homogenous and inhomogeneous materials (Robinson et al., 2003). They all have in common, that their measurement results are limited to the mean value of the permittivity of the respective material of interest.

Due to the existing relation between the permittivity and the moisture of a material probe, permittivity measurements became a well established method for the determination of the water content of soils (Kupfer, 2005; Kupfer et al., 2000; Hoekstra and Delaney, 1974). By the use of capacitive sensors or time delay measurements it is possible to determine the mean value of the water content along a sensor which is penetrating into the medium.

This contribution deals with an improved measurement method, capable to determine the distribution of the permittivity of soils with an adjustable spatial resolution along a sensor, which is penetrating into the medium. By this, the method may be a solution for the upcoming needs to determine the moisture distribution inside soils for observing e.g. the infiltration of moisture on waste dumps and dikes and for solving the increasing numbers of industrial applications like the characterization of emulsions.

The conventional TDR (Time Domain Reflectometry) technique serves as the basis for the presented measurement principle, which itself can be subdivided into three alternative solutions: the reflection measurement, the transmission measurement and the double transmission measurement. All solutions make use of the delay time measurement of the transmitted or reflected signal, which is correlated to the permittivity of the material surrounding the sensor. The main difference between the three alternative solutions are the configurations of the measurement ports and the implementation of the local interaction between the electromagnetic signal travelling along the probe and the surrounding material. The following considerations focus on the double transmission concept, which is expected to be the most robust and reliable concept of the three mentioned alternatives.

2 Basics of the double transmission measurement

The basic functional principle of the double transmission measurement system is illustrated in Fig. 1. Figure 1a illustrates the setup for the reference measurement, while b, c, d illustrate an arbitrary measurement situation to determine the permittivity of the material located around the probe. The measurement setup itself mainly consists of the following components:

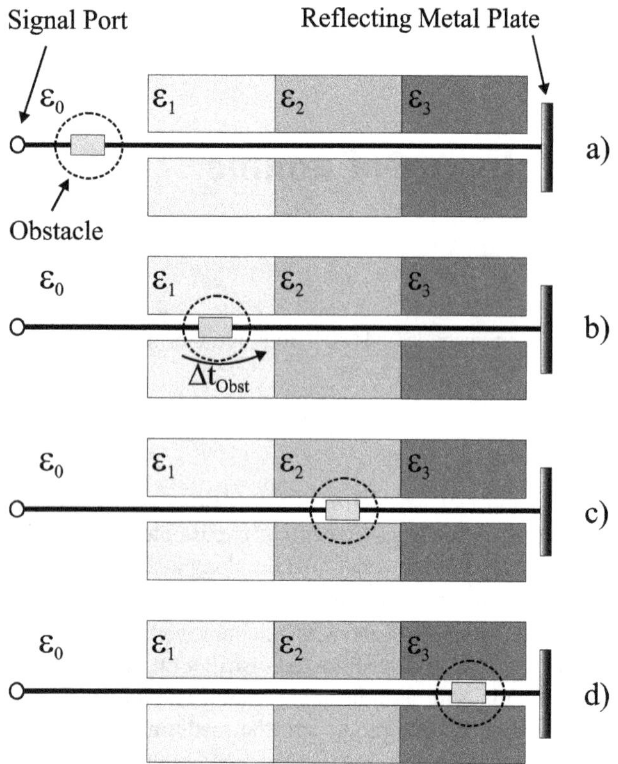

Fig. 1. Functional principle of the double transmission measurement: **(a)** Obstacle in reference position; **(b)**, **(c)**, **(d)** Obstacle in different measurement positions.

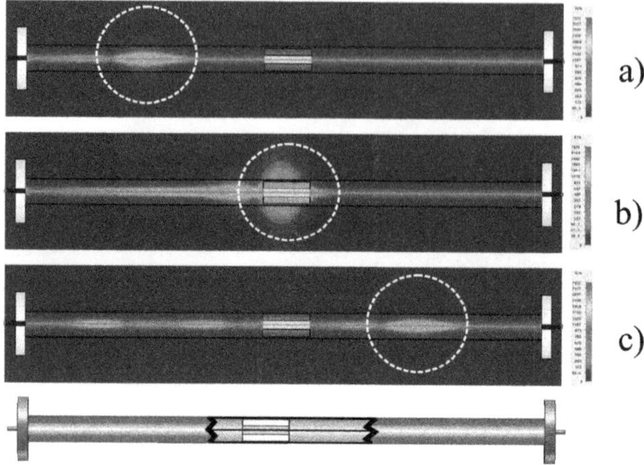

Fig. 2. EM-simulation illustrating the movement of the electromagnetic pulse along the probe: **(a)**, **(c)** The EM-field is concentrically concentrated around the waveguide; **(b)** The EM-field penetrates into the surrounding of the probe.

- A conductive single rod or cable, acting as a transmission line.
- The so called "obstacle", a dielectric cylinder, which can mechanically be moved along the waveguide, to provide a movable local electromagnetic disturbance.
- A metal plate at the end of the electromagnetic waveguide, reflecting the electromagnetic wave

The transmission line and the obstacle are covered by a thin-walled plastic tube for practical reasons. By this, the complete probe can directly be inserted into the medium while the space for the obstacle movement is still guaranteed by the plastic tube. Finally, the complete setup is connected to the measurement electronic via one signal port at the beginning of the probe. Thus, the signal flow is basically compatible with a conventional industrial baseband TDR-System.

Generally, the measurement principle is based on the propagation delay time of a transmitted electromagnetic pulse. The signal propagates along the probe, guided by the single rod inside the plastic tube. Excluding the volume in the vicinity of the obstacle, the electromagnetic field is concentrically concentrated around the transmission line with hardly any field components outside the plastic tube, as it is illustrated in Fig. 2a and c. At the end of the probe, the signal is reflected by the metallic plate and travels back in the opposite direction. Following the signal path, it is obvious that the signal is transmitted twice through the region of interest, illustrating the name of the introduced "Double Transmission" principle. At the obstacle, the electromagnetic field penetrates into the surrounding of the probe and thus penetrating into the material, which has to be characterized (Figs. 1b, 2b). This displacement of the electromagnetic field and the penetration of the outer material increases the propagation delay by Δt_{obst} and gives a measure of its permittivity. By mechanically moving the obstacle along the probe it is possible to characterize the region of interest at nearly any arbitrary position, as illustrated in Fig. 1b, c, d.

The displacement of the electromagnetic field into the surrounding, caused by the obstacle, has been validated via simulations and is due to the difference between the permittivity inside the plastic tube, which is close to the permittivity in free-space and the comparable high permittivity of the dielectric obstacle. The dielectric obstacle itself consists of a ceramic material ($\varepsilon_r \approx 15$). As a matter of fact, the electromagnetic wave takes the way providing the highest possible propagation speed. Estimating a permittivity of the area surrounding the obstacle of less than $\varepsilon_r \approx 15$, this forces the electromagnetic field to penetrate the outer medium, following the relation $c = c_0/\sqrt{\varepsilon_{r\mathrm{eff}}}$.

Thus, the propagation speed and the corresponding signal delay time of the transmitted signal is a measure for the relative effective permittivity of the surrounding material.

Initially, the obstacle is placed in its reference position outside the material under test, as illustrated in Fig. 1a, in order to achieve the overall signal delay time Δt_{ref} as a reference value:

$$\Delta t_{\mathrm{ref}} = \Delta \tilde{t}_{\mathrm{obst}} + \Delta t_{\mathrm{probe}} \tag{1}$$

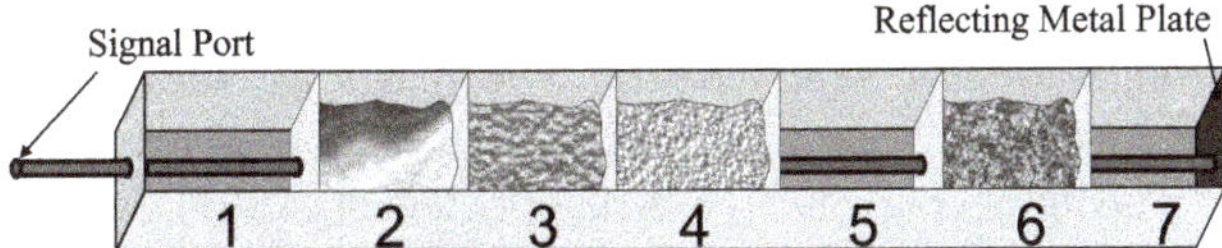

Fig. 3. Simplified schematic overview of the double transmission measurement setup which is used for the exemplary measurement results.

$\Delta\tilde{t}_{\text{obst}}$ represents the delay time, caused by the obstacle in its reference position and Δt_{probe} implies the signal delay time along the rest of the probe. The reference position of the obstacle is nearly arbitrary, as long as the obstacle is located outside the material under test.
In general, the delay time caused by the obstacle can be described as follows:

$$\Delta t_{\text{obst}} = 2 \cdot l_{\text{obst}} \frac{\sqrt{\varepsilon_{r_{\text{eff}}}}}{c_0} \tag{2}$$

The effective permittivity $\varepsilon_{r_{\text{eff}}}$ mainly comprises the permittivity of the plastic tube and of the outer material, which is penetrated by the electromagnetic field. The permittivity of the obstacle itself has nearly no influence on the effective permittivity $\varepsilon_{r_{\text{eff}}}$, as it has been verified by simulations, due to the effect, that the electromagnetic field is mainly not penetrating into the obstacle. In fact, the electromagnetic field is mainly pushed into the surrounding of the obstacle. During the reference measurement, the obstacle should preferably be covered by air, so that $\varepsilon_{r_{\text{eff}}}$ is mainly defined by the permittivity of air ($\varepsilon_r \approx 1$). Thus the effective permittivity $\tilde{\varepsilon}_{r_{\text{eff}}}$ in Eq. (2) of the active surrounding material during the reference measurement can be approximated to be close to 1, leading to:

$$\begin{aligned}\Delta\tilde{t}_{\text{obst}} &= 2 \cdot l_{\text{obst}} \frac{\sqrt{\tilde{\varepsilon}_{r_{\text{eff}}}}}{c_0} \\ &\approx 2 \cdot l_{\text{obst}} \frac{1}{c_0}\end{aligned} \tag{3}$$

l_{obst} is the mechanical length of the obstacle. Due to the double transmission principle, the transmitted signal passes the obstacle twice, which is taken into account by the factor of 2 in Eq. (3). Δt_{probe} describes the signal delay time, caused by the rest of the probe, excluding the obstacle. In a first approximation Δt_{probe} can be assumed to be independent of the obstacle position. This leads to the following equation for an arbitrary obstacle position (index i) as illustrated in Fig. 1b, c, d:

$$\Delta t(i) = \Delta t_{\text{obst}}(i) + \Delta t_{\text{probe}} \tag{4}$$

Defining the difference $\Delta T(i)$ between $\Delta t(i)$ and Δt_{ref} to be a measure for the effective permittivity of the material at the obstacles position, this yield:

$$\begin{aligned}\Delta T(i) &= \Delta t(i) - \Delta t_{\text{ref}} \\ &= \Delta t_{\text{obst}}(i) - \Delta\tilde{t}_{\text{obst}} \\ &= 2 \cdot \frac{l_{\text{obst}}}{c_0}\left(\sqrt{\varepsilon_{r_{\text{eff}}}(i)} - \sqrt{\tilde{\varepsilon}_{r_{\text{eff}}}}\right)\end{aligned} \tag{5}$$

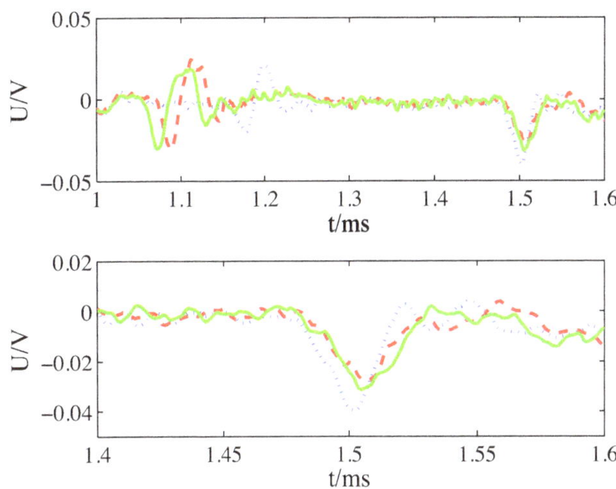

Fig. 4. Qualitative illustration of the different propagation delay times, caused by different materials under test aligned at three different positions along the probe.

Simplifying the equation by the use of the relation from Eq. (3) results in:

$$\varepsilon_{r_{\text{eff}}}(i) = \left(\frac{1}{2}\frac{\Delta T(i) c_0}{l_{\text{obst}}} + 1\right)^2 \tag{6}$$

Thus, it is possible to determine the effective permittivity $\varepsilon_{r_{\text{eff}}}(i)$ in the immediate vicinity of each arbitrary obstacle position. This yields a measurement system for the determination of dielectric profiles along the probe, e.g. by moving the obstacle stepwise along known positions. Furthermore it should be remarked, that the clearly layered measurement scenarios in Fig. 1 are simplified examples, only. The functionality of the introduced double transmission principle exceeds these examples and is also capable of handling mixed layers and irregular layered materials.

3 Measurements

The capability of the double transmission principle is verified by the following exemplary measurement results: the basis for these measurement is a test fixture, similar to the illustration in Fig. 3. According to the most probable application, the probe will be used in a vertical position, but for practical reasons, the test fixture is mounted in a horizontal position. This has no negative effect on the measurement principle and furthermore allows an easy placement of well defined test scenarios. The probe itself is fixed in the center of a horizontally placed box, which acts as a container for the material under test. The box itself is subdivided into single

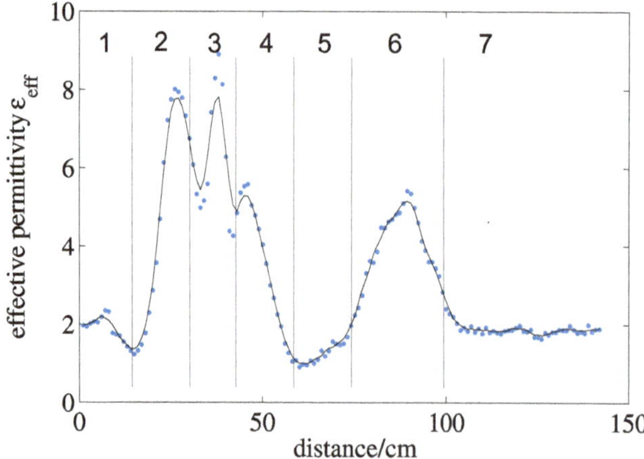

Fig. 5. Calculated effective permittivity based on measurement results corresponding to the measurement setup, illustrated in Fig. 3. (The step width of the obstacle movement is 1 cm. The solid line is the 5-point moving average curve of all single datapoints (dotted line).)

cavities, allowing the alignment of different "discrete" material layers for testing purposes. The signal port is located on the left side of the test fixture and the reflecting metal plate is located at the right side of the test fixture, directly at the end of the probe. According to the exemplary measurement scenario, the illustration in Fig. 3 shows four of the seven tests volumes filled with sand, each having a different humidity.

The propagation delay time of the transmitted signal measured at each obstacle position is a measure for the effective permittivity of the material along the probe position. The effect of the surrounding of the obstacle on the propagation delay time is qualitatively shown in Fig. 4. The curves show the resulting time-domain signals for arbitrary obstacle positions. The pulse signals around the particular time of 1.5 ms show the resulting time delay, which is used for calculating the effective permittivity by the use of Eq. (6).

In order to determine a permittivity profile, the obstacle stepwise has to be moved along the probe. The step width is a user-defined parameter and has a direct influence on the spatial resolution of the final permittivity profile. Here the step width is chosen to be 1 cm. Moving the obstacle along the complete probe and characterizing the test fixture as it is illustrated in Fig. 3 yields the permittivity profile in Fig. 5. As it can clearly be seen, the determined effective permittivity of the surrounding of the probe is always greater than 1, even if the test volume in three of the seven cavities is air. This is due to the mechanical design of the probe and has to be taken into account when finally calculating the absolute permittivity of the surrounding material under test.

Even if the test fixture gives the impression, that the presented principle is limited to well layered materials only, it has to be remarked that especially the double transmission method can handle smooth and homogeneous material mixtures, because the principle does not rely on the reflection of the separation layer between two materials, as it is the case by already known measuring principles (Huebner et al., 2007; Kupfer et al., 2007; Scheuermann et al., 2007). In case of the double transmission principle, the reflection of the separation layer and the reflection of the obstacle itself could be used to have some additional information to increase the reliability and the robustness of the system, but so far it has not been taken into account.

4 Conclusion

The so-called double transmission principle is introduced as an advanced and advantageous technique for the determination of the water content of soil along a probe. The basis of the advanced TDR technique is a probe, realized by a combination of a waveguide and a dielectric obstacle, which can mechanically be moved along the probe acting as the key component for the allocation between the measuring value and the place where it is taken. This leads to the system capability of characterizing permittivity profiles with an arbitrary step size. The step size is directly related to the resulting spatial resolution of the profile and can be adapted problem specific to the individual needs. Furthermore the system electronic is based on the well known TDR technique and can easily be adapted to existing industrial TDR-measurement systems as they are used in level measurements, for example.

References

Hoekstra, P. and Delaney, A.: Dielectric properties of soils at UHF and microwave frequencies, J. Geophys. Res, 79, 1699–1708, 1974.

Huebner, C., Schlaeger, S., and Kupfer, K.: Spatial Water Content Measurement with Time-Domain Reflectometry, tm-Tech. Mess., 74, 316–326, 2007.

Kupfer, K.: Electromagnetic Aquametry: Electromagnetic Wave Interaction With Water And Moist Substances, Springer, 2005.

Kupfer, K., Kraszewski, A., and Knöchel, R.: RF and Microwave Sensing of Moist Materials, Food and Other Dielectrics, Wiley-VCH, 2000.

Kupfer, K., Trinks, E., Schafer, T., Wagner, N., and Hubner, C.: Determination of Moisture and Density Distributions using TDR-Sensors, tm-Tech. Mess., 74, 298–307, 2007.

Robinson, D., Jones, S., Wraith, J., Or, D., and Friedman, S.: A Review of Advances in Dielectric and Electrical Conductivity Measurement in Soils Using Time Domain Reflectometry, Vadose Zone J., 2, 444–475, 2003.

Scheuermann, A., Bieberstein, A., Schlaeger, S., and Becker, R.: Optimized Sensor Design for the Determination of the Spatial Moisture Distribution in Electrical Lossy Soils, tm-Tech. Mess., 74, 308–315, 2007.

11

A comparison of software- and hardware-gating techniques applied to near-field antenna measurements

M. M. Leibfritz, M. D. Blech, F. M. Landstorfer, and T. F. Eibert

Institute of Radio Frequency Technology (IHF), Universität Stuttgart, Germany

Abstract. It is well-known that antenna measurements are error prone with respect to reflections within an antenna measurements test facility. The influence on near-field (NF) measurements with subsequent NF to far-field (FF) transformation can be significantly reduced applying soft- or hard-gating techniques. Hard-gating systems are often used in compact range facilities employing fast PIN-diode switches (Hartmann, 2000) whereas soft-gating systems utilize a network analyzer to gather frequency samples and eliminate objectionable distortions in the time-domain by means of Fourier-transformation techniques. Near-field (NF) antenna measurements are known to be sensitive to various errors concerning the measurement setup as there have to be mentioned the accuracy of the positioner, the measurement instruments or the quality of the anechoic chamber itself. Two different approaches employing soft- and hard-gating techniques are discussed with respect to practical applications. Signal generation for the antenna under test (AUT) is implemented using a newly developed hard-gating system based on digital signal synthesis allowing gate-widths of 250 ps to 10 ns. Measurement results obtained from a Yagi-Uda antenna under test (AUT) and a dual polarized open-ended waveguide used as probe antenna are presented for the GSM 1800 frequency range.

1 Introduction

Antenna measurements require free space conditions. However, in practice they are carried out in anechoic chambers equipped with absorbing material. These absorbers still cause reflections although the reflection level is significantly reduced as compared to a non-anechoic test site. They lead to incorrect amplitude and phase values of the acquired electromagnetic fields.

To counteract these errors it is possible to gate out reflected multipath components either by conventional software-gating or hardware-gating techniques. Both methods avail themselves of the time difference of arrival (TDOA) of line of sight and multipath signal components. These techniques are presented in Sect. 2.

Multipath propagation errors are mitigated applying gating techniques to measured near-field data. After that a NF to FF transformation is performed and a comparison of results obtained by either gating-technique and an ungated measurement are presented in Sect. 3. It is investigated how marginal errors of maximal 1 dB of the NF raw data effect the transformed FF as the error propagation through the transformation algorithm is not quite obvious.

The paper closes with a discussion of the measurement results and an evaluation of the novel gating method proposed.

2 Gating fundamentals

2.1 Software-gating

The software-gating principle is the most commonly applied gating technique. In this technique frequency samples acquired by a vector network analyzer (VNA) are postprocessed by means of a discrete Fourier-transformation (DFT). For this method no special RF measurement instruments are needed. As the postprocessing of the measurement data requires low computational effort an ordinary desktop PC is sufficient.

N_M equidistant frequency samples of the scattering parameter $S_{21}(n_M F_S)$ ($n_M = 0, 1, .., N_M - 1$) have to be measured over a wide frequency range. This data obtained at positive frequencies has to be zero-padded by at least $N_M - 1$ more samples yielding

$$S_{21,\text{padded}}(nF_S) = \begin{cases} S_{21}(nF_S) & , n \le N_M - 1 \\ 0 & , n > N_M - 1 \end{cases} \qquad (1)$$

with n=0, 1, .., $N-1$ before beeing transformed to the discrete analytic time-domain signal $s_{21,a}(kT_S)$ (k=0, 1, ..,

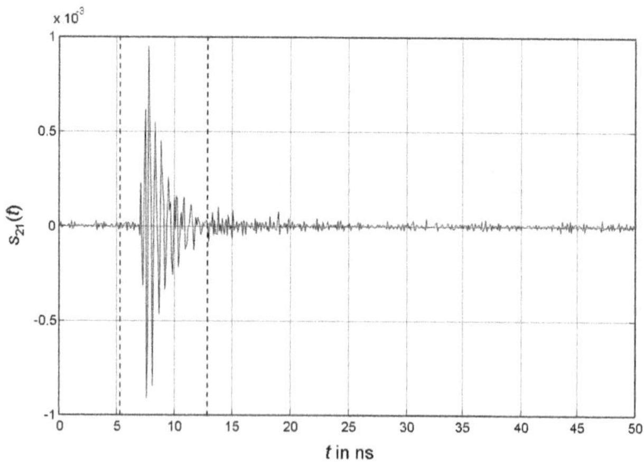

Fig. 1. Impulse response of the transmission link between AUT and probe antenna (black line) and possible time window (dashed lines).

$N-1$) (Oppenheim et al., 2004) by the inverse discrete Fourier-transform (IDFT)

$$s_{21,\mathrm{a}}(kT_\mathrm{S}) = \frac{2}{N}\sum_{n=0}^{N-1} S_{21,\mathrm{padded}}(nF_\mathrm{S})\ \mathrm{e}^{\mathrm{j}\frac{2\pi}{N}kn}. \tag{2}$$

In order to get a fine interpolation in the time-domain signal it is adequate to pad ten times $N_\mathrm{M}-1$) zeros.

The sampling interval in the frequency-domain is denoted by F_S, the time step in the time-domain by T_S. It has to be pointed out that the overall bandwidth $B_\mathrm{LP}=F_\mathrm{S}(N-1)$ of the N discrete frequencies determines the time step T_S between adjacent samples in the time-domain

$$T_\mathrm{S} = \frac{1}{B_\mathrm{LP}}, \tag{3}$$

whereas the sampling intervall F_S defines the unambigous range in the time-domain

$$T_\mathrm{unambiguous} = \frac{1}{F_\mathrm{S}}. \tag{4}$$

The real-valued impulse response

$$s_{21}(kT_\mathrm{S}) = \mathrm{Re}\left\{s_{21,\mathrm{a}}(kT_\mathrm{S})\right\} \tag{5}$$

is expected to be similar to the one depicted in Fig. 1. Signal components which have been received after the main peak due to multipath propagation in the anechoic chamber can be eliminated by applying a time window $w(kT_\mathrm{S})$ drawn in by dashed lines which gates out the ringing of the impulse

$$s_{21,\mathrm{gated}}(kT_\mathrm{S}) = s_{21}(kT_\mathrm{S})\ w(kT_\mathrm{S}). \tag{6}$$

This window should have a smooth amplitude tapering in order to avoid leakage effekts in the frequency-domain (Harris, 1978). In the measurements presented here a Hann window has been applied. After that the analytic signal $s_{21,\mathrm{gated,a}}(kT_\mathrm{S})$ of the gated signal $s_{21,\mathrm{gated}}(kT_\mathrm{S})$ has to be computed by means of discrete Fourier-transform as described in Marple (1999) (see Appendix A) before being transformed back to the frequency-domain by applying the discrete fourier transform (DFT)

$$S_{21,\mathrm{gated}}(nF_\mathrm{S}) = \sum_{k=0}^{N-1} s_{21,\mathrm{gated,a}}(kT_\mathrm{S})\ \mathrm{e}^{-\mathrm{j}\frac{2\pi}{N}nk}. \tag{7}$$

This gating procedure has to be applied to all measurement points on the spherical grid before being able to postprocess the data by a NF to FF transformation. The transformation can individually be done for each frequency sample of $S_{21,\mathrm{gated}}(nF_\mathrm{S})$.

2.2 Hardware-Gating

The principle of the new hardware gating technique developed at the IHF is to evaluate the measured sinusoidal test signal in the time interval after the envelope has reached its steady state and before the first reflected multipath component of the signal occurs. For a typical near-field measurement setup within the GSM 1800 range this is approximately 3–4 ns depending on the settling time of the AUT and the probe, respectively.

In order to evaluate the measured signal properly it is necessary that at least 1.5 ns of the sinusoidal impulse remain free of multipath components. This time windows corresponds to the interval marked by dashed lines in Fig. 2. As the settling time of the signal transmitted through the bandpass system consisting of both broadband antennas is 4 ns and the TDOA of the line of sight signal and the first multipath component is approximately 7 ns (see Fig. 2) there are 3 ns left for a proper measurement. In this interval the switched carrier has already reached its steady state and multipath components corrupting the received signals have not occured yet.

It should be noted that the bandwidth B of a system and its according rise time T_r are inverse proportional:

$$BT_\mathrm{r} = C,\ \ C = \mathrm{const.}\,. \tag{8}$$

For the acquisition of these short transient signals a fast real time scope or a broadband sampling scope is needed. With a sampling scope a higher dynamic range as well as a better resolution can be obtained.

For the evaluation of the envelope of a received signal a CW-signal with a rise time in the sub-ns region is required. This is necessary to ensure that the sinusoidal carrier has reached its steady state before multi-path components corrupt the signal. PIN-diode switches which are commonly used to gate out objectionable signal components in far-field measurement setups (Hartmann, 2000) cannot be used here as they are not available with a switching time below 1 ns. Therefore a custom signal generator was developed (Leibfritz, 2006) that synthesizes a sinusoidal impulse

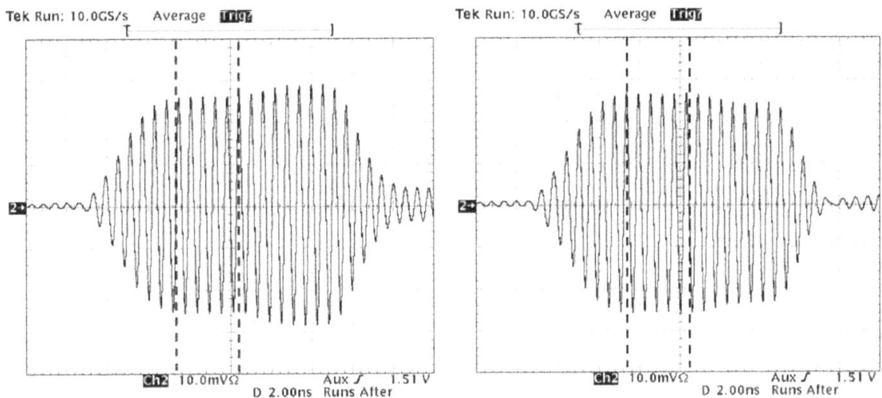

Fig. 2. Constructive (left) and destructive (right) interference caused by multipath propagation in the anechoic chamber and possible time window.

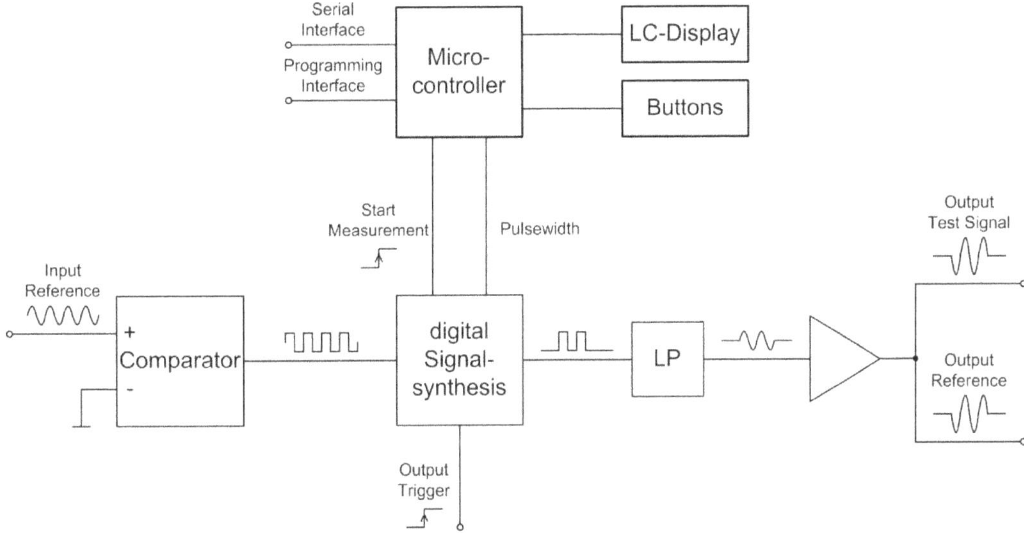

Fig. 3. Block diagram of the impulse generator employed for hardgating measurements.

by means of digital signal processing and subsequent low pass filtering. In this way a rise time of 150 ps is obtained. The block diagram including a microcontroller based interface and a broadband gain stage providing an output level of 12 dBm is depicted in Fig. 3.

3 Measurements

To evaluate the influence of the proposed gating techniques on the quality of a measured radiation pattern of an AUT a broadband two element yagi antenna is taken as test antenna (Fig. 4). This custom-made antenna has a center frequency of 1.7 GHz, a 10 dB bandwidth of approximately 500 MHz and thus exhibits a suitable rise time of the envelope for the hardgating system. It is fed by a coaxial $\frac{\lambda}{4}$-balun and its boom is made of polyvinyl chloride (PVC).

The size of this single test antenna allows far field measurements in a distance of a few meteres, but very often much larger arrays like base station antennas have to be measured requiring a minimum distance of more than 10 m to ensure far-field conditions. This and the fact that decreasing measurement frequencies result in an increasing distance of the Fraunhofer-region lead to near-field measurements as the preferred universal measurement technique.
Reference measurements are carried out on an optimized GSM 1800 test facility of a well known manufacturer of mobile phones. Formerly a division of Siemens. These measurement results are assumed to be error-free as the probe is mounted on an arc orbiting around the AUT causing hardly any distortions of the measurement signals. Thus all measurements namely an ungated measurement in the frequency-domain and a soft- as well as a hard-gated measurement carried out in the anechoic chamber of the Institute of Radio

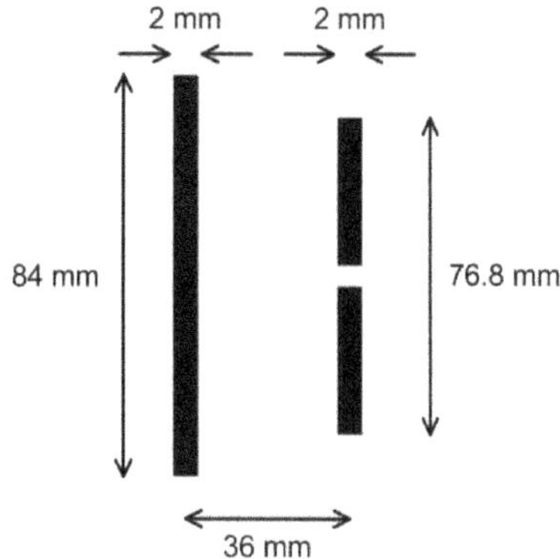

Fig. 4. Dimensions of the investigated AUT.

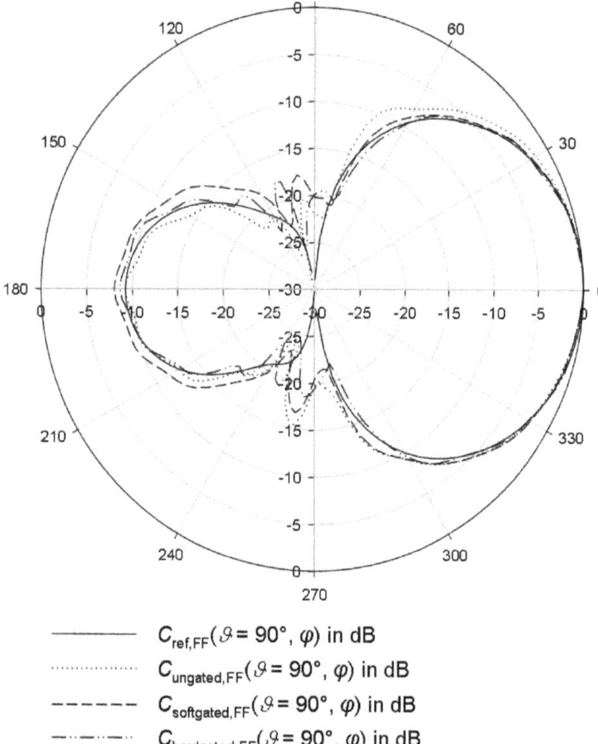

Fig. 5. Computed far-field patterns of the AUT (E-plane).

Frequency Technology at the Universität Stuttgart are compared to them. All frequency-domain measurements conducted there are made with a Wiltron 360B VNA. The time-domain signal acquisition of the hardgating system is carried out by a Tektronix TDS694C real time scope.

The spherical electric near-field is measured with an elevation over azimuth positioner from Orbit and a custom-made dual polarized open-ended waveguide used as probe antenna. This dual polarized probe helps to save time for the long-lasting measurements as it does not have to be positioned and all data can be acquired in one run. The probe antenna

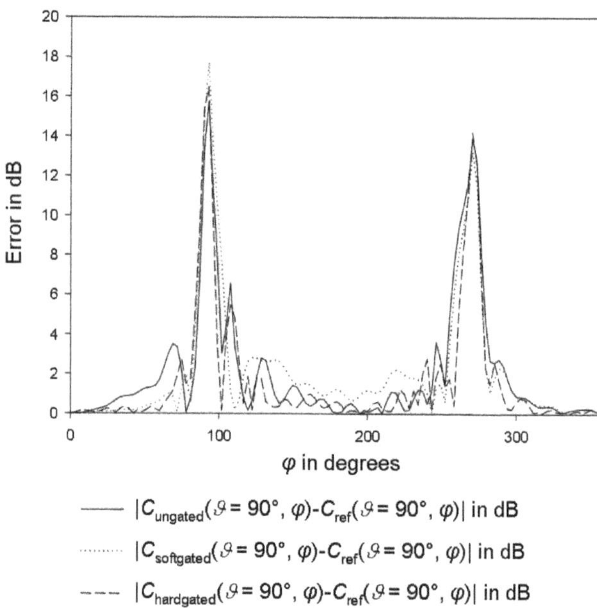

Fig. 6. Absolute error of computed far-field patterns.

is mounted at a distance of 1.7 m from the AUT whereas the angular measurement grid is 3° in ϑ and φ.

After the gating procedure the near-field data is transformed to the far field by an algorithm (Christ, 1995) also developed at the Universität Stuttgart using first order probe correction (Hansen, 1988).

As the near-field to far-field transformation algorithm is sensitive to amplitude and phase the near-field data should be acquired as accurate as possible. Otherwise small NF errors can lead to even worse distortions in the FF transormation results. Figure 5 shows the measured and computed far-field radiation patterns of the AUT in the E-plane. In Fig. 6 the absolute error of the computed far-field in the E-plane is shown. It can be seen that the error in the near- and far-field has its maximum close to the zeros of the reference measurement pattern. Because of this error peak of almost 20 dB this cannot be caused by the measurement instruments themselves even if the signals were slightly above the noise floor the measurement error cannot be more than 6 dB. In fact these errors arise from the elevation positioner the AUT is attached to which has a diameter of 30 cm. Hence the AUT suffers from the shadowing effect due to the dimensions of this obstacle when it moves in between the yagi and the probe. In this case neither of both gating techniques can improve the measurements, because the positioner is placed right in the line of sight between the two antennas and only reflected non line of sight components exist. The influence of the azimuth positioner can be neglected as it is located 2.5 m below the AUT and completely covered with absorbers.

Therefore further error analysis is only carried out for the main beam. In Fig. 6 no distinct improvement by either of

Table 1. Absolute mean error of ungated and gated far-field patterns after NF to FF transformation.

Angular Range	Measurement	$\lvert\overline{E}\rvert$ (NF)	$\lvert\overline{E}\rvert$ (FF)
−60°..+60°	ungated	0.42 dB	0.49 dB
−60°..+60°	software-gating	0.35 dB	0.25 dB
−60°..+60°	hardware-gating	0.42 dB	0.17 dB
−75°..+75°	ungated	1.10 dB	0.88 dB
−75°..+75°	software-gating	1.12 dB	0.46 dB
−75°..+75°	hardware-gating	1.10 dB	0.38 dB

Table 2. Absolute mean error of ungated and hardgated FF measurements.

Angular Range	Measurement	$\lvert\overline{E}\rvert$ (FF)
−60°..+60°	ungated	0.25 dB
−60°..+60°	hardware-gating	0.21 dB
−75°..+75°	ungated	0.28 dB
−75°..+75°	hardware-gating	0.25 dB
0°..360°	ungated	0.93 dB
0°..360°	hardware-gating	0.60 dB

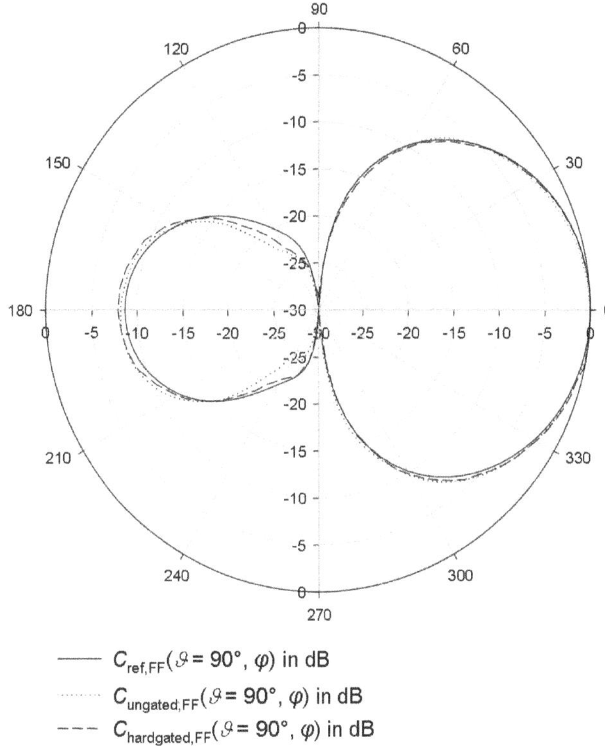

Fig. 7. Measured far-field patterns of the AUT (E-plane).

the gating techniques can be seen. Therefore in the following the absolute mean error (9) between ungated, soft- and hardgated measurements and reference data is evaluated in order to evaluate the different measurement results,

$$\lvert\overline{E}\rvert = \frac{\sum_{n=N_{\text{Start}}}^{N_{\text{Stop}}} \lvert C(n\Delta\varphi) - C_{\text{ref}}(n\Delta\varphi)\rvert}{\left(N_{\text{Start}} - N_{\text{Stop}} + 1\right)} \tag{9}$$

where $\Delta\varphi$ is the angle increment of the measurement grid. N_{Start} and N_{Stop} are the indices of of the first and the last sample in the regarded interval, respectively. $C_{\text{ref}}(n\Delta\varphi)$ is the radiation characteristic taken as reference whereas $C(n\Delta\varphi)$ is the radiation characteristic being erroneous.

As described earlier it is only reasonable to evaluate the main lobe of the AUT. For the comparison shown in Table 1 the maximum angular range of the radiation pattern of the two element yagi in the E-plane is limited to ±75°. As can be seen there the absolute value is evaluated in the near-field as well as in the far-field. The absolute mean error is reduced by both software- and hardware-gating techniques. Regarding the main lobe one can note that the ungated measurement yields a half-power beam width which has an error of 5° whereas the gated measurements coincide with the reference pattern.

The influence of the roll axis of the positioner can be shown if this axis is removed and a far-field pattern of the E-plane of the AUT is measured with the azimuth rotor only. Then the distortions caused by the positioner should vanish. The corresponding patterns for an ungated and a hardgated far-field measurement are compared to the reference pattern determined earlier (Fig. 7). It shows that there are no major distortions in the vicinity of the zeros any more. So the sidelobes in the transformed far-field are traced back to the incorrect near-field in some regions of the sphere caused by the disadvantageous dimension of the roll axis as proposed. The resulting absolute errors of this far-field measurement are plotted in Fig. 8. Table 2 shows a comparison of the absolute mean error. In this case an evaluation of the whole circular range is possible, significantly reducing the errorlevel as compared to the ungated measurement.

4 Conclusions

The influence of software- and hardware-gating techniques on erroneous near-field measurements data and subsequent near- to far-field transformation have been investigated. It has been shown that the resulting main lobe accuracy of the far-field pattern in the E-plane of a two element yagi antenna used as AUT can be improved by 0.5 dB using the novel hardware-gating system.

For a far-field measurement the improvement in the whole E-plane is 0.33 dB as compared to the reference pattern.

Further investigations will have to be carried out using UWB antennas as well as an elevation positioner with

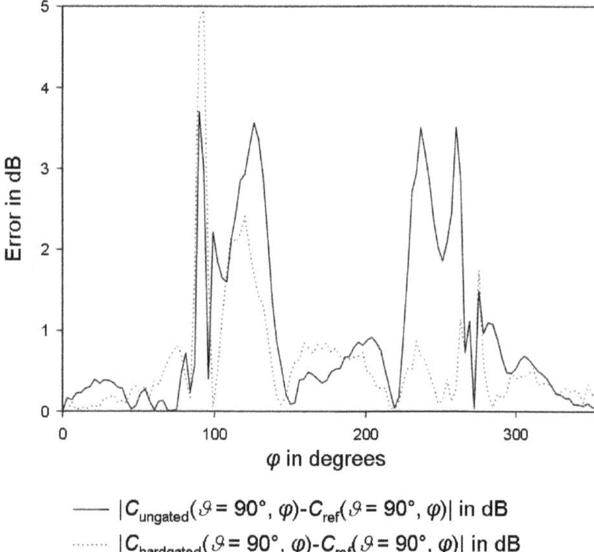

Fig. 8. Absolute error of measured far-field patterns.

reduced size. Employing state of the art measurement instruments with higher dynamic range than the ones used for the measurements presented here is supposed to lead to even better results.

Appendix A

Computation of the discrete time analytic signal

In Sect. 2.1 it is necessary to compute the discrete time analytic signal of the time-domain signal $s_{21,\text{gated}}(kT_S)$ in order to obtain a one sided spectrum. Therefor $S_{\text{aux},1}(nF_S)$, the DFT of $s_{21,\text{gated}}(kT_S)$, is computed in a first step

$$S_{\text{aux},1}(nF_S) = \text{DFT}\left\{s_{21,\text{gated}}(kT_S)\right\}. \tag{A1}$$

Then the auxiliary signal $S_{\text{aux},2}(nF_S)$ has to be created according to

$$S_{\text{aux},2}(nF_S) = \begin{cases} S_{\text{aux},1}(0) & ,n=0 \\ 2S_{\text{aux},1}(nF_S) & ,1 \le n \le \frac{N}{2}-1 \\ S_{\text{aux},1}\left(\frac{N}{2}F_S\right) & ,n=\frac{N}{2} \\ 0 & ,\frac{N}{2}+1 \le n \le N-1. \end{cases} \tag{A2}$$

In a final step the analytic time-domain signal $s_{21,\text{gated,a}}(kT_S)$ is obtained by transforming $S_{\text{aux},2}(nF_S)$ back to the time-domain by applying the IDFT

$$s_{21,\text{gated,a}}(kT_S) = \text{IDFT}\left\{S_{\text{aux},2}(nF_S)\right\}. \tag{A3}$$

The original time-domain signal $s_{21,\text{gated}}(kT_S)$ can be obtained by taking the real part of $s_{21,\text{gated,a}}(kT_S)$, whereas the absolute value of $s_{21,\text{gated,a}}(kT_S)$ provides its envelope.

Acknowledgements. The authors wish to thank the Hittite Microwave Corporation and ON Semiconductor for providing samples of their integrated circuits for the development of the hardgating system.

References

Christ, J.: Korrektur prinzipbedingter und durch die Meßumgebung verursachter Fehler bei der Nahfeld-Fernfeld-Transformation, Dissertation, Universität Stuttgart, Shaker-Verlag, Aachen, 1995.

Hansen, J. E.: Spherical Near-Field Antenna Measurements, Institution of Electrical Engineers, London, 1988.

Harris, F. J.: On the Use of Windows for Harmonic Analysis with the Discrete Fourier Transform, Proceedings of the IEEE, vol. 66, no. 1, January, 1978.

Hartmann, J.: Grenzen der Störstrahlungsunterdrückung bei der Vermessung von Mikrowellen- und Millimeterwellenantennen in kompensierten Doppelspiegel-Compact-Range-Meßanlagen, Dissertation, Universität der Bundeswehr, München, 2000.

Leibfritz, M. M.: Effiziente Methoden zur Diagnose bei Gruppenantennen unter Verwendung der sphärischen Nahfeldmesstechnik, Fortschrittsbericht, Universität Stuttgart, 2006.

Marple Jr., S. L.: Computing the Discrete-Time "Analytic" Signal via FFT, IEEE Transactions on Signal Processing, vol. 47, no. 9, September, 1999.

Oppenheim, A. V., Schafer, R. W., and Buck, J. R.: Zeitdiskrete Signalverarbeitung, Pearson Studium, München, 2004.

12

Development of a low cost robot system for autonomous measuring of spatial field distributions

B. Schetelig[1], S. Parr[1], S. Potthast[2], and S. Dickmann[1]

[1]Faculty of Electrical Engineering, Helmut-Schmidt-University/University of the Federal Armed Forces Hamburg, Germany
[2]Bundeswehr Research Institute for Protective Technologies and NBC Protection (WIS) Munster, Germany

Correspondence to: B. Schetelig (schetelig@hsu-hh.de)

Abstract. A new kind of a modular multi-purpose robot system is developed to measure the spatial field distributions of very large as well as of small and crowded areas. The probe is automatically placed at a number of pre-defined positions where measurements are carried out. The advantages of this system are its very low influence on the measured field as well as its wide area of possible applications. In addition, the initial costs are quite low. In this paper the theory underlying the measurement principle is explained. The accuracy is analyzed and sample measurements are presented.

1 Introduction

In the context of electromagnetic compatibility, the knowledge of the electromagnetic field distribution in a certain plane or volume is often required. This information can be used to verify the efficiency of shielding measures or to calibrate field simulators, for example. During such measurement campaigns, usually a large amount of data is collected for a lot of positions in the test area. Taking these measurements manually can be very time-consuming. To speed-up the process, the field probe can be placed at the designated positions by a robot. There are several specialized automated measurement systems, designed for special applications (e.g. Haake, 2010). Most of these systems cover a given maximum volume depending on its skeleton size. The probe manipulation is usually done with rectangularly arranged wooden or plastic arms.

The disadvantage of these systems is that due to their fixed-size arms and skeleton they are limited to applications matching their size. If the volume under test is too small, the robot cannot be placed inside. If it is much larger than the scanning area of the robot, the total measurement has to be divided into several parts. In addition, such systems often are too big to speak of an easily movable measuring device.

To overcome these disadvantages, the system presented in this paper uses a different approach. To achieve mobility and to cover measurement sites of very different sizes, the measurement probe is positioned at the measurement points using two belts (Fig. 1). By manipulating the lengths of these two belts, it is possible to move the probe to any requested position. By using belts with adequate lengths, it is possible to vary the size of the covered area very easily. In spite of this variability, the total system mainly consists of two actuator boxes and two belts. In the presented setup, rubber belts are used which are armed with fibreglass to ensure both tensile strength and a very small influence of the measurement system on the field to be measured.

2 Theory

The field probe is placed at the requested position $(x_\mathrm{m}, y_\mathrm{m})$ in the measurement plane by pulling and releasing the belts. The necessary lengths of the belts for a placement can be calculated by

$$l_i = \sqrt{(x_\mathrm{m} - x_i)^2 + (y_\mathrm{m} - y_i)^2}\,, \tag{1}$$

with: l_i: length of the uncoiled parts of the belts, (x_i, y_i): positions of the actuator boxes.

To enhance the precision, one has to take into account that the belts are not ideal strings. The belts get stretched a little bit when attaching heavy sensors. This influence can usually be ignored if using belts that are oversized with respect to the expected loads. Another influence comes from the weight of the belts itself. The belts do not connect their fixing points in straight ways but are sagging the more the sensor weight is getting smaller in relation to the weight of the belts. From the balance of forces at every position x of

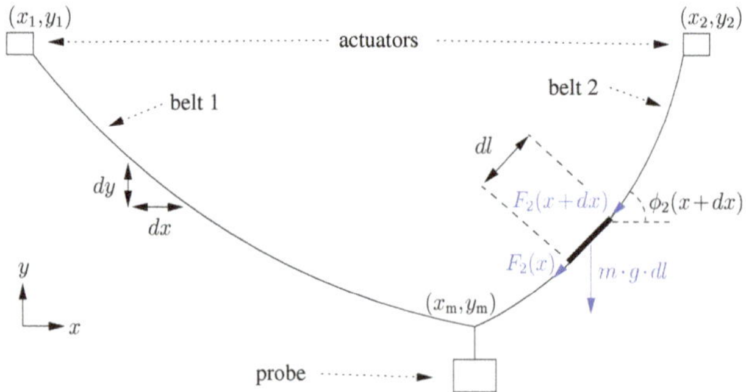

Fig. 1. Positioning a measurement probe in an electromagnetic field using belts.

each belt (Fig. 1), a non-linear secondary order differential equation can be derived (Hagedorn, 2006):

$$\frac{\mathrm{d}^2 y}{\mathrm{d}x^2} = \frac{m \cdot g}{F_\mathrm{x}} \cdot \sqrt{1 + \left(\frac{\mathrm{d}y}{\mathrm{d}x}\right)^2}\,, \tag{2}$$

with: m: weight of the belts per element $\mathrm{d}l$, g: gravity acceleration. $F_\mathrm{x} = F \cdot \cos(\phi)$ is the horizontal force at each belt.

The solution of this differential equation is the so-called catenary equation:

$$y(x, x_0, y_0, k) = \frac{1}{k}\cosh(k(x - x_0)) - \frac{1}{k} + y_0\,, \tag{3}$$

with: (x_0, y_0): shift of the catenary and:

$$k := \frac{m \cdot g}{F_x}\,. \tag{4}$$

Equation (3) must be applied separately to the left and the right belt, as they usually have different shapes (Fig. 1). The balance of forces at the positions of the actuators (x_1, y_1), (x_2, y_2) and of the probe $(x_\mathrm{m}, y_\mathrm{m})$ allow to derive a set of equations which can be used to calculate the shift-parameters x_0, y_0 for both belts $(x_{10}, y_{10}, x_{20}, y_{20})$.

The requested lengths of both belts can be calculated as

$$\begin{aligned} l_{1,2} &= \pm \int_{x_{1,2}}^{x_\mathrm{m}} \mathrm{d}l = \pm \int_{x_{1,2}}^{x_\mathrm{m}} \sqrt{1 + \left(\frac{\mathrm{d}y}{\mathrm{d}x}\right)^2}\,\mathrm{d}x \\ &= \pm \int_{x_{1,2}}^{x_\mathrm{m}} \frac{1}{k}\frac{\mathrm{d}^2 y}{\mathrm{d}x^2}\,\mathrm{d}x = \pm \frac{1}{k} \cdot \frac{\mathrm{d}y}{\mathrm{d}x}\bigg|_{x_{1,2}}^{x_\mathrm{m}}\,, \end{aligned} \tag{5}$$

whereas the derivation of Eq. (3) is used:

$$\begin{aligned} l_1 &= \frac{\sinh(k \cdot (x_\mathrm{m} - x_{10})) - \sinh(k \cdot (x_1 - x_{10}))}{k}\,, \\ l_2 &= \frac{\sinh(k \cdot (x_2 - x_{20})) - \sinh(k \cdot (x_\mathrm{m} - x_{20}))}{k}\,. \end{aligned} \tag{6}$$

The quantity k can be calculated according to Eq. (4). As the horizontal force F_x is not exactly known a priori, a first estimation value has to be calculated assuming that the per unit weight of the belts is zero. In this borderline case, F_x can be calculated from the geometry as:

Fig. 2. Measurement setup in the DIES EMP simulator Munster. The actuators are mounted on a mobile wooden frame.

$$F_\mathrm{x} = \frac{\cos(\phi_1) \cdot \cos(\phi_2)}{\sin(\phi_2 - \phi_1)} \cdot M \cdot g\,, \tag{7}$$

with: M: weight of the sensor and $\phi_{1,2}$: angles between horizon and the belts.

The final value of k is determined by an iterative approach by:

$$k_\mathrm{new} = [\sinh(k \cdot (x_\mathrm{m} - x_{20})) - \sinh(k \cdot (x_\mathrm{m} - x_{10}))] \cdot \frac{m}{M}\,. \tag{8}$$

The precision of placing the probe depends on its absolute position in the scanning area. Horizontally in the middle between the actuators at the bottom of the scanning area, the positioning precision is the highest with a given pulling precision of the stepping motor. Moving the probe up towards its highest possible position, the accuracy is decreasing. The same way, the force F_i to be applied on the belts is getting larger and larger. It can be derived as

$$F_{1,2} = \frac{M \cdot g \cdot \cos(\phi_{2,1})}{\cos(\phi_1) \cdot \sin(\phi_2) - \sin(\phi_1) \cdot \cos(\phi_2)}\,, \tag{9}$$

if the weight of the belts is neglected for simplification.

3 Applied mechanics

Previous work (Haake and ter Haseborg, 2008) has shown that the type of rope or belt chosen is critical for the precision of the system. A rope is not adequate for precise positioning. The main problem is that it is not possible to measure the length of the uncoiled rope in a sufficient precise and reliable way. One reason is that the rope is always slipping a bit at the cylinder, which is counting the length of the uncoiled rope. In addition, even high-performance Dyneema ropes stretch too

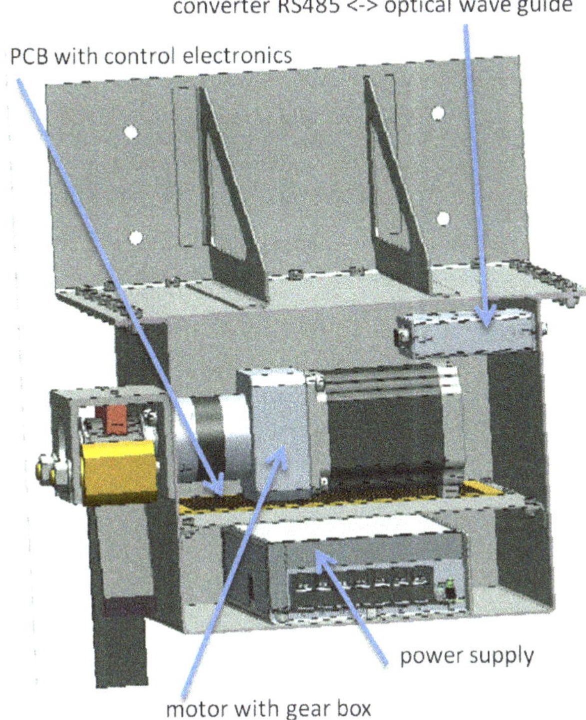

Fig. 3. CAD drawing of one of the actuator boxes.

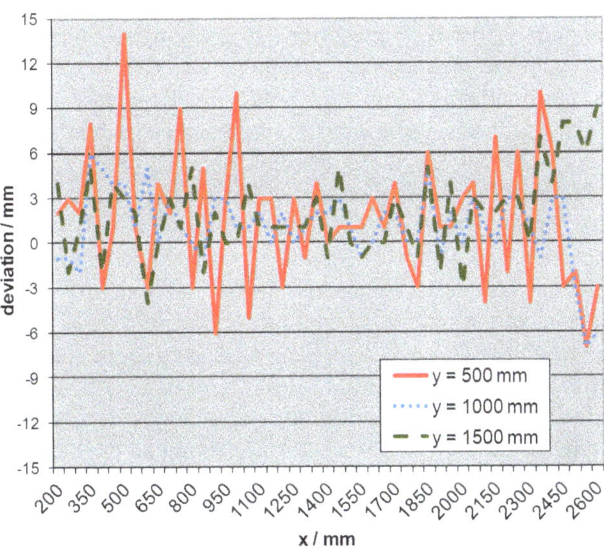

Fig. 4. Positioning deviations in the horizontal direction.

much, depending on the position of the sensor, its weight and even the actual kind of coiling operation: The rope usually runs over some deflection rollers. In this area, the lengthening of the rope depends on whether the motor winds the rope up (more stretching) or unwinds it (less stretching).

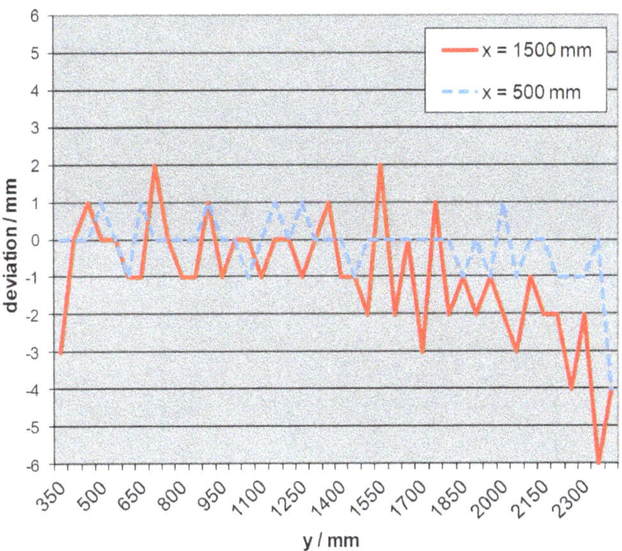

Fig. 5. Positioning deviations in the vertical direction.

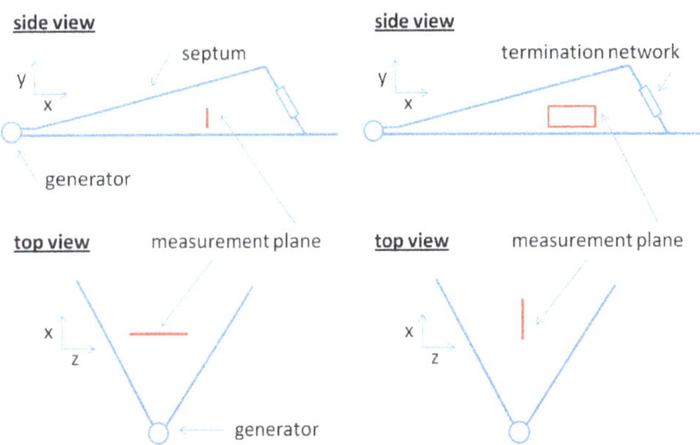

Fig. 6. Measurement setups in the EMP simulator. Left pictures: transversal measurement, right pictures: length-wise measurement.

These difficulties can be overcome by using a cam belt. It cannot slip and in combination with a toothed wheel it is perfect for a precise calculation of the actual length. In addition, there are several fibreglass-armed cam belts available that have almost no lengthening when loaded with the sensor weight.

The mechanical setup is simplified, too. Stepping motors have a fourth state (besides "off", "right", "left"): it can act as a brake and fix the sensor at the desired positions. So, no additional elements are needed. In Fig. 3, the inside of one of the actuator boxes is presented. The belt can be seen on the left, moved by the motor on the inside. Below, you find the PCB controlling the motor. It consists of a power electronics area and a section realizing the communication with the personal computer, which controls the measurement devices.

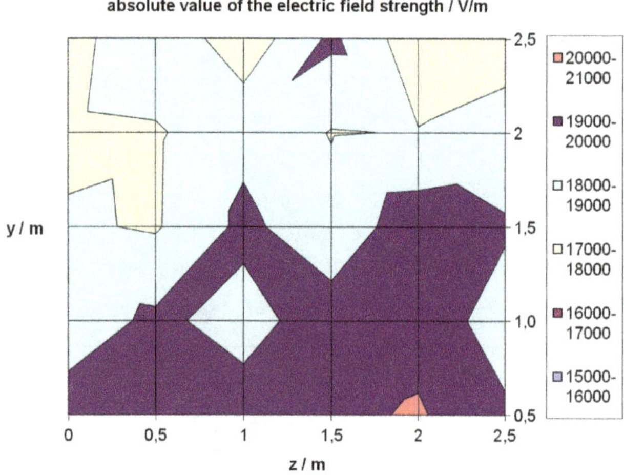

Fig. 7. Vertical component of the spatial field distribution in the DIES EMP simulator (transversal plane).

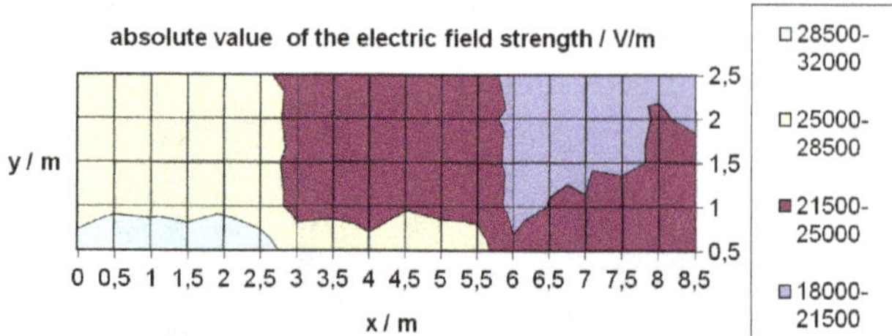

Fig. 8. Vertical component of the spatial field distribution in the DIES EMP simulator (length-wise plane).

Data enters the box via a full duplex optical data link. The small box on the upper right is the appropriate converter.

The entire system is designed to work with a drag force F_i of up to 44 N at each cam belt. The maximum weight of the sensor depends on this limit and the size of the respective measurement area. The conveying velocity of the belt is $21.4\,\mathrm{mm\,s^{-1}}$.

4 Control electronics and software

In each of the actuator boxes, identical PCBs can be found. They are equipped with Atmel ATMega32 microcontrollers for communication with the personal computer and to control the power electronics sections. This section of the PCB mainly consists of the STMicroelectronics devices L297, L298N and the belonging external circuit elements. These devices are used to easily drive the stepping motor using the microcontroller. The microcontrollers communicate with the personal computer via a full duplex RS-485 interface (MAX488, MAX489). This kind of bus system was chosen as its differential wires help to withstand the influence of external electromagnetic field. The bus cables are connected to a PC interface which translates the RS-485 signals to USB. The interface is detected by the computer as a virtual COM port.

To reduce the vulnerability even more, the RS-485 copper wires connecting the actuator boxes with the interface can be replaced by fibre optical wires. Additional converters in the boxes and the interface execute a code-transparent translation.

To ensure the electromagnetic compatibility, the electronics and the motor are encapsulated in a metal box that is sealed by a tight row of screws. All feed-throughs are manufactured very carefully. The data connection is realized with a fibre optical link. Power supply uses shielded cables and connectors. In addition, filters are applied at the ports of the PCB and the power supply. These measures allow operation up to an electric field strength of at least $32\,\mathrm{kV\,m^{-1}}$.

The entire system is controlled by a Matlab-based computer programme. Nevertheless, the open documentation of all available instructions, which can be sent to the actuator box microcontrollers, allow to connect the hardware to any other programming environment that is able to communicate via a serial computer port.

5 Precision of positioning

The precision of sensor positioning depends on the accuracy of the rotational movement of the motor and the modelling accuracy of the belt catenaries. It is also quite important to set up the measurement site properly. That means that the positioning can only work sufficient, if the vertical and horizontal distances between the actuator boxes are known exactly and this data is fed into the control software.

In the presented setup, a step of the motor equals 0.45 degrees at the gear shaft. One degree at the shaft results in a belt drive of 0.2 mm. This allows to coil up and uncoil the belt very precisely. From this follows that the positioning faults mainly result from a poor measurement of the distance between the boxes and inaccurate modelling of the shape of the belts.

In Figs. 4 and 5 the positioning accuracy is measured in a typical measurement setup for horizontal as well as vertical movement. The boxes were mounted on a wooden frame (cf. Fig. 2). The deviations are measured in a range of 2.4 m (horizontally) and 1.95 m (vertically). Figure 4 shows that the horizontal precision is best in the middle between the boxes and decreases quite symmetrically towards the left and the right end. To enhance the area of high precision in the middle, the distance between the mounting positions of the boxes can be increased. The deviations in the vertical direction are limited to 6 mm in this setup.

The precision of positioning of the field probe is sufficient for the intended use of measuring the spatial field distribution of an EMP simulator.

6 Field measurements in the EMP simulator

Test measurements were carried out at the DIES simulator ("Deutsches Impulserzeugendes System zur EMP-Simulation") in Munster in two measurement planes. Their orientations are sketched in Fig. 6. The electromagnetic pulse is generated at the left end (side view) with a Marx generator and propagates along the expanding transmission line. At the right end, the line ends in a termination network.

Figure 7 shows the spatial distribution of the electric field with a maximum field strength of $20\,\mathrm{kV\,m^{-1}}$ in the transversal plane. The larger field strength in the lower area can be traced back to the fact that only the vertical component is measured by the sensor. In the upper area, the field lines are slightly bent and are not longer totally vertical as they have to be rectangular to the septum.

The length-wise measurement plane is shown in Fig. 8. The generator is placed at the left end. As expected, the field strength decreases with growing x-values. This can be explained with the increasing distance between generator and measurement point and with the growing distance between floor and septum.

7 Conclusions

In this paper we presented a multi-purpose robot system for autonomous field measurements. Its concept allows the application in very large as well as in very narrow and crowded environments. Due to its modularity, it can be easily moved to other test sites. To minimize the distortion of the measured field caused by the system, fibreglass-armed belts are used to position the sensor. Summarized, it is a very easy to handle, reliable and yet economic system.

References

Haake, K.: Automatisierte Messsysteme für elektromagnetische Felder komplexer Strahlungsquellen, Ph.D. Thesis, Technische Universität Hamburg-Harburg, 2010.

Haake, K. and ter Haseborg, J. L.: Development of a modular low cost robot for scanning the electromagnetic field within very large arbitrary areas or volumes, Serbian Journal of Electrical Engineering, 5, 49–56, 2008.

Hagedorn, P.: Technische Mechanik, Band 1 – Statik, Harri Deutsch Verlag, 2006.

13

Benefits of on-wafer calibration standards fabricated in membrane technology

M. Rohland[1,2], U. Arz[1], and S. Büttgenbach[2]

[1]Physikalisch-Technische Bundesanstalt, 38116 Braunschweig, Germany
[2]Institut für Mikrotechnik, Langer Kamp 8, 38106 Braunschweig, Germany

Abstract. In this work we compare on-wafer calibration standards fabricated in membrane technology with standards built in conventional thin-film technology. We perform this comparison by investigating the propagation of uncertainties in the geometry and material properties to the broadband electrical properties of the standards. For coplanar waveguides used as line standards the analysis based on Monte Carlo simulations demonstrates an up to tenfold reduction in uncertainty depending on the electromagnetic waveguide property we look at.

1 Introduction

Traceability for scattering parameter measurements back to SI units has long been possible for coaxial waveguides. The characteristic impedance can be calculated very accurately from the cross-sectional dimensions of the coaxial geometry. It is in principle also possible to determine the properties of coplanar waveguides to a high degree of accuracy.

However, uncertainties in the complex permittivity of the dielectric substrate can have a substantial effect on the propagation characteristics of the coplanar waveguide (CPW) as recent studys (Arz et al., 2008a) have shown. Even when using state-of-the-art measurement techniques (Arz et al., 2008b) to accurately capture the frequency-dependent permittivity and loss tangent of the substrate, the remaining uncertainties in the waveguide properties are still much higher than in the coaxial airline case.

A carrier substrate beneath the coplanar lines with the effective dielectric constant value close to the ideal value of 1 provides one possible solution of this problem. Such a carrier substrate would lead to excellent dispersion properties and also prevent the excitation of undesired surface modes. A good approximation of these ideal structures are CPWs fabricated on thin insulating carrier membranes. These coplanar waveguides have been built successfully. The applied technologies are normally used for fabricating microelectromechanical systems (MEMS).

2 CPW fabrication

The technology for fabricating membrane-supported CPWs is currently being developed in a collaboration with the Institute for Microtechnology in Braunschweig, Germany. The cross section of a coplanar airline built in membrane technology together with its geometrical and material parameters is shown in Fig. 1. The CPW is supported by a thin film stretched across a silicon frame. In order to meet the requirements, the layer must possess good electrical and mechanical properties. The thin film must have low losses at microwave and millimeter-wave frequencies in order to achieve superior electrical performance, as well as be compatible with semiconducting and conducting materials. Reduced sensitivity to applied pressure and temperature variations, along with increased membrane sizes must be considered for optimization of the mechanical properties. Therefore, we use a thin film of silicon nitride produced by plasma enhanced chemical vapor deposition (PECVD) at multi-frequency (mf-nitride), made available by Rohde & Schwarz, to process membrane supported CPWs.

In contrary to the well-known three-layer approaches described in Dib et al. (1991); Weller et al. (1994); Katehi and Rebeiz (1996), our approach consists of one layer. The three-layer membranes in Dib et al. (1991); Weller et al. (1994); Katehi and Rebeiz (1996) use three process steps to deposit one layer of thermal oxide, one layer of low pressure chemical vapor deposition (LPCVD) silicon nitride and one layer of LPCVD silicon oxide.

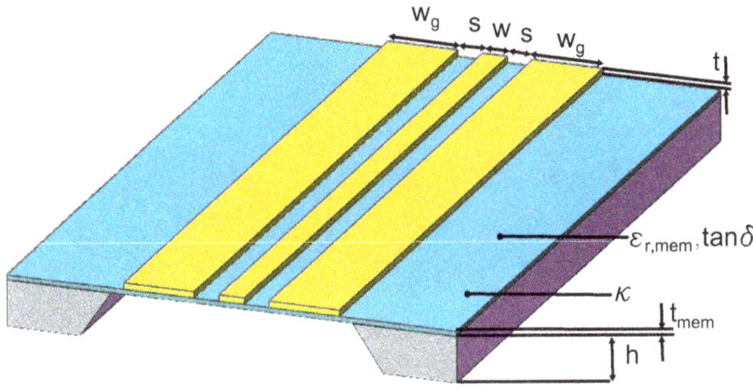

Fig. 1. Cross section of CPW based on membrane technology.

The mf-nitride used in this work is one dielectric layer which is deposited in one step on a silicon wafer. The conducting lines are constructed by applying and structuring a gold layer on the mf-nitride. To selectively remove silicon, wet chemical etching is used. This fabrication step opens the membrane windows on which the conducting lines are carried.

3 CPW modelling

The model presented in (Heinrich, 1993) was applied to investigate the electromagnetic propagation properties of the CPWs. The model is based on the assumption of an infinitely expanded dielectric substrate beneath the coplanar metal lines. This assumption is well met for the case of conventional, several hundred µm thick substrates.

For the CPW fabricated in membrane technology, however, this assumption is not valid. Instead of a several hundred µm thick substrate we have a $t_{\mathrm{mem}} = 1$ µm thick dielectric membrane over air. To calculate the effect of the membrane, we use an equivalent permittivity value $\varepsilon_{\mathrm{r,LHP}}$ of a fictitious infinitely-expanded lower half-plane allowing to capture the dielectric influence of the thin membrane material on the equivalent lower half-plane in the model from (Heinrich, 1993).

To calculate $\varepsilon_{\mathrm{r,LHP}}$, we first computed the capacitance per unit length C'_{mem} of the membrane-based CPW with high-precision 2D-FEM simulations (QuickField™, 5.5). We then used a simple analytical relationship between $\varepsilon_{\mathrm{r,LHP}}$ and C'_{mem}, which can be derived from (Heinrich, 1993):

$$\varepsilon_{\mathrm{r,LHP}} = \frac{C'_{\mathrm{mem}}}{2\varepsilon_0 F_{\mathrm{low}}} - \frac{F_{\mathrm{up}}}{F_{\mathrm{low}}} \qquad (1)$$

The expressions for F_{low} and F_{up} are given in (Heinrich, 1993), they depend only on the geometry of the CPW's cross section.

Table 1. Membrane CPW parameters and associated uncertainties.

Parameter	value	half-width of pdf interval
w_g	258 µm	0.25 µm
w	166 µm	0.25 µm
s	10 µm	0.25 µm
t	1 µm	0.03 µm
κ	27 MS/m	1 MS/m
$\varepsilon_{\mathrm{r,mem}}$	4.1	0.2
$\varepsilon_{\mathrm{r,LHP}}$	1.2707	0.0148
$\tan\delta$	0.0001	$5\cdot10^{-5}$

Table 2. Virtual CPW parameters and associated uncertainties.

Parameter	value	half-width of pdf interval
w_g	258 µm	0.25 µm
w	166 µm	0.25 µm
s	10 µm	0.25 µm
t	1 µm	0.03 µm
κ	27 MS/m	1 MS/m
ε_{r}	4.1	0.2
$\tan\delta$	0.0001	$5\cdot10^{-5}$

4 CPW parameter sets

The performance of three different CPWs against each other is compared in this paper. To calculate the broadband electrical properties of these CPWs, the model from (Heinrich, 1993) was used in the frequency range 1–110 GHz in all three cases. The input quantities used in the CPW model are the width of the ground planes w_g, the center conductor width w, the width s of the slot between center line and ground planes, the relative permittivity $\varepsilon_{\mathrm{r,mem}}$, the dielectric loss tangent of the substrate $\tan\delta$, the thickness t and the conductivity κ of the metal layer, respectively (see Fig. 1). Tables 1–3 contain the parameters and associated uncertainties of the three CPWs under investigation. The assumptions regarding the propability density functions (pdf) will be discussed in the next section.

The parameters of the first CPW (see Table 1), constructed in membrane technology, are such that an impedance level near 50 Ω is provided. 50 Ω is the reference impedance of most instruments and connectors used in microwave scattering-parameter measurements. As a consequence of this requirement, the slot width is comparatively narrow with $s = 10$ µm. Otherwise, the impedance level would be much higher than in conventional thin-film technology on account of the absent substantial dielectric substrate. The dielectric constant $\varepsilon_{\mathrm{r,LHP}}$ of the CPW in membrane technology is effective in the lower half plane of the model from (Heinrich, 1993). Using the equation specified in the previous section this effective dielectric constant was calculated to

Table 3. 50 Ω-CPW parameters and associated uncertainties.

Parameter	value	half-width of pdf interval
w_g	280 μm	0.5 μm
w	50 μm	0.5 μm
s	23.5 μm	0.5 μm
t	3.161 μm	0.25 μm
κ	26.5 MS/m	1 MS/m
ε_r	10.2	0.2
$\tan\delta$	0.001	$5 \cdot 10^{-5}$

$\varepsilon_{\mathrm{r,LHP}} = 1.2707 \pm 0.0148$. This already points to one significant advantage of using membrane technology: the CPW is much less sensitive to uncertainties in the dielectric constant of the membrane.

The parameters of the second CPW (see Table 2) differ with regard to the first CPW in only one aspect: a dielectric bulk substrate replaces the dielectric membrane. The material parameters of this bulk substrate are identical to the supporting membrane (ε_r in Table 2 equal to $\varepsilon_{\mathrm{r,mem}}$ in Table 1). In order to apply the model from (Heinrich, 1993), the thickness of this bulk substrate is assumed to be large enough. This parameter set was chosen deliberately to illustrate the difference between the dielectric setup used in membrane and conventional thin-film technology, leaving the rest of the cross-sectional parameters unchanged. The parameter set of this CPW does not necessarily correspond to a CPW one would design in practice. Therefore, we called it a virtual CPW.

Finally, the parameters of the third CPW (see Table 3) were chosen from a realistic example of a 50 Ω-CPW fabricated on an alumina substrate. The dimensions of this CPW are close to the dimensions one can find on a commercial impedance standard substrate (ISS).

Offering the possibility to produce metallic structures with a very smooth surface and an almost perfect edge definition, an evaporation process for the metal lines of the first two CPWs is assumed. The metallic structures of the third CPW are produced by electroplating. This leads to higher fabrication deviations, but a more substantial metal thickness can be achieved.

Before investigating the uncertainties of the different CPW cases, we first examined the nominal electrical properties over a frequency range of 110 GHz. Fig. 2 presents the three data sets for the real and imaginary part of the propagation constant. The phase constant β is normalized to its free-space value β_0. Apparently, the propagation characteristics of the membrane-supported CPW are better than for the other two CPWs discussed: the normalized phase constant is close to the ideal value of 1, and the attenuation constant is lowest over frequency.

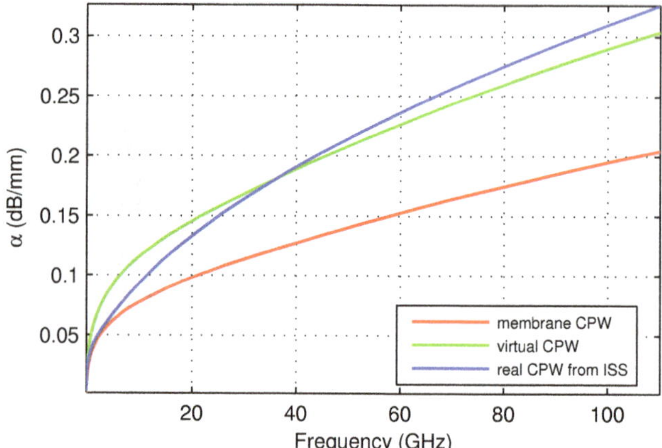

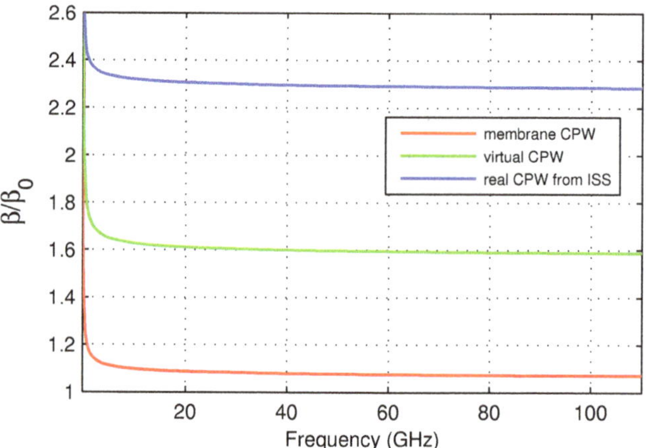

Fig. 2. Nominal values of propagation constant for all three CPW types discussed.

5 Comparison of uncertainty distributions

To study the propagation of uncertainties we use the Monte Carlo method as recommended in the *Guide to the Expression of Uncertainty in Measurement* (GUM), Supplement 1 (BIPM, 2008). The input quantities are assumed to be independent of each other as well as to have an uniform probability density function. Thus, the values of all input quantities are assumed to lie within an interval $[a,b]$ including a lower limit a and an upper limit b. In Tables 1–3 the half-width of this interval $\frac{b-a}{2}$ is indicated for all input quantities.

As the output quantities of interest we investigated the real and imaginary part of the propagation constant as well as the magnitude and phase of the characteristic impedance at nine frequencies from 0.1 to 110 GHz. We first explored the composition of the uncertainty budgets over frequency in order to better understand the uncertainty mechanisms leading to the total observed uncertainty.

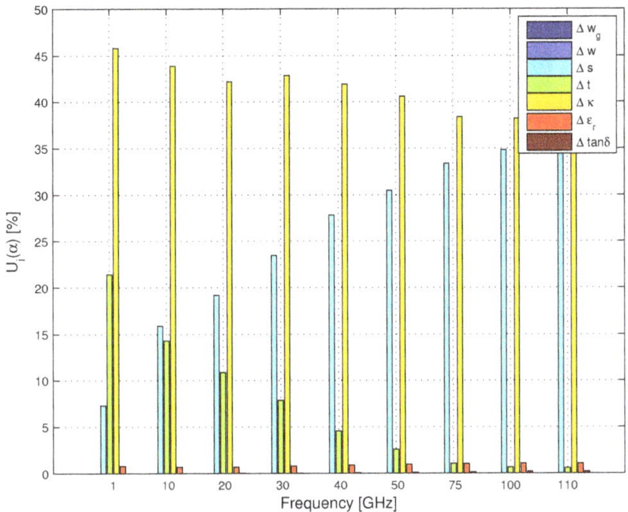

Fig. 3. Distribution of the uncertainties in the attenuation constant α for the membrane CPW (see Table 1).

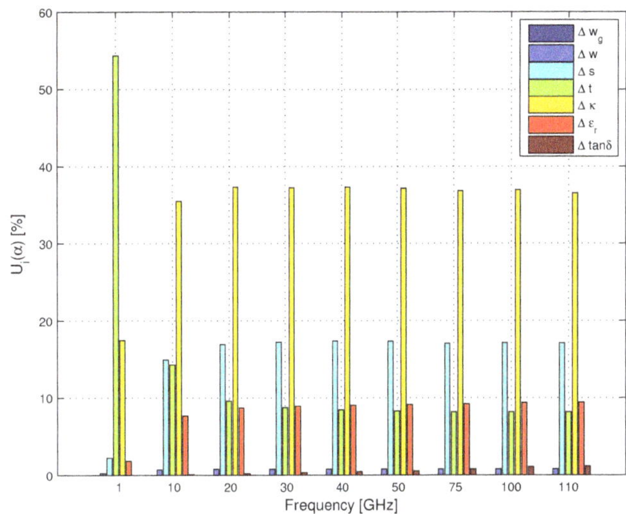

Fig. 5. Distribution of the uncertainties in the attenuation constant α for the 50-Ω CPW (see Table 3).

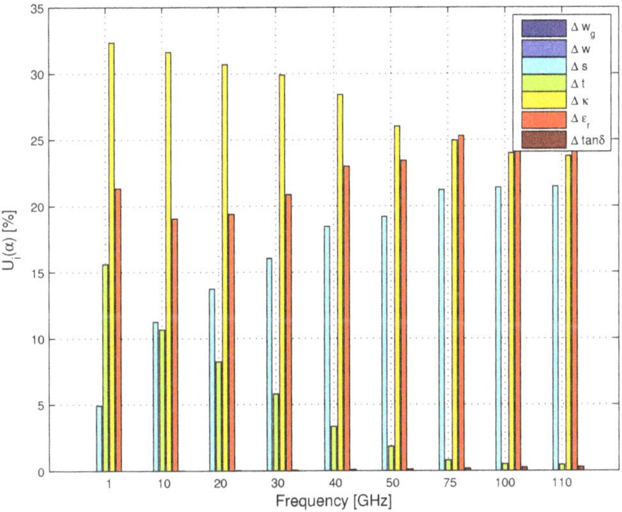

Fig. 4. Distribution of the uncertainties in the attenuation constant α for the virtual CPW (see Table 2).

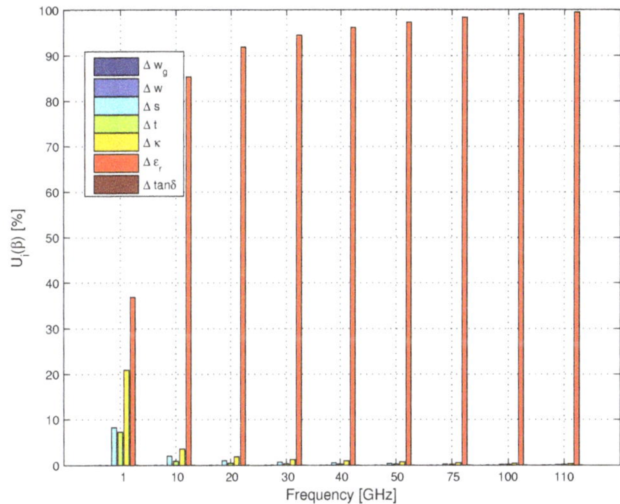

Fig. 6. Distribution of the uncertainties in the normalized phase constant β for the membrane CPW (see Table 1).

The percental contribution of each individual input quantity to the total standard deviation of the attenuation constant α is shown in Figs. 3–5. The total standard deviation was calculated using the adaptive Monte Carlo method of (BIPM, 2008) with statistics of the input quantities according to Tables 1–3. Then the percental contribution was calculated as the ratio of the squared uncertainty contribution of the individual input quantity to the square of the total standard deviation. This ratio is sometimes termed uncertainty index U_i. In Figs. 3–5 each colored bar corresponds to the uncertainty index $U_i(\alpha)$, the contribution of a given input quantity (see legend) to the uncertainty in α at a given frequency.

For the virtual CPW and the 50 Ω-CPW (Figs. 4 and 5), at almost all frequencies the uncertainty in the bulk substrate's dielectric constant is an important contributor to the total uncertainty in α. By contrast, the influence of the membrane's dielectric on the uncertainty in α can be neglected for the membrane supported CPW, illustrated in Fig. 3. The dominant influences in Fig. 3 stem from the thickness and the conductivity of the metal lines, with an increasing contribution of the gap definition at higher frequencies.

The cross-section's geometrical uncertainty is assumed to be rather low for the first and second CPW (within ± 0.25 μm for w_g, w and s). Figures 3 and 4 visualize the domination of the uncertainty budgets by the remaining input parameters. For the second CPW the uncertainties in the conductivity, the

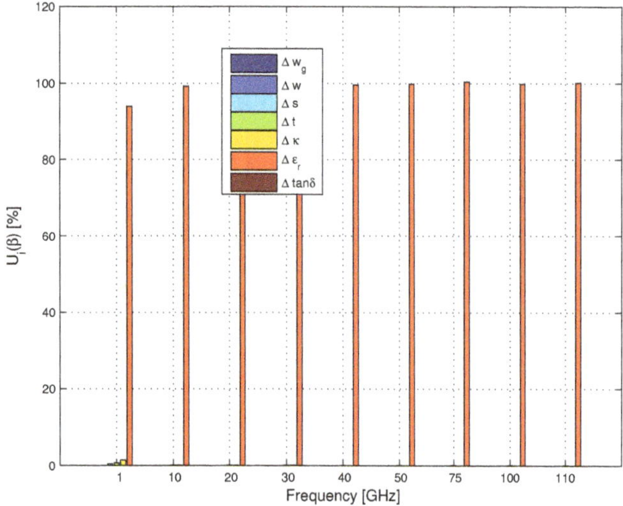

Fig. 7. Distribution of the uncertainties in the normalized phase constant β for the virtual CPW (see Table 2).

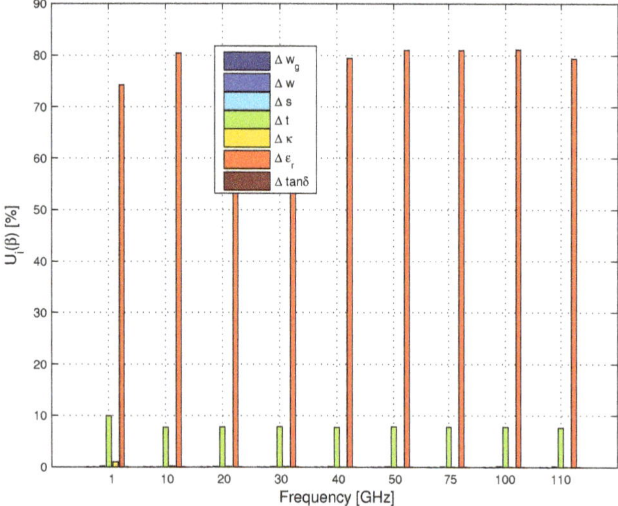

Fig. 8. Distribution of the uncertainties in the normalized phase constant β for the 50-Ω CPW (see Table 3).

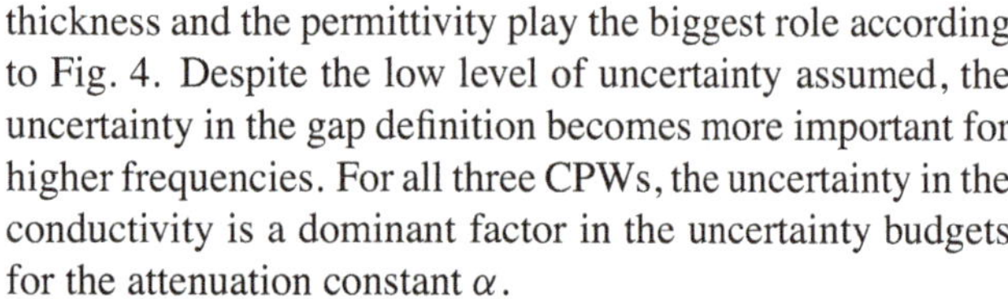
thickness and the permittivity play the biggest role according to Fig. 4. Despite the low level of uncertainty assumed, the uncertainty in the gap definition becomes more important for higher frequencies. For all three CPWs, the uncertainty in the conductivity is a dominant factor in the uncertainty budgets for the attenuation constant α.

The uncertainty budgets over frequency for the phase constant normalized to the free-space value are represented in Figs. 6–8. The normalized phase constant is denoted with β in this paper. We know from Arz et al., 2010 that the uncertainties in β are dominated by the bulk substrate's dielectric properties for conventionally fabricated CPW and microstrip lines (MSL). Figure 6 demonstrates that this is also

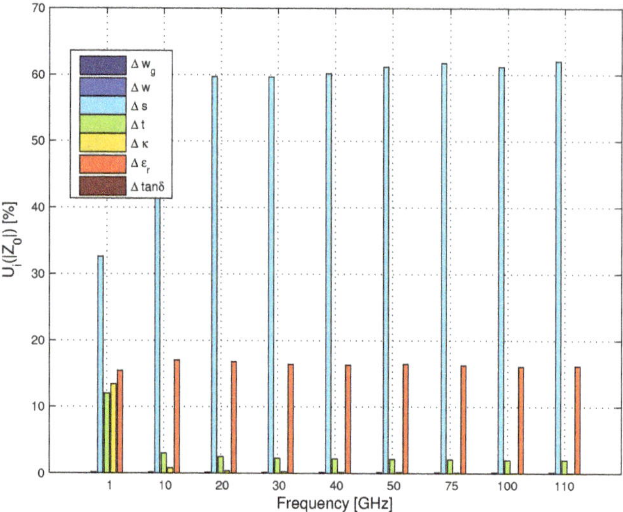

Fig. 9. Distribution of the uncertainties in the magnitude of the characteristic impedance $|Z_0|$ for the membrane CPW (see Table 1).

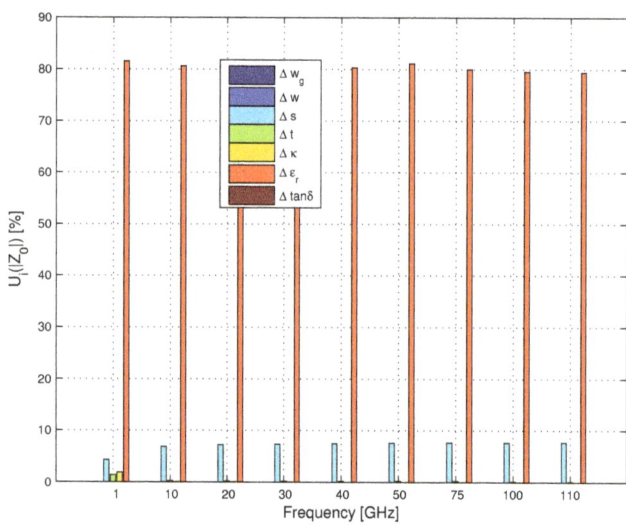

Fig. 10. Distribution of the uncertainties in the magnitude of the characteristic impedance $|Z_0|$ for the virtual CPW (see Table 2).

true of membrane-supported CPWs, even though the thickness of the membrane we assumed in our study was only $t_{mem} = 1$ μm.

The uncertainty budgets over frequency for the magnitude of the characteristic impedance Z_0 are shown in Figs. 9–11. While the biggest contributor for the uncertainty budgets of the CPWs with a bulk substrate is the uncertainty in ε_r (Figs. 10–11), the uncertainty budget of the membrane-supported CPW is dominated by the uncertainty in the width s of the gap (Fig. 9).

Finally, Figs. 12–14 demonstrate the evolution over frequency of the uncertainty contributions to the phase of the characteristic impedance. For all three CPWs, the uncer-

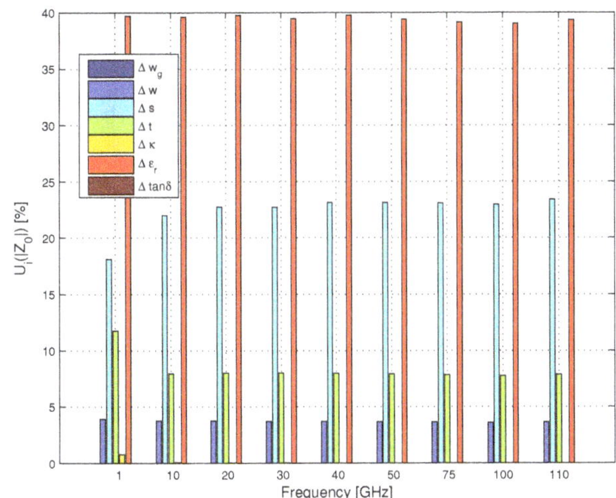

Fig. 11. Distribution of the uncertainties in the magnitude of the characteristic impedance $|Z_0|$ for the 50-Ω CPW (see Table 3).

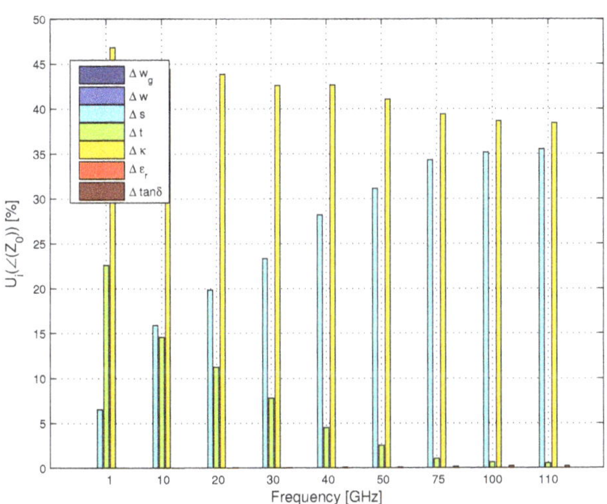

Fig. 12. Distribution of the uncertainties in the phase of the characteristic impedance for the membrane CPW (see Table 1).

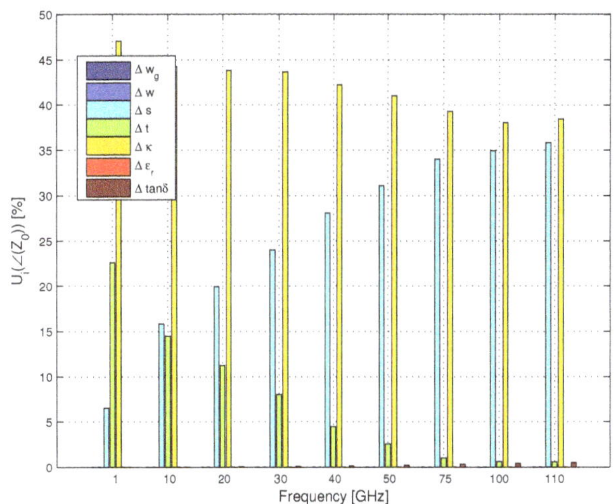

Fig. 13. Distribution of the uncertainties in the phase of the characteristic impedance for the virtual CPW (see Table 2).

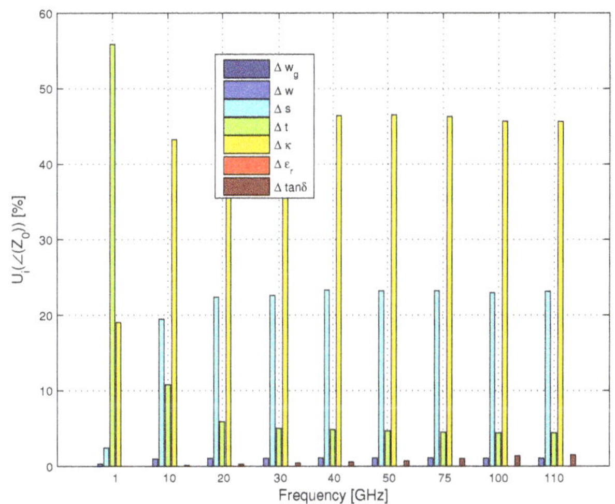

Fig. 14. Distribution of the uncertainties in the phase of the characteristic impedance for the 50-Ω CPW (see Table 3).

tainty distributions look qualitatively quite similar. In all cases, the contributions of the uncertainty in the dielectric constant is negligible. From Figs. 12-14 one can conclude that, in terms of the uncertainty in $\angle(Z_0)$, the use of membrane technology does not offer any advantages.

6 Comparison of total expanded uncertainties

The absolute values of the uncertainties are not shown in Figs. 3–14. Therefore, we have summarized the total expanded uncertainties for all output quantities of the membrane-supported CPW in Table 4. We assumed a coverage factor of $k = 2$. The ratio of the total uncertainties of either the first CPW to the second CPW (blue bars) or of the first CPW to the third CPW (brown bars) is plotted in Figs. 15–18. This is equivalent to the reduction of the uncertainties when using membrane technology instead of conventional technology. The total uncertainties for the second and third CPW can be also derived from Table 4 together with Figs. 15–18.

The results for the uncertainties of the real and imaginary part of the propagation constant are illustrated in Figs. 15–16. The uncertainty in α for the membrane-supported CPW is less than 70% of the corresponding uncertainty of either CPW with a bulk substrate for higher frequencies. The uncertainty in β is reduced to values between 10 and 15%. This is equivalent to an almost tenfold reduction of the uncertainty in β by the use of membrane technology.

Figures 17–18 present the results for the uncertainties of the magnitude and phase of the characteristic impedance. For the magnitude of Z_0, the membrane technology reduces the

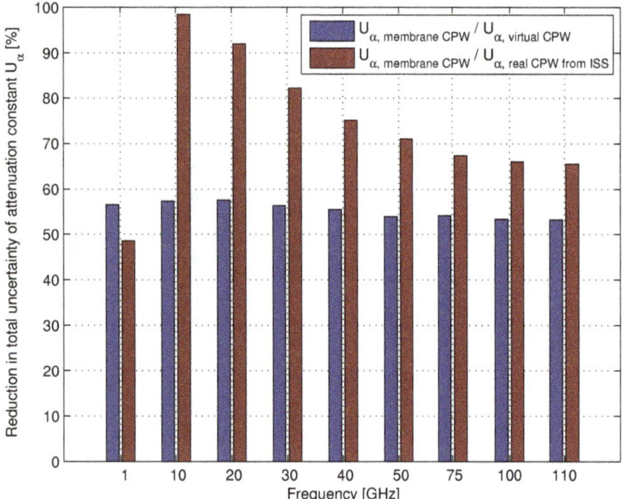

Fig. 15. Reduction of the uncertainties in α when using membrane instead of conventional technology.

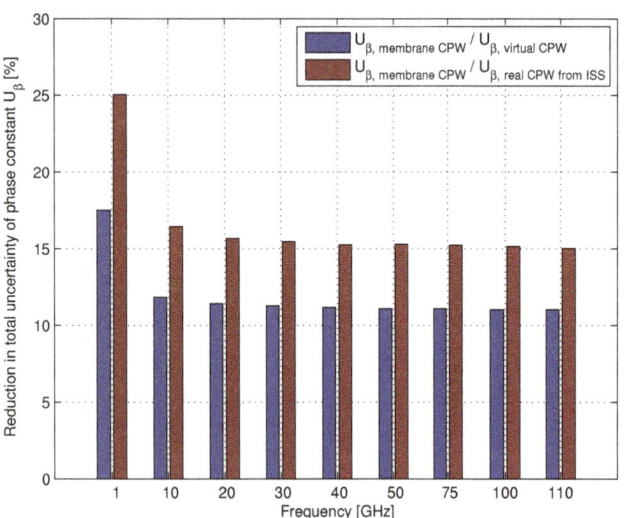

Fig. 16. Reduction of the uncertainties in β when using membrane instead of conventional technology.

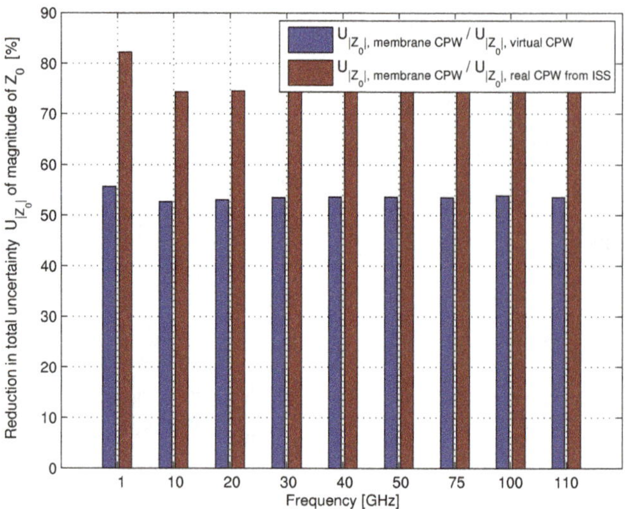

Fig. 17. Reduction of the uncertainties in $|Z_0|$ when using membrane instead of conventional technology.

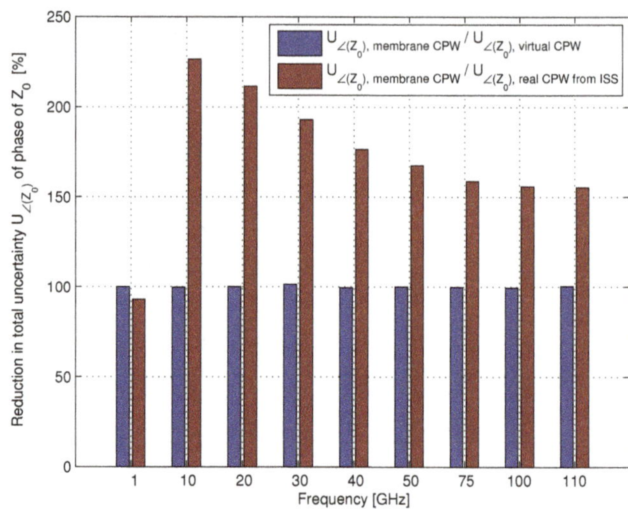

Fig. 18. Reduction of the uncertainties in phase of Z_0 when using membrane instead of conventional technology.

uncertainty to values between ~55% (blue bars) and ~75% (brown bars). No reduction is observed concerning the phase of Z_0. The comparison of the first and second CPW shows identical values, while the total uncertainty in $\angle(Z_0)$ for the 50-Ω CPW is even lower than for the membrane-supported CPW. This is consistent with the conclusions drawn from Figs. 12–14: the uncertainty in the phase of Z_0 does not depend on the uncertainty in ε_r.

7 Conclusions

In this work, we presented a closer examination of the uncertainties in the electrical properties of a CPW fabricated in membrane technology compared to CPWs built on several hundred μm thick dielectric substrates. An expression was derived for the dielectric constant of an equivalent half" space beneath the signal conductors. This was used to describe the dielectric effect of the supporting membrane, in order to be able to use a broadband analytic CPW model for the membrane-supported CPW.

The sensitivities of the propagation constant and the characteristic impedance were analyzed while allowing for tolerances in the cross-sectional CPW parameters for frequencies

Table 4. Total expanded uncertainties for membrane-supported CPW.

f [GHz]	U_α[dB/mm]	U_β	$U_{\|Z_0\|}[\Omega]$	$U_{\angle Z_0}$[rad]
1	0.0012	0.0058	0.55	0.0048
10	0.0028	0.0035	0.48	0.0013
20	0.0035	0.0034	0.47	0.0008
30	0.0038	0.0033	0.48	0.0006
40	0.0040	0.0033	0.48	0.0005
50	0.0042	0.0032	0.47	0.0004
75	0.0050	0.0032	0.48	0.0003
100	0.0056	0.0032	0.48	0.0003
110	0.0059	0.0032	0.48	0.0003

from 1 to 110 GHz. We were able to demonstrate that, for the CPW parameter sets chosen in this study, an almost tenfold reduction in the uncertainty of the phase constant β can be achieved through the use of membrane technology. Moderate improvements of the uncertainty can be obtained for the attenuation constant and the magnitude of the characteristic impedance. Membrane technology does not offer any improvements for output quantities that do not depend on the dielectric constant of the substrate, like e.g. the phase of the characteristic impedance.

Acknowledgements. The authors would like to thank the Test and Measurement Division, Rohde & Schwarz GmbH & Co. KG, München, Germany for providing the membrane material.

References

Arz, U., Leinhos, J., and Janezic, M. D.: Effect of Material Properties on Broadband Electrical Behavior of Coplanar Waveguides, Conference on Precision Electromagnetics CPEM, Broomfield, Colorado, 8–13 June 2008, 470–471, doi:10.1109/CPEM.2008.4574857, 2008a.

Arz, U., Leinhos, J., and Janezic, M. D.: Broadband Dielectric Material Characterization: A Comparison of On-Wafer and Split-Cylinder Measurements, 38th European Microwave Conference EuMC, 27–31 October 2008 , Amsterdam, Netherlands, 913–916, doi:10.1109/EUMC.2008.4751602, 2008b.

Arz, U. and Kuhlmann, K.: Uncertainties in Coplanar Waveguide and Microstrip Line Standards for On-Wafer Thru-Reflect-Line Calibrations, 75th ARFTG Microwave Measurement Conference, 28–28 May 2010 , Anaheim, California, 1–5, doi:10.1109/ARFTG.2010.5496328, 2010.

BIPM, IEC, IFCC, ISO, IUPAC, IUPAP, and OIML: Evaluation of measurement data – Supplement 1 to the "Guide to the expression of uncertainty in measurement": Propagation of distributions using a Monte Carlo method, 1, BIPM Joint Committee for Guides in Metrology, JCGM 101:2008, Paris, 2008.

Dib, N., Harokopus Jr., W. P., Katehi, L. P., Ling, C. C., and Rebeiz, G. M.: Study of a Novel Planar Transmission Line, IEEE MTT-S, 10–14 June 1991, Boston, 2, 623–626, doi:10.1109/MWSYM.1991.147080, 1991.

Heinrich, W.: Quasi-TEM description of MMIC coplanar lines including conductor-loss effects, IEEE T MICROW THEORY, 41, 45–52, doi:10.1109/22.210228, 1993.

Katehi, L. P. and Rebeiz, G. M.: Novel micromachined approaches to MMICs using low-parasitic, high-performance transmission media and environments, IEEE MTT-S, 17–21 June 1996, San Francisco, California, 2, 1145–1148, doi:10.1109/MWSYM.1996.511232, 1996.

QuickField™ 5.5, Tera Analysis Ltd..

Weller, T. M., Katehi, L. P., Herman, M. I., and Wamhof, P. D.: Membrane technology (MIST-T) applied to microstrip: a 33 GHz Wilkinson power divider, IEEE MTT-S, 23–27 May 1994, San Diego, California, 2, 911–914, doi:10.1109/MWSYM.1994.335209, 1994.

14

Verification of scattering parameter measurements in waveguides up to 325 GHz including highly-reflective devices

T. Schrader[1], K. Kuhlmann[1], R. Dickhoff[1], J. Dittmer[1], and M. Hiebel[2]

[1]Physikalisch-Technische Bundesanstalt, Braunschweig, Germany
[2]Rohde & Schwarz, Munich, Germany

Abstract. Radio-frequency (RF) scattering parameters (S-parameters) play an important role to characterise RF signal transmission and reflection of active and passive devices such as transmission lines, components, and small-signal amplifiers. Vector network analysers (VNAs) are employed as instrumentation for such measurements. During the last years, the upper frequency limit of this instrumentation has been extended up to several hundreds of GHz for waveguide measurements. Calibration and verification procedures are obligatory prior to the VNA measurement to achieve accurate results and/or to obtain traceability to the International System of Units (SI). Usually, verification is performed by measuring well-matched devices with known S-parameters such as attenuators or short precision waveguide sections (shims). In waveguides, especially above 110 GHz, such devices may not exist and/or are not traceably calibrated. In some cases, e.g. filter networks, the devices under test (DUT) are partly highly reflective. This paper describes the dependency of the S-parameters a) on the calibration procedure, b) on the applied torque to the flange screws during the mating process of the single waveguide elements. It describes further c) how highly-reflective devices (HRD) can be used to verify a calibrated VNA, and d) how a measured attenuation at several hundreds of GHz can be substituted by a well-known coaxial attenuation at 279 MHz, the intermediate frequency (IF) of the VNA, to verify the linearity. This work is a contribution towards traceability and to obtain knowledge about the measurement uncertainty of VNA instrumentation in the millimetre-wave range.

1 Introduction

With the development of millimetre-wave instrumentation e.g. VNA mainframes and waveguide frequency extensions (Fig. 1), more and more insight is gained in network analysis at frequencies beyond 110 GHz (Adamson et al., 2009). In general, network analysis is a necessary tool for development and testing, also at millimeter-wave frequencies. Many applications benefit from the development of this versatile instrumentation. Some practical examples are the characterisation of passive and active devices, calibration of antennas, systems, and material properties. The waveguide VNA technology is applied e.g. to characterise signal propagation to develop future ultrafast and high-bitrate data transfer systems and communication links (Jastrow et al., 2010; Priebe et al., 2010), to characterise absorber material or anechoic environments using time-domain techniques (Schrader et al., 2010) or to measure the response of field probes at millimetre wavelengths (Salhi et al., 2010).

Before usage, any VNA has to be calibrated with respect to the measurement ports at the reference plane. Typical calibration procedures such as TMSO, TRL, and LRL are applicable also in the waveguide system (T: through, M: match, S: short, O: open – here offset short, L: line, R: reflect).

With increasing frequency it becomes more difficult to design a well-matched load or an open standard. Hence, TRL, LRL, and offset-short calibrations become more important at higher frequencies. Furthermore, with increasing frequency losses due to the skin effect become more significant. Moreover, surface roughness becomes a major factor with respect to losses and wave propagation (Hoffmann, 2009). In general, mechanical dimensions and their deviations from the ideal geometry become more critical. Therefore, high precision manufacturing of waveguides and components becomes a crucial factor with respect to the measurement uncertainty.

Fig. 1. VNA mainframe and sets of frequency converters covering the WR15, WR10, WR6, WR5, and WR3 waveguide bands.

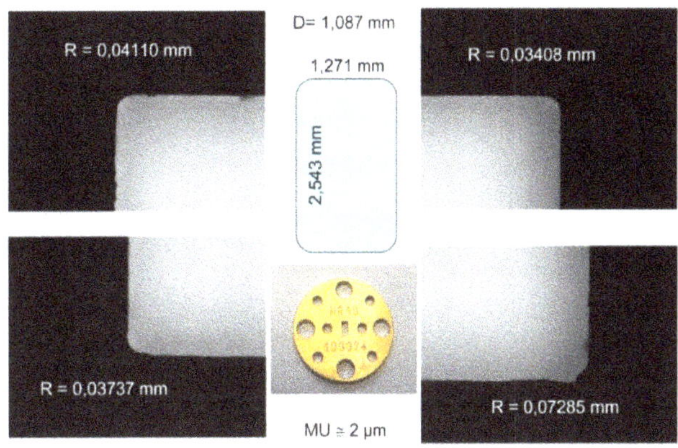

Fig. 2. Zoom images of the edges of the WR10 waveguide shim. The radii R and the waveguide square dimensions, the thickness of the shim, and the measurement uncertainty MU are given.

During calibration, several standards are connected to each VNA measurement test port. In addition, a through and – depending on the used calibration method – a waveguide of short length (shim) or shims with different line lengths are inserted between the measurement ports. Mating of the flanges is supported by four alignment pins, and for precision flanges by two additional dowel pins. By this the remaining uncertainty due to the waveguide mating process is reduced to a minimum but still leads to irregularities along the line (Stumper, 2001; Lau, 2010; Lok et al., 2009), again increasing with frequency. After mating of the flanges the connection is fixed by four screws around the flange. Only precision flanges have an additional "anti-cocking" rim to avoid air gaps between the mating planes or mechanical misalignment due to non-uniform screw torque. As waveguide flanges may be manufactured from thin or weak materials, the applied screw torque should be well-defined using a torque wrench. This tool also ensures a flange mating with a repeatable force both at calibration and measurements.

Typically, after calibration a known device is measured as verification standard (Clarke et al., 2009; Ridler et al., 2010a, 2010b; Horibe et al., 2010). Its properties may be known from prior and other calibration measurements, mechanical measurements, analytical calculations, or other methods of characterisation. Precision gold-plated shims provide a quarter-wavelength transmission phase shift at the band mid frequency. The input reflection coefficient of the shim as thru device is dominated by a) the waveguide cross-sectional mechanical dimensions and b) by the effective load match of the output port (Williams, 2010), (Judaschke, 2011). If the thru device is highly reflective instead of being a perfect waveguide also the effective source match generates an input standing wave. Furthermore, the effective directivity at both test ports has significant influence on the measurement result. Both effects of effective source and load match and effective directivity play an important role determining the S-parameters of DUTs, especially for highly reflective devices (HRD), which belong to the most challenging type of DUTs. Such a HRD could be e.g. a filter network offside its pass-band. A well-matched attenuator as a typical verification device does not reveal the effects of effective directivity and/or effective source and load match. Usually, a precision waveguide section, a short and a match are used to apply the classical ripple method, thus identifying the effective directivity and source and load match. The methods described here are an alternative to the classical ripple method as precision waveguide sections are not available yet.

Measurements of filter networks typically require a high dynamic range of the instrumentation. As VNAs offer a high dynamic range, linearity has to be guaranteed for precise high attenuation measurements and therefore to be verified experimentally. To investigate the VNA linearity, we compare a known RF attenuation inserted between the test ports with a calibrated coaxial attenuator (at 279.28 MHz) inserted in the "meas input" path of the VNA.

2 The equipment

2.1 Instrumentation

We have used a 4-port vector network analyser R&S ZVA 50 GHz with frequency converters (type ZVA-Z, cp. Fig. 1) covering the frequency range of the corresponding waveguide bands 50–75 GHz (WR15), 75–110 GHz (WR10), 110–170 GHz (WR6), 140–220 GHz (WR5), and 220–325 GHz (WR3). The waveguide calibration kits (R&S ZV-WR) are corresponding to the waveguide bands above. We have used the kits with fixed loads. An adjustable WR10 precision rotary attenuator (0–60 dB, Flann 27110) and PC3.5 mm coaxial precision fixed attenuators (Rosenberger 03AS102, with 3 dB, 6 dB, 10 dB, 20 dB, and three 30 dB attenuators) were

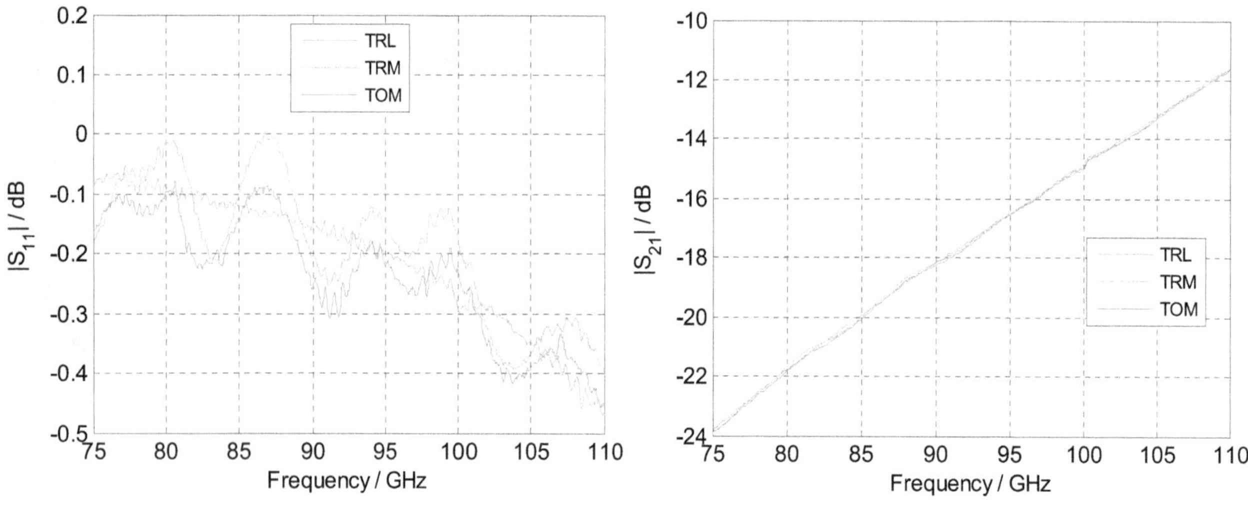

Fig. 3. Measured S-parameters of the HRD type "90° shim" for different calibrations schemes.

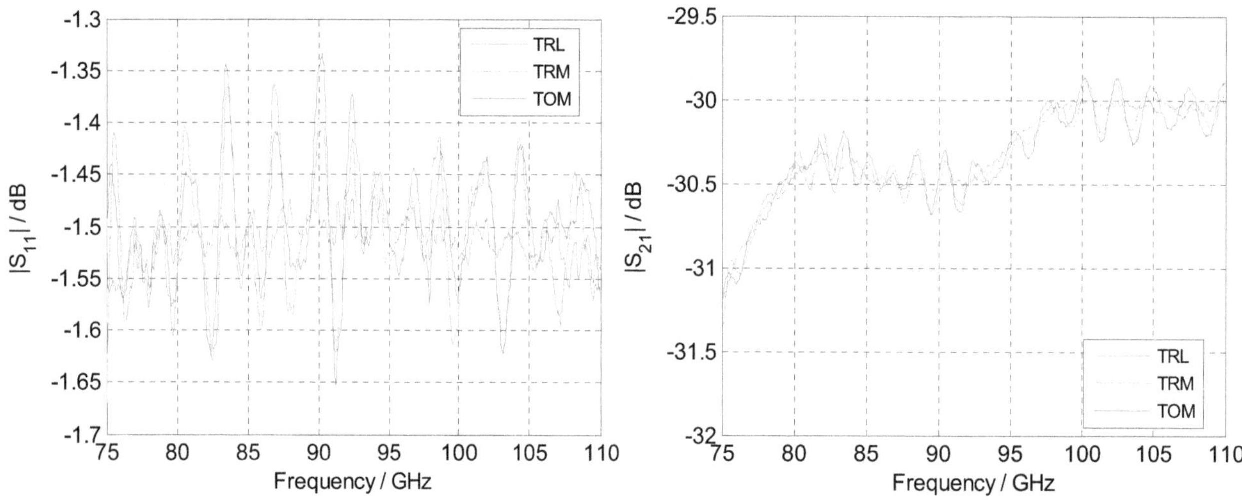

Fig. 4. Measured S-parameters of the HRD type "thin foil" obtained by applying different calibration schemes.

used for the linearity measurements. The higher values 40 dB, 50 dB, and 60 dB were assembled from (10+30) dB, (20+30) dB and (30+30) dB attenuators.

2.2 Design and characterisation of highly reflective devices (HRD)

We have used three different kinds of HRDs. First, a 125 μm polyimide foil (Goodfellow) coated with a thin one-sided thermal deposition Titanium layer of about 60 nm was assembled between the test port waveguide flanges. Therefore, ohmic contact to the flanges is only possible at one side. The Titanium layer is very stable with respect to the mechanical stability and oxidation. As the thickness of the metal layer is kept 10 times below the smallest skin depth of about 550 nm, the resulting attenuation does not change with frequency. We obtained a nearly constant frequency response in all used waveguide bands (cp. Sect. 2.1) from 50 GHz up to 325 GHz.

The second HRD was a regular shim as part of the waveguide calibration kit, that was rotated by 90 degrees before mounting. Thereby the shim changes its behaviour from a matched line section to a waveguide below cut-off. The evanescent mode in the rotated shim provides a low transmission. The attenuation only depends on the mechanical dimensions while surface roughness and skin effect losses can be neglected due to the short line length. Mechanical measurements on an optical coordinate measuring machine provided the exact shim dimensions, as shown in Fig. 2 for the WR10 shim. Here, the pixel resolution of the optical image was about one μm, the measurement uncertainty was 2 μm. The thickness D of the shim was determined to be

Table 1. Contributions to the measurement uncertainty of reflection and transmission (in dB) for a HRD and a temperature range of 22.7 °C to 24.2 °C.

	torque		Drift 17 h		model simulation		calibration		attenuation	repro-ducibility	
in dB	$\lvert S_{ii}\rvert$	$\lvert S_{ij}\rvert$	$\lvert S_{ii}\rvert$	$\lvert S_{ij}\rvert$	$\lvert S_{ii}\rvert$	$\lvert S_{ij}\rvert$	$\lvert S_{ii}\rvert$	$\lvert S_{ij}\rvert$	$\lvert S_{ij}\rvert$	$\lvert S_{ii}\rvert$	$\lvert S_{ij}\rvert$
WR15	0.1	0.05	0.01	0.02	0.1	0.1	n.a.		n.a.	0.03	0.05
WR10	0.1	0.1	0.02	0.05	0.1	0.1	0.15	0.25	0.1	0.05	0.05
WR6	0.25	0.2	0.05	0.05	0.1	0.1	n.a.		n.a.	0.05	0.05
WR5	0.25	0.2	0.05	0.05	0.15	0.5	n.a.		n.a.	0.05	0.05
WR3	1.0	0.25	0.1	0.3	0.5	0.5	1.0	1.0	n.a.	0.1	0.5

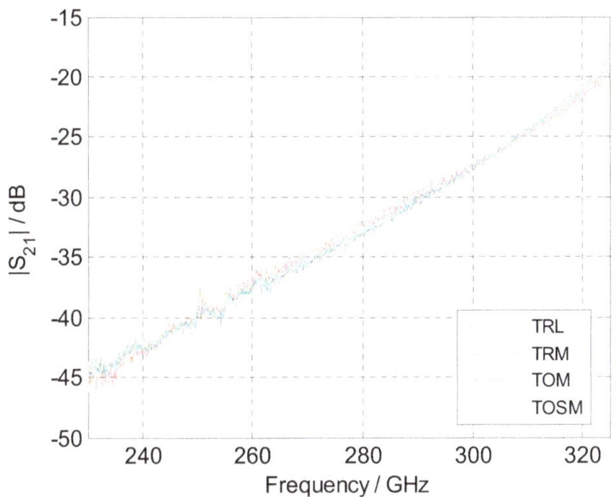

Fig. 5. Measured S-parameters of the HRD type "shim with cylindrical hole" obtained by applying different calibration schemes.

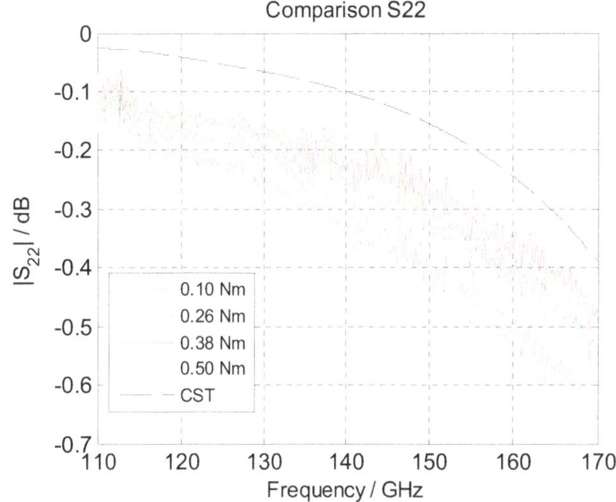

Fig. 6. Measurement of S_{22} of the HRD type "90° shim" in WR6 waveguide using a torque wrench compared with CST simulations. The waveguide connection was opened and completely re-assembled for each torque setting.

1.087 mm, the waveguide cross-sectional dimensions were 2.543 mm×1.271 mm.

Again, this HRD type was used in all waveguide bands (cp. Sect. 2.1).

The third HRD was a shim of 1 mm thickness made of brass with a cylindrical hole of 500 μm diameter manufactured with a precision reamer. Here, we have used this HRD only in the WR3 waveguide band, by choosing the diameter accordingly, it can be used for the lower bands as well.

The numerical computations were performed using CST Microwave Studio (Computer Simulation Technology, www.cst.com, Darmstadt, Germany). To model the thin metal foil a conductive non-metallic material has to be chosen instead of a lossy metal. In a second step, the material was assigned the conductivity of Ti ($\sigma = 2.56 \cdot 10^6$ S/m).

2.3 Assembly

In order to determine the dependency of the scattering parameters S_{ii} and S_{ij} on the mechanical torque applied to the mounting screws at the waveguide flanges we used straight and 90° offset torque wrenches.

3 Measurement results

For VNA calibration we used the standard firmware procedures. The measurements were performed in a laboratory environment without air-conditioning. The temperature was monitored in the range between 22.7 °C and 24.2 °C. We also measured the hardware drift during 17 h. All VNA measurements were performed with 801 frequency points and a resolution bandwidth of 100 Hz. We applied 5 averages per frequency point.

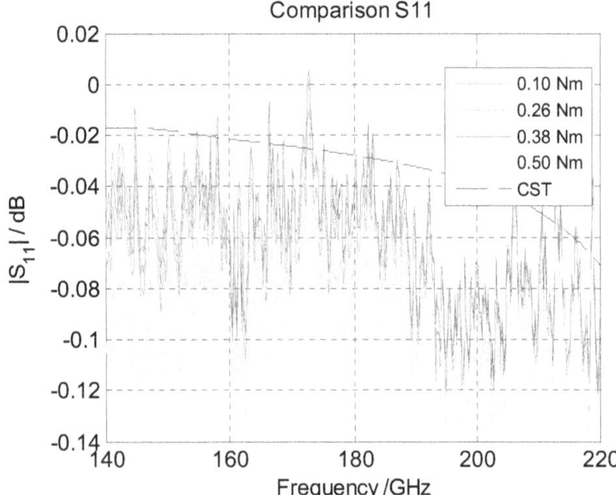

Fig. 7. Measured and simulated magnitude of S_{11} of the HRD type "90° shim" in WR5 waveguide using a torque wrench.

3.1 Dependency of scattering parameters on the calibration scheme

To investigate the influence of the calibration method on the measured S-parameters, we have performed different calibrations. Subsequently measurements of the HRD (90° shim) have been performed, as can be seen in Fig. 3. The abbreviations of the calibration standards are T: through, R: reflect (short), L: line (shim in regular orientation), M: broadband match, O: offset short (built from shim and short of the calibration kit).

Secondly, we used the HRD type "thin foil" as DUT. The results are shown in Fig. 4. Obviously, the deviations of S-parameters due to different calibration schemes depend on the DUT (and its length). For the HRD type "90° shim", S_{11} varies slowly with frequency, but for the HRD type "thin foil" S_{11} changes much faster with frequency, but with almost similar amplitude. The TRL scheme shows the best performance as the frequency response is rather flat compared to TRM and TOM. The variations of $|S_{21}|$ in Figs. 3 and 4 show the same behaviour and frequency response. For the WR3 waveguide and HRD type "shim with cylindrical hole" we obtained a continuously increasing frequency response with superimposed oscillations in the order of 1 dB, depicted in Fig. 5.

All results of the dependency of S-parameters on the calibration scheme are summarised in Table 1.

3.2 Influence of the fastening torque on S-Parameters

A parametric study was performed in order to identify the influence of the fastening torque of the flange screws to assemble the waveguide devices. The torque was increased from 0.1 nm to 0.5 nm in steps of 0.04 nm using the same torque wrench (cp. Figs. 6–8).

It can be noted that the S-parameters are significantly depending on the applied torque. The reflection parameters S_{ii} values show a much higher sensitivity on the torque than the transmission parameters S_{ij}. Increasing the torque from 0.1 nm to 0.5 nm brings S_{22} closer to the theoretical reflection. The increasing torque reduces the ohmic contact resistance between waveguide and shim by reducing the remaining gaps between the flanges. As the axes of DUTs and test ports are not perfectly in line, these investigations are only possible, if the frequency converters are not fixed on the workbench.

It is evident in Figs. 6 and 8, that for the WR6 and WR3 waveguides increasing the torque matches simulation and measurement better, whereas in Fig. 7 the reproducibility limit for the WR5 waveguide is reached. The initial alignment of the WR5 waveguide was better than that of WR6. For a precision measurement it is crucial, that mating of the flanges is done without mechanical stress due to axial misalignment. The reproducibility of flange mating was tested by assembling, opening and re-assembling one waveguide connection ten times using 0.3 nm for all waveguide bands. The results are given in Table 1. It can be seen, that reproducibility and hardware drift are not the major contributions to uncertainty. The influence of the measurement bandwidth on the S-parameters was not tested yet.

3.3 Measurement and simulation of S-parameters of highly reflective devices (waveguide shims)

Usually, well-matched attenuators are used for verification of calibrated VNAs. However, in the millimetre-wave range, especially beyond 110 GHz, a traceable calibration of such devices is not available yet. To make progress in that regard, we have used regular waveguide shims according to Sect. 2.2 for that purpose. They are mechanically stable devices and can be characterised by their mechanical dimensions. Instead of mounting them correctly we have rotated them by 90° and inserted them as iris into the waveguide setup. Due to the increasing coupling with frequency, the response of S_{21} covers about 10 dB dynamic range within the WR10 waveguide band as shown in Fig. 9 and about 23 dB in the WR6 waveguide band, respectively. The resulting differences between simulation and measurement are listed in Table 1. It can be seen that numerical simulations predict the measurement results quite closely, especially for the HRD 90°.

3.4 Verification of linearity by substituting the RF attenuation by an IF attenuation

3.4.1 DUT A: precision WR10 rotary attenuator

In order to compare a known RF attenuation with traceably calibrated coaxial attenuators in the MHz range we used a precision WR10 rotary attenuator (PRA) having a maximum attenuation of 60 dB. As the frequency response of the PRA

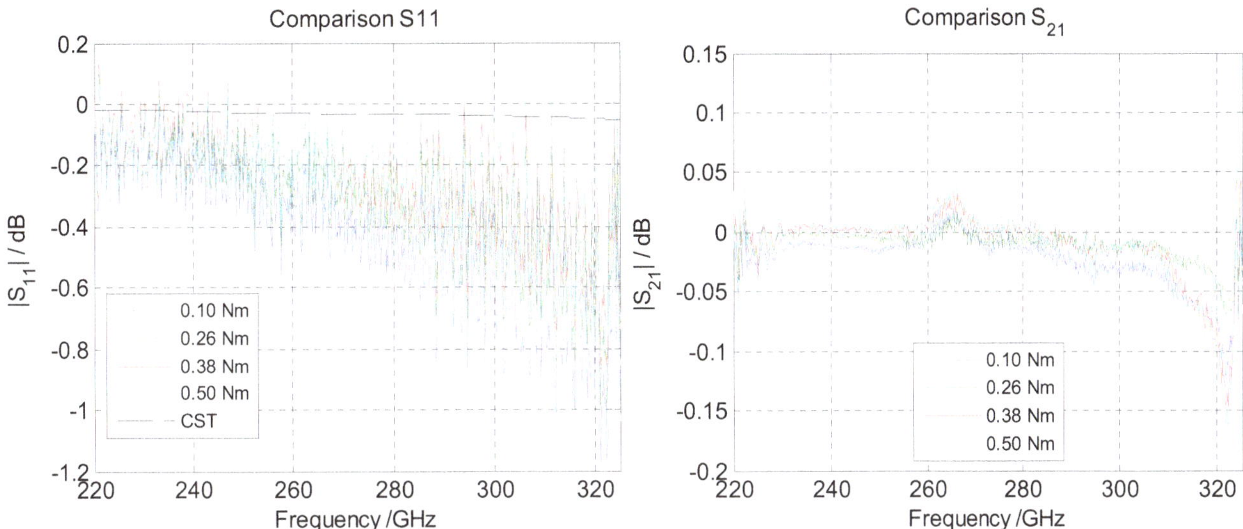

Fig. 8. Comparison of measured S-parameters of the HRD type "90° shim" using shim #2 (the WR3 calibration kit contains two shims with different length) and of the through connection using a torque wrench. The connection between the WR3 waveguides was opened and completely re-assembled for each torque setting.

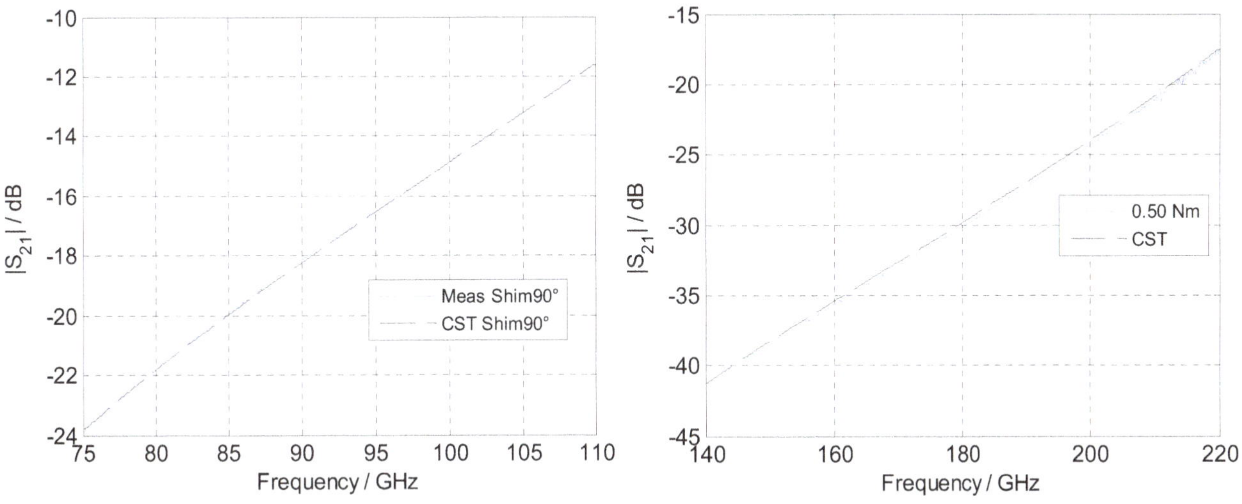

Fig. 9. Numerical simulation (CST) and measurement of S_{21} of a WR10 and a WR6 shim each rotated by 90°.

is nearly independent of frequency and the attenuation setting, we can compare a presumably known RF attenuation inserted between the test ports with a coaxial attenuator which is precisely calibrated at 279.28 MHz (the IF of the VNA) and inserted in the "meas input" path of the VNA (Fig. 10). Under the assumption that input and output reflection coefficient of the hardware of the "meas input" path of the VNA are low, inserting the coaxial attenuators does not substantially raise the mismatch uncertainty in the "meas in" path. The traceably calibrated attenuation values of the coaxial attenuators can therefore be used for comparison. We have chosen combinations of such coaxial attenuators to cover the dynamic range. Furthermore, we have assumed that the small reflection coefficients at the in- and output ports of the attenuators do not show a significant influence on the total attenuation due to the VSWR when mounting the attenuators in series.

The regular signal at 279.28 MHz as sent from the frequency converter to the "meas in" port is a measure for the attenuation A_{GHz} of the attenuator in the GHz range inserted between the test ports, leading to an indication of $|S_{21\mathrm{GHz}}|$. Then, the attenuator is replaced by a thru. Subsequently, we insert a well-known attenuation (A_{MHz}) into the "meas in" path and read the indication $|S_{21,\mathrm{MHz}}|$. Ideally,

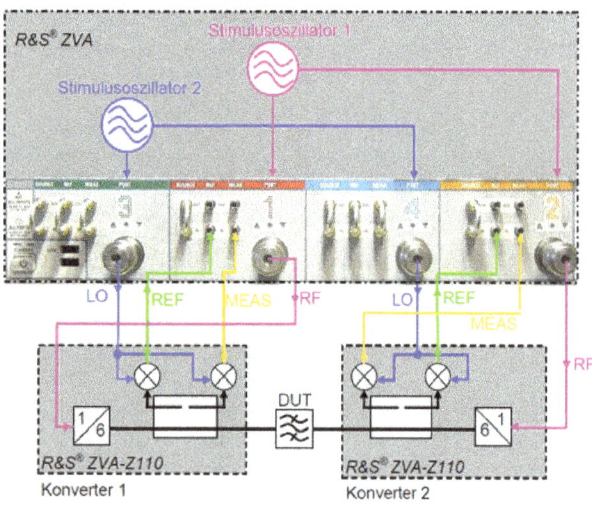

Fig. 10. Setup of VNA mainframe and frequency converters. The coaxial attenuator is inserted into the "Meas" connection at port 2.

$D_S = |S_{21,\text{GHz}}| - |S_{21,\text{MHz}}| = 0$. If A_{GHz} and A_{MHz} were *both* traceable quantities, D_S would indicate the deviation from linearity of the instrumentation (VNA hardware). Here, we only assume an attenuation A_{GHz} of the PRA, but we have used a traceably defined A_{MHz}.

We applied the following procedure: We have measured the frequency response for a 0 dB setting of the PRA and observed a typical flat frequency response with a dip at 88 GHz (see Fig. 11) and a low floor attenuation, even for a 0 dB setting. The frequency response did not change regardless of the attenuation setting (cp. Fig. 11). The floor residual attenuation was measured at 100 GHz each time the PRA was set to 0 dB. From 10 measurements we obtained 1.25 ± 0.02 dB. We have then taken the attenuation value from the calibration certificate, e.g. 9.918 dB for the nominal 10 dB attenuator, and adjusted the PRA to the calibrated value of the attenuator (in this example 9.9 dB). We took the marker value at 100 GHz again which was now -11.15 dB. After we have taken the S-parameter measurements, we set the PRA back to zero attenuation and inserted the attenuator into the coaxial "meas in" branch of the VNA. We took again the S-parameter measurements and the marker value at 100 GHz which was -11.17 dB. The differences of the marker readings with either the PRA or the coaxial attenuator in place were in the order of 0.01 dB up to 0.05 dB. These results show on the one hand that the setting of the PRA (indication to actual value) is very accurate (but not traceable by itself) and on the other hand that the linearity of the frequency converter is excellent (Fig. 11).

3.4.2 DUT B: highly reflective device (90° shim)

As a second device for linearity verification we have used the HRD 90° shim. Again, we have chosen combinations of the coaxial attenuators to cover the dynamic range. The theoretical value of the A_{GHz} attenuation of the 90° shim is obtained from numerical simulations using CST (cp. Sect. 2.2). Its mechanical data were traceably measured (cp. Sect. 2.2). The black circles in Fig. 12 indicate the intercept points of the GHz attenuation of the HRD 90° shim and the A_{MHz} attenuation in the "meas in" path overlap. Table 1 indicates a maximum deviation of S_{21} between simulation and measurement of 0.1 dB for WR10. As shims are part of standard calibration kits they are standard equipment. In combination with traceably calibrated coaxial attenuators they provide a convenient method for VNA verification up to the highest frequencies of interest.

4 Conclusions

Verification of vector network analysers in the millimetre- and sub-millimeter wave range can not yet make use of traceably calibrated attenuators at these high frequencies due to the lack of calibration facilities. Often, devices under test (DUT) exhibit a purely reflective behaviour, then a prior verification of the VNA using only low-reflective calibrated attenuators would not reveal effects such as effective load and source match errors or effective directivity, but the measured S-parameters of the DUT would induce these effects.

Therefore, we have introduced three different types of highly reflective devices (HRD) being suitable for verification in the millimetre- and submillimeter-wave bands including their mechanical and computational characterisation. We have shown that measurements of S-parameters of these HRD can depend strongly on the calibration procedure. Assembling of waveguides should only be done using torque wrenches as the S-parameters depend on the mating forces of the flanges.

Furthermore we have shown that a known millimetre-wave attenuation e.g. from a 90° rotated waveguide shim can be substituted by a traceably and precisely calibrated coaxial PC3.5 mm attenuator being inserted in the measurement path between the frequency converter and the VNA.

Acknowledgements. The authors thank their colleagues at PTB, Dr. Ulrich Neuschäfer-Rube and Dieter Schulz ("Multisensor Coordinate Metrology Group") for the mechanical measurements, Dirk Schubert ("High-frequency Measuring Techniques Group") for the precision RF calibration of the attenuators, and Daniel Hagedorn ("Surface Technology Group") for coating polyimide foils with thermal deposition of Titanium.

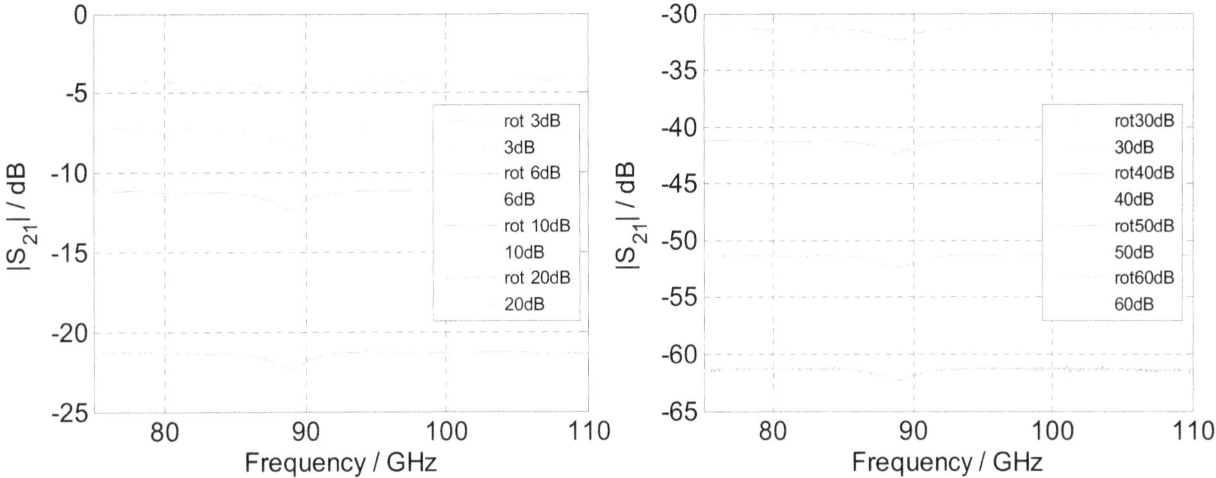

Fig. 11. Substitution of GHz attenuation provided by a precision rotary attenuator with traceably calibrated coaxial PC3.5mm attenuators at 279.28 MHz. Rot3dB indicates measurements with the PRA at a 3 dB setting, 3 dB indicates the substitution with a coaxial 3-dB-attenuator.

References

Adamson, D., Ridler, N., and Howes, J.: , Recent and future developments in millimetre and sub-millimeter wavelength standards at NPL, Proc. of the Joint 5th ESA workshop on millimetre-wave Technology and applications and 31th ESA Antenna Workshop, 463–467, Nooordwijk, The Netherlands, 2009.

Clarke, R., Pollard, R., Ridler, N., and Salter, M.: Traceability to national standards for S-parameter measurements of waveguide devices from 110 GHz to 170 GHz, Microwave Measurement Conference, June 2009, 73rd ARFTG, Boston, USA, 2009.

Hoffmann, J.: Traceable S-Parameter Measurements in Coaxial Transmission Lines up to 70 GHz, Dissertation ETH Zürich, Switzerland, 2009.

Horibe, M., Kishikawa, R., and Shida, M.: Complete Characterization of Rectangular Waveguide Measurement Standards for Vector Network Analyzer in the range of Millimeter and Submillimeter wave frequencies, Microwave Measurements Conference (ARFTG), 76th ARFTG, December 2010, Clearwater Beach, USA, 2010.

Jastrow, C., Priebe, S., Spitschan, B., Hartmann, J., Jacob, M., Kürner, T., Schrader, T., and Kleine-Ostmann, T.: Wireless digital data transmission at 300 GHz, Electron. Lett. 46, 661–663, 2010.

Judaschke, R.: Second-Order Waveguide Calibration of a One-Port Vector Network Analyzer, to be published 77th ARFTG, 2011.

Lau, Y.: Understanding the residual waveguide interface variations on millimeter wave calibration, Microwave Measurements Conference (ARFTG), 76th ARFTG, December 2010, Clearwater Beach, USA, 2010.

Lok, L. B., Singh, S., Wilson, A., and Elgaid, K.: Impact of waveguide aperture dimensions and misalignment on the calibrated performance of a network analyzer from 140 to 325 GHz, Microwave Measurement Conference, 73rd ARFTG, June 2009, Boston USA, 1–4, 2009.

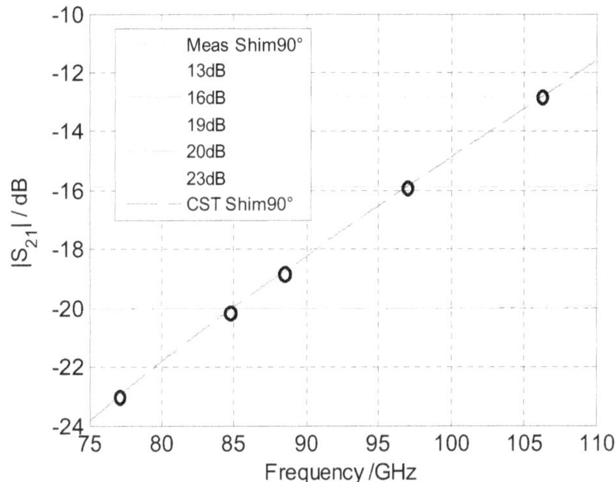

Fig. 12. Comparison of numerical simulation (CST), measurement of S_{21} of a WR10 shim rotated by 90° (HRD 90° shim), and "measurement of S_{21}" with a well-defined attenuation using coaxial attenuators inserted in the "meas in" path of the VNA covering the same dynamic range.

Priebe, S., Jastrow, C., Jacob, M., Kleine-Ostmann, T., Schrader, T., and Kürner, T.: A Measurement System for Propagation Measurements at 300 GHz, Conf. Proc. PIERS 2010, Cambridge, USA, 5–8 July, 2010

Ridler, N., Salter, M., Goy, P., Caroopen, S., Watts, J., Clarke, R., Yuenie Lau, Linton, D., Dickie, R., Huggard, P., Henry, M., Hesler, J., Barker, S., and Stanec, J.: Inter-laboratory comparison of reflection and transmission measurements in WR-06 waveguide (110 GHz to 170 GHz), Microwave Measurements Conference (ARFTG), 75th ARFTG, May 2010, Anaheim, USA, 2010a.

Ridler, N., Clarke, R., Salter, M., and Wilson, A.: Traceability to National Standards for S-parameter Measurements in Waveguide at Frequencies from 140 GHz to 220 GHz, Microwave Measurements Conference (ARFTG), 76th ARFTG, December 2010, Clearwater Beach, USA, 2010b.

Salhi, M., Kleine-Ostmann, T., and Schrader, T.: Broadband Electromagnetic Field Strength Sensors for 40-300 GHz Based on Planar Log Per Antennas and High-Speed Schottky Diodes, Conf. Proc. APMC 2010, Yokohama, Japan, 7–10 Dezember 2010.

Schrader, T., Baaske, K., Salhi, M., and Kleine-Ostmann, T.: Time-Domain Evaluation of Anechoic Environments up to 325 GHz, Conf. Proc. of 2010 Asia-Pacific Symposium on EM Compatibility, Beijing, China, 12–16 April, 2010, 778–781, doi 10.1109/APEMC.2010.5475831, 2010.

Stumper, U.: Extended Cross-Ratio Reflection Correction at Microwave Frequencies Using Waveguide Air-Lines, IEEE Transactions on Instrumentation and Measurement, 50(2), 364–367, 2001.

Williams, D. F.: Rectangular-Waveguide Vector-Network-Analyzer Calibrations With Imperfect Test Ports, Microwave Measurements Conference (ARFTG), 76th ARFTG, December 2010, Clearwater Beach, USA, 2010.

Evaluation of a concept for density measurement of solid particle flows in pneumatic conveying systems with microwaves (8–12 GHz)

C. Baer[1], T. Musch[1], M. Gerding[1], and M. Vogt[2]

[1]Ruhr-Universität Bochum, Institute for Electronic Circuits, 44780 Bochum, Germany
[2]Ruhr-Universität Bochum, High Frequency Engineering Research Group, 44780 Bochum, Germany

Abstract. This paper presents a novel density measuring concept for gas/coal particle compositions in pneumatic conveying systems. The proposed monitoring system uses horn antennas to perform complex electromagnetic transmittance measurements through the cross section of the conveying tube. The phase of the complex transmittance gives information about the effective permittivity, which is related to the mean volume fraction of the coal. Electromagnetic field simulations have been performed for the evaluation of the concept and the performance of the designed setup. A test stand for measurements on coal dust under reproducible conditions and with well-defined particle concentrations has been developed and implemented. The test tube has a diameter of 200 mm and a length of 400 mm. Coal particles with a diameter between 20 μm and 100 μm have been dispersed by injecting nitrogen gas inside the test tube. Complex transmission measurements are performed in the implemented setup with a calibrated vector network analyzer within a frequency range of 8–12 GHz. Results of the conceptual evaluation by measurements with different concentrations of coal particles are presented and discussed.

1 Introduction

Pneumatic conveying is of interest for the handling of particulate bulk materials like grain, pellets, etc., and pulverized fuel like coal dust. Mass or volume flow rate are parameters to quantify the mass flow (Miller et al., 2000). Common methods for the measurement of the particles velocity are correlation and Doppler techniques. Besides the flow velocity, the mass density and volume concentration of the solid particles inside the conveying tube are additional parameters to be assessed. Available techniques for the monitoring of the volume concentration are based on cut-off frequency measurements (Conrads, 1996), determination of backscatter coefficients (Happel, 2006), different resonator properties (Penirschke and Jakoby, 2010), and various optical, acoustical, and electro-statical concepts. In this contribution, concepts for the measurement of the volume concentration by means of microwaves are discussed and evaluated. The proposed concept is based on transmission measurements through the cross-section of a conveying tube. The complex transmittance between the antennas is measured in order to determine electrical material properties of the gas/dust composition, which are representative for the mean volume fraction of the pulverized coal. The applied wide frequency range (8–12 GHz) permits a robust and precise analysis.

2 Monitoring sytem

2.1 Concept

The most important parameter in the described application is the mass flow $\dot{m}$ of the coal particles, which is defined as:

$$\dot{m} = \frac{\partial m}{\partial t} = \rho \cdot \dot{V} \tag{1}$$

In Eq. (1), ρ is the mass density of the coal. The volume flow $\dot{V}$ of the coal dust is given as follows:

$$\dot{V} = \frac{\partial V}{\partial t} = \overline{v} \cdot A \cdot \overline{\zeta} \tag{2}$$

In Eq. (2), $\overline{v}$ is the mean particle flow velocity over the cross-section of the tube, A the cross-section of the tube and $\overline{\zeta}$ the mean volume fraction of the coal dust through the cross-section. By means of the subsequently described measurement concept, the mean volume fraction $\overline{\zeta}$ is measured. Figure 1 schematically shows the setup of the implemented monitoring system. Two horn antennas are arranged opposite to each other with the line-of-sight being oriented perpendicularly to the tube axis. By measuring the complex transmittance $S_{21}(\omega)$, a projection along the tube's cross-section of

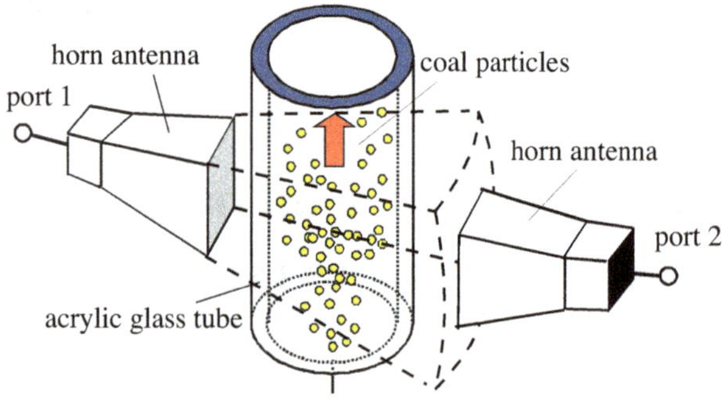

Fig. 1: Schematic of setup with conveying tube: measurement tube between two rectangular horn antennas.

the particles is obtained. With the high propagation speed of the electromagnetic waves in relation to the particle flow velocity, the particle density within the measuring tube can be assumed as static.

In order to obtain a robust model, various material properties of the coal particles have to be taken into consideration. Because the maximum diameter K of the particles is less than $100\,\mu$m, it follows:

$$K << \lambda_{min} \tag{3}$$

In Eq. (3), λ_{min} is the smallest wavelength in the used frequency band i.e. at 12 GHz. In addition, coal can be described as a dielectric medium with a relative permittivity $\varepsilon_{r,coal}$ in a range (Ruhrkohlenhandbuch, 1984):

$$\varepsilon_{r,coal} = 2.4 \ \ldots \ 2.8 \tag{4}$$

The conductivity σ of coal is negligibly small in this application in relation to the product of angular frequency and permittivity. Consequently, the coal/gas composition can be described as a quasi-homogeneous dielectric medium described by a single effective permittivity $\varepsilon_{r,eff}$ which is directly related to the mean volume fraction $\overline{\zeta}$ of the coal. Based on Sihvola (2000) the effective permittivity is given as follows:

$$\varepsilon_{r,eff} = \varepsilon_{r,gas} + \frac{3 \cdot \overline{\zeta} \cdot \varepsilon_{r,gas} \cdot \left(\varepsilon_{r,coal} - \varepsilon_{r,gas}\right)}{\varepsilon_{r,coal} + 2 \cdot \varepsilon_{r,gas} - \overline{\zeta} \cdot \left(\varepsilon_{r,coal} - \varepsilon_{r,gas}\right)} \tag{5}$$

By measuring the effective permittivity $\varepsilon_{r,eff}$, the mean volume fraction $\overline{\zeta}$ of the coal can be obtained from Eq. (5) by taking the known permittivities into consideration. Formula (5) is known as the Maxwell-Garnett mixing formula for spherical particles (Sihvola, 2000), which is a good approximation for the description of the behavior of the coal particles.

2.2 Measurement approach

Under the assumption, that standing waves and multiple reflections in the system can be neglected, the phase $\phi(\omega)$ of the complex transmittance $S_{21}(\omega)$ is given as the line-integral along the transmission path between the two ports, see Fig. 1, as follows:

$$\begin{aligned} \phi(\omega) &= \int_s \frac{\omega}{v_{ph}(s)} ds + \phi_0(\omega) \\ &= \frac{\omega}{c_0} \cdot \int_s \sqrt{\varepsilon_r'(s)}\ ds + \phi_0(\omega) \\ &= \frac{\omega}{c_0} \cdot \sqrt{\varepsilon_{r,eff}} \cdot D + \phi_0(\omega) \end{aligned} \tag{6}$$

By integrating the effective permittivity over the cross section a closed expression for the phase relation can be derived. In Eq. (6), D is the diameter of the measuring tube, ω the angular frequency and c_0 the speed of light. The frequency-dependent phase $\phi_0(\omega)$ denotes the wave propagation inside the antennas and feeding networks. By means of a reference measurement without any particles inside the measurement tube, the phase term $\phi_0(\omega)$ given in Eq. (6) can be determined:

$$\phi_{ref}(\omega) = \frac{\omega}{c_0} \sqrt{\varepsilon_{r,gas}} \cdot D + \phi_0(\omega) \tag{7}$$

The measured phase $\phi_{meas}(\omega)$ obtained in a measurement can be eliminated for $\phi_0(\omega)$ as follows, by use of Eqs. (6) and (7):

$$\begin{aligned} \Delta\phi(\omega) &= \phi_{meas}(\omega) - \phi_{ref}(\omega) \\ &= \frac{\omega}{c_0} \cdot \left(\sqrt{\varepsilon_{r,eff}} - \sqrt{\varepsilon_{r,gas}}\right) \cdot D \end{aligned} \tag{8}$$

The phase $\Delta\phi(\omega)$ in Eq. (8) is a linear function of $\sqrt{\varepsilon_{r,eff}}$, and the mean volume fraction $\overline{\zeta}$ can be assessed by means of Eq. (5). Based on a linear regression fit of the function $\Delta\phi(\omega)$, the measurement is very robust.

2.3 Prototype system

A test tube for measurements on coal dust under reproducible conditions and well-defined particle concentrations has been developed and implemented. Coal particles are dispersed inside the test tube by injecting nitrogen gas.

Figure 2 shows a photo of the test stand consisting of an acrylic glass tube (I), two rectangular horn antennas (II), a nitrogen gas inlet pipe (III), and a filtering exhaust (IV). The coal particles are dispersed with a gas pressure of 5 bar. The exhausted gas is cleaned in a gas scrubbing system in order to avoid contamination. The whole test stand has a height of 80 cm and a diameter of 20 cm. The complex transmittance is measured with a Rohde & Schwarz ZVK two-port vector network analyzer.

3 Simulations and exemplary measurements

For the verification of the measurement concept, various simulations with a 3-D-field simulator were performed.

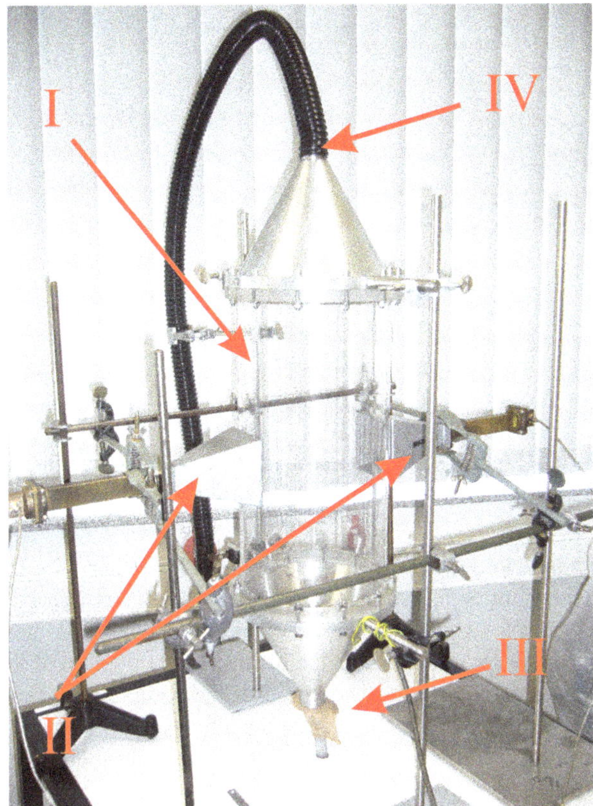

Fig. 2: Photo of the implemented test stand. **I**: Acrylic glass cylinder, **II**: Rectangular horn antennas, **III**: Nitrogen gas inlet pipe, **IV**: Filtering exhaust.

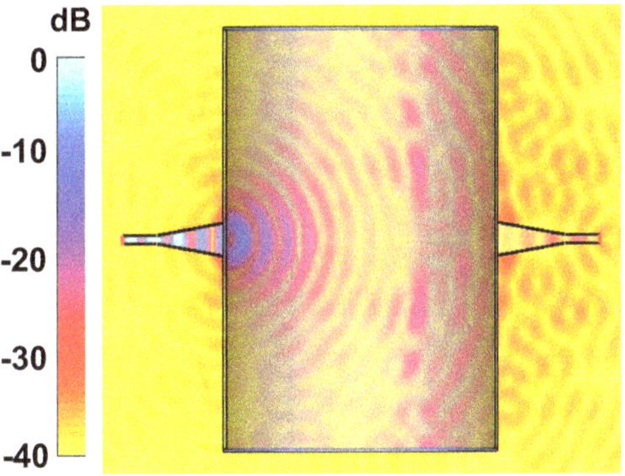

(a) 3-D simulation, magnitude of the E-field at a frequency of 10 GHz with a homogeneous relative permittivity of $\varepsilon_{\mathrm{r,eff}} = 1$.

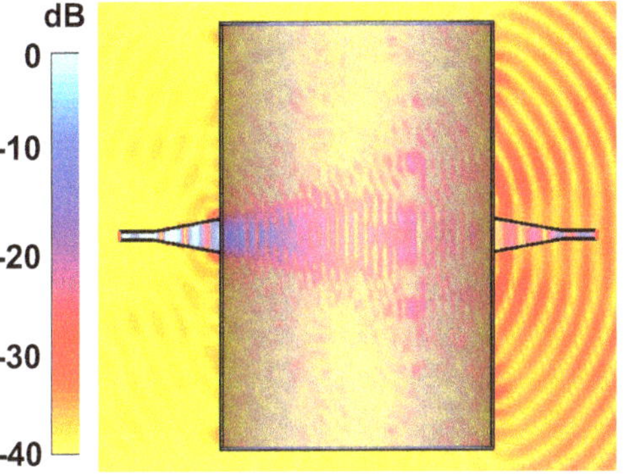

(b) 3-D simulation, magnitude of the E-field at a frequency of 10 GHz with a homogeneous relative permittivity $\varepsilon_{\mathrm{r,eff}}$=4.

Fig. 3: Screenshots of transient 3-D field simulations at constant frequency for changing permittivity, CST Microwave Studio 2010.

3.1 Simulations

Figure 3 shows simulation results for two different setups of the measuring tube. In both configurations, the measuring tube is placed in center, while the two horn antennas are arranged side wise, opposite to each other. The permittivity of the inner volume of the tube has been varied. A transient solver was used for the following simulations. While the field monitor's frequency was set to 10 GHz in all cases, relative permittivities $\varepsilon_{\mathrm{r,eff}} = 1$ and $\varepsilon_{\mathrm{r,eff}} = 4$ have been used for the results in Fig. 3a and Fig. 3b, respectively. A comparison shows that the phase fronts in Fig. 3b are tighter than the phase fronts in Fig. 3a. This is in agreement with the assumption that the slope of the phase line depends on the permittivity. Figures 3a and 3b also show that multiple reflections, effected by the acrylic glass cylinder with a relative permittivity $\varepsilon_{\mathrm{r,acryl}} = 3$, are negligible as they are more than 40 dB smaller compared to the maximum field strength.

Figure 4 shows the simulated phase $\Delta\phi$ versus frequency f, for increasing permittivities $\varepsilon_{\mathrm{r}} = 1$ to $\varepsilon_{\mathrm{r}} = 1.8$ of the inner volume. Comparing the effective permittivities from the slopes of the linear functions in Fig. 4 to the permittivities used for the simulation, an evaluation error of about 1% is obtained. This confirms the applicability of the proposed concept and model. Due to the linear run of the phases in Fig. 4 it can be assumed, that standing waves are also negligible as they would cause nonlinear phase runs.

3.2 Measurements

In a next step, measurements with different homogeneous materials with known permittivities have been performed. Two phantoms have been made of polyethylene (PE, relative permittivity $\varepsilon_{\mathrm{r,eff,PE}} = 1.4$) and polyoxymethylene (POM, relative permittivity $\varepsilon_{\mathrm{r,eff,POM}} = 3.8$). Figure 5 shows the measured phase $\Delta\phi$ versus frequency f for both phantoms. Like in the simulation, the slope of the phase increases with increasing permittivity. The permittivites calculated from the slopes show a relative measurement error of 9.2% for PE and 0.7 % for POM, respectively.

Both, simulation and phantom measurement results, confirm the validity of the proposed method.

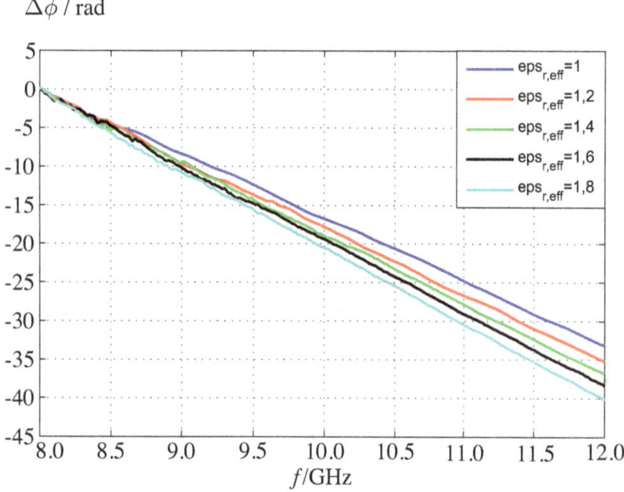

Fig. 4: Simulated phase differences $\Delta\phi$ for different homogenous permittivities.

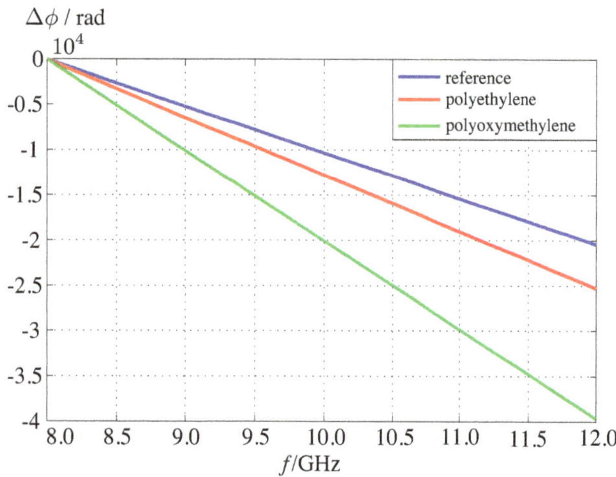

Fig. 5: Measured phase: polyethylene and polyoxymethylene phantom.

Measurements on a coal/gas composition have been performed by inserting different amounts of coal particles with well known weight into the tube. Since the total volume of the measurement tube and the density of the coal is known, measurements results can be compared to the known volume fractions. Figure 6 shows the measured total coal mass versus the actually inserted and dispersed coal mass. The relative measurement error is approximately 13%.

4 Conclusions

In this paper, a novel density measuring method for pulverized fuels in pneumatic conveying systems is proposed. The concept is based on electromagnetic transmission measurements along the cross-section of a conveying tube consisting of acrylic glass. In the implemented setup, two

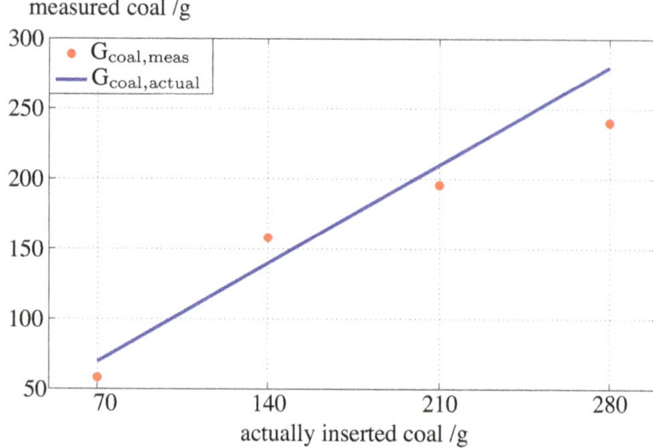

Fig. 6: Comparison between mass of coal measured inside the measurement tube and actually inserted coal mass.

horn-antennas are arranged opposite to each other with the line-of-sight being oriented perpendicularly to the tube axis. The complex transmittance between both antennas is measured in order to determine the mean volume fraction of the gas/coal particle flow. The phase of the complex transmittance contains information about the effective permittivity of the quasi-continuous medium. The phase contributions of the antennas and feeding networks are compensated by taking a reference measurement with the empty tube into account. Furthermore, electromagnetic field simulations have been performed to evaluate the concept and the performance of the method. Concerning the permittivity measurements, the evaluation error is smaller than 1%. The error of measurement with two homogenous phantoms with different permittivities are 0.7% and 9.7%, respectively. Different tests on dispersed coal particles revealed an error of 13%. The proposed monitoring concept has been verified based on various electromagnetic simulations and measurements. The goal of future developments is to improve the measurement accuracy by employing more sophisticated calibration techniques.

References

Conrads, H. G.: Method and device for a contact-free measurement of the mass flow rate in a two phase pneumativ transport using microwaves, Eur ER Patent 07176269 A2, 1996.

Happel, J.: Method and device for measuring a mass flow, US Patent 7102133, 2006.

Miller, D., Baimbridge, P., and Eyre, D.: Technology status of PF flow measurement and control methods for utility boilers, Report No. COAL R201 DTI/PUB 00/1445, Crown Copyright 2000.

Penirschke, A. and Jakoby, R.: Design of a Moisture Independent Microwave Mass Flow Detector for Particulate Solids, GEMIC 2010.

Ruhrkohlenhandbuch: Glueckaufverlag GmbH, Essen, 1984.

Sihvola, A.: Mixing Rules with complex Dielectric Coefficients, Subsurface Sensing Technologies and Applications Vol. 1, No. 4, 2000.

16

Computation of electrostatic fields in anisotropic human tissues using the Finite Integration Technique (FIT)

V. C. Motrescu and U. van Rienen

Institute of General Electrical Engineering, University of Rostock, Germany

Abstract. The exposure of human body to electromagnetic fields has in the recent years become a matter of great interest for scientists working in the area of biology and biomedicine. Due to the difficulty of performing measurements, accurate models of the human body, in the form of a computer data set, are used for computations of the fields inside the body by employing numerical methods such as the method used for our calculations, namely the Finite Integration Technique (FIT). A fact that has to be taken into account when computing electromagnetic fields in the human body is that some tissue classes, i.e. cardiac and skeletal muscles, have higher electrical conductivity and permittivity along fibers rather than across them. This property leads to diagonal conductivity and permittivity tensors only when expressing them in a local coordinate system while in a global coordinate system they become full tensors. The Finite Integration Technique (FIT) in its classical form can handle diagonally anisotropic materials quite effectively but it needed an extension for handling fully anisotropic materials. New electric voltages were placed on the grid and a new averaging method of conductivity and permittivity on the grid was found. In this paper, we present results from electrostatic computations performed with the extended version of FIT for fully anisotropic materials.

1 Introduction

With continuously increasing numbers of electrical and electronic devices being used both in households and for communication purposes, special concerns arise regarding the possible adverse biological effects of high or low frequency electromagnetic fields on the human body. The difficulties in directly measuring the fields' induced currents or energy inside the body have resulted in different maximum values within the safety guidelines for limiting the effects of human exposure to non-ionizing electromagnetic radiation. In such conditions, the estimation of electromagnetic fields inside the body using numerical methods adequate for computer code implementations, represents a good alternative to the experimental measurements, especially, because the computational capacities are currently able to respond to such complex tasks and continue to have a quick evolution. Another argument for using computer based numerical methods is the availability of realistic human body models of high resolution based on anatomical data in the form of a computer data set. Numerical methods such as the Finite Element Method (FEM), the Finite Difference Method (FD), the Boundary Element Method (BEM), and the Finite Integration Technique (FIT) have already been used in computations of electromagnetic fields in the human body. Our computations are based on an extended version of the Finite Integration Technique (FIT) which allows us to account for the anisotropic character of some tissue classes.

2 Human body model

The human body model used in our simulations was created by a working group at the Institute of Biomedical Engineering, University of Karlsruhe, Germany, who applied strategies of image processing (Sachse et al., 1996a; 1996b) to a set of anatomical data obtained from Computed Tomography (CT) and Magnetic Resonance Tomography (MRT) provided by the National Library of Medicine, Maryland, USA. The images were first assembled in a 3D anatomical model consisting of cubic voxels each voxel having been assigned one of 40 different tissue types. To every tissue class was assigned a frequency dependent electrical conductivity and permittivity in order to obtain a dielectric model which can be used for electromagnetic simulations in computerized applications. The dielectric properties of tissues are known from measurements (Gabriel et al., 1996). The model, known as HUGO, is available in different resolutions, from 8 mm^3 to 1 mm^3. Furthermore, this model was extended with the

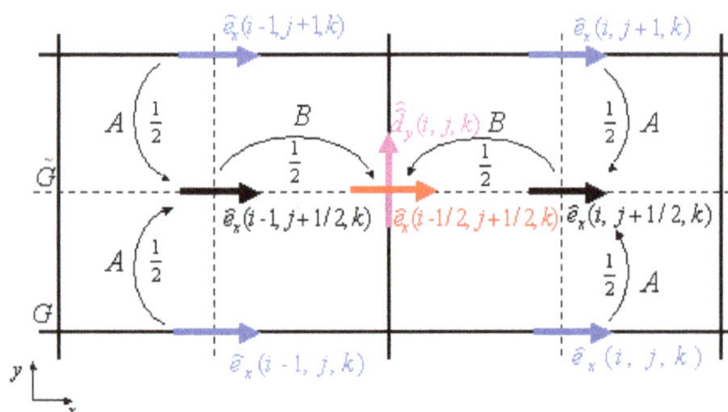

Fig. 1. Local interpolation scheme of a grid voltage in x-direction at the location of a grid flux in y-direction.

Orientation Data Set which gives the direction of fibers by providing two angles for every voxel containing muscle tissue (Sachse et al., 1998).

3 Electromagnetic properties of biological tissues

It is known that biological tissue is non-magnetic i.e. the permeability of biological tissue is equal to that of free space (Durney et al., 1986). The permittivity and conductivity vary with temperature but a stronger variance is experienced with frequency. While permittivity of biological tissue generally decreases with frequency, its conductivity generally increases.

Some tissue types which present a fiber structure (e.g. skeletal and cardiac muscles) are anisotropic having higher conductivity and permittivity in the longitudinal direction of the fibers than on the perpendicular direction to the fibers (Sachse et al., 1997). Shortly, these tissues present a transversely isotropic anisotropy with regard to their dielectric properties.

(Note: In compliance with the subject matter, the remainder of this paper will refer only to permittivity, though, it should be noted that conductivity can be treated in exactly the same way.)

In a local coordinate system, the permittivity of muscle tissue is described by a diagonal tensor of rank two:

$$\mathbf{T}_{\varepsilon_\mathrm{l}}=\begin{pmatrix}\varepsilon_l & 0 & 0\\ 0 & \varepsilon_p & 0\\ 0 & 0 & \varepsilon_p\end{pmatrix}, \tag{1}$$

where ε_l is the longitudinal permittivity and ε_p is the perpendicular permittivity, relative to the fiber's direction.

Since muscle fibers are miscellaneously oriented in the body, for computational reasons, it is useful to express this tensor in a global coordinate system. In this respect, the local tensor has to be rotated according to the following equation:

$$\mathbf{T}_{\varepsilon_\mathrm{G}}=\mathbf{R}\mathbf{T}_{\varepsilon_\mathrm{L}}\mathbf{R}^{-1}, \tag{2}$$

where the rotation matrix $\mathbf{R}$ is the product of two other rotation matrices: $\mathbf{R}=\mathbf{R}_{xy}\mathbf{R}_{xz}$, with $\mathbf{R}_{xy}$ and $\mathbf{R}_{xz}$ given below.

$$\mathbf{R}_{xy}=\begin{pmatrix}\cos\phi & -\sin\phi & 0\\ \sin\phi & \cos\phi & 0\\ 0 & 0 & 1\end{pmatrix} \quad \mathbf{R}_{xz}=\begin{pmatrix}\sin\theta & 0 & \cos\theta\\ 0 & 1 & 0\\ \cos\theta & 0 & -\sin\theta\end{pmatrix}. \tag{3}$$

The angles ϕ and θ are the rotation angles about the z- and y-axes, respectively, provided by the Orientation Data Set. After the rotation in Eq. (2) we obtain a full and symmetric tensor which expresses the permittivity of muscle tissue in a global coordinate system.

4 Considerations concerning the classical Finite Integration Technique (FIT)

Since a lot of literature has already been published about the FIT, only a short introduction is provided here concerning the allocation of some grid-state variables and material treatment related to the subject of this paper. For more general details about FIT, we recommend you see van Rienen (2001), meanwhile, for more information concerning the subject matter of this paper, see van Rienen et al. (2003).

The Finite Integration Technique (FIT) was first published by Weiland (1977) and developed as a numerical method which discretizes the Maxwell's equations on a grid pair preserving the analytical properties of the original equations (van Rienen, 2001). On the FIT's grid doublet, in connection with Electrostatics, the following variables are defined:

- the electric potentials (denoted with φ_n), allocated in every mesh node belonging to the primary grid;
- the electric grid voltages (denoted with $\hat{e}_n$), allocated in the middle of the edges belonging to the primary grid and calculated as the integral of the electric field along primary edges;
- the electric grid flux densities (denoted with $\hat{\hat{d}}_n$), normal in the middle of the surfaces belonging to the dual grid and calculated as the integral of the electric flux on dual surfaces.

In a 3D Cartesian coordinate system, because the dual grid is shifted with half an edge length in all positive directions with respect to the primary grid, the primary edges intersect dual surfaces on the normal direction in the middle. This means that an electric grid voltage $\hat{e}_n$ from the primary grid is allocated in the same point with the corresponding electric grid flux $\hat{\hat{d}}_n$ in the same direction, from the dual grid.

The electric fluxes and voltages with coinciding both, locations and orientations, are related to each other by the permittivity according to the following equation:

$$\frac{\hat{\hat{d}}_n}{\hat{e}_n}=\frac{\int\int_{\tilde{A}_n}\mathbf{D}\cdot d\mathbf{A}}{\int_{L_n}\mathbf{E}\cdot d\mathbf{s}}=\frac{\int\int_{\tilde{A}_n}\varepsilon dA+O(\Delta^{\kappa+1})}{\int_{L_n}ds+O(\Delta^{\kappa})}$$
$$\approx\frac{\overline{\varepsilon}\int\int_{\tilde{A}_n}dA}{\int_{L_n}ds}+O(\Delta^{\kappa})\approx\overline{\varepsilon}\frac{|\tilde{A}_n|}{|L_n|}=[M_\varepsilon]_{n,m}, \tag{4}$$

where κ takes values between $\kappa=2$ for varying permittivity or non-uniform step size and $\kappa=3$ otherwise. The symbol $\overline{\varepsilon}$ denotes a weighted average of the permittivity on a dual surface from four possibly different values belonging to the primary grid cells intersected by that dual surface. For the entire grid, the point-wise relation in Eq. (4) becomes:

$$\hat{\hat{d}}=\tilde{\mathbf{D}}_{\mathbf{A}}\mathbf{D}_{\varepsilon}\mathbf{D}_{\mathbf{S}}^{-1}\hat{e}=\mathbf{M}_{\varepsilon}\hat{e}, \tag{5}$$

where: $\tilde{\mathbf{D}}_{\mathbf{A}}$, $\mathbf{D}_{\varepsilon}$ and $\mathbf{D}_{\mathbf{S}}^{-1}$ are diagonal matrices containing the areas of the dual surfaces, the permittivity of the primary grid cells averaged on the dual surfaces and the inverse of the primary edge lengths, respectively. The vectors $\hat{\hat{d}}$ and $\hat{e}$ contain the electric grid fluxes allocated in the middle of dual surfaces and the electric grid voltages along the primary edges. $\mathbf{M}_{\varepsilon}$ is the material operator which decomposed along the axes of a Cartesian coordinate system has the following diagonal form:

$$\mathbf{M}_{\varepsilon}=\begin{pmatrix} \tilde{\mathbf{D}}_{\mathbf{A}_{yz}}\mathbf{D}_{\varepsilon_{xx}}\mathbf{D}_{\mathbf{S}_x}^{-1} & 0 & 0 \\ 0 & \tilde{\mathbf{D}}_{\mathbf{A}_{xz}}\mathbf{D}_{\varepsilon_{yy}}\mathbf{D}_{\mathbf{S}_y}^{-1} & 0 \\ 0 & 0 & \tilde{\mathbf{D}}_{\mathbf{A}_{xy}}\mathbf{D}_{\varepsilon_{zz}}\mathbf{D}_{\mathbf{S}_z}^{-1} \end{pmatrix} \tag{6}$$

In its classical form, FIT allows the presence of diagonally anisotropic material on the grid but, as it was shown, this allowance is not enough for computing muscle tissues which are fully anisotropic in a global coordinate system.

5 Extension of the Finite Integration Technique for computing anisotropic tissues

To deal with anisotropic tissues, we follow the idea from (Krüger, 2000) where gyrotropic materials were treated in time domain.

When the diagonal matrix $\mathbf{D}_{\varepsilon}$ in Eq. (5) is replaced with a full one, the material operator in Eq. (6) becomes:

$$\mathbf{M}_{\varepsilon}=\begin{pmatrix} \tilde{\mathbf{D}}_{\mathbf{A}_{yz}}\mathbf{D}_{\varepsilon_{xx}}\mathbf{D}_{\mathbf{S}_x}^{-1} & \tilde{\mathbf{D}}_{\mathbf{A}_{yz}}\mathbf{D}_{\varepsilon_{xy}}\mathbf{D}_{\mathbf{S}_y}^{-1} & \tilde{\mathbf{D}}_{\mathbf{A}_{yz}}\mathbf{D}_{\varepsilon_{xz}}\mathbf{D}_{\mathbf{S}_z}^{-1} \\ \tilde{\mathbf{D}}_{\mathbf{A}_{xz}}\mathbf{D}_{\varepsilon_{yx}}\mathbf{D}_{\mathbf{S}_x}^{-1} & \tilde{\mathbf{D}}_{\mathbf{A}_{xz}}\mathbf{D}_{\varepsilon_{yy}}\mathbf{D}_{\mathbf{S}_y}^{-1} & \tilde{\mathbf{D}}_{\mathbf{A}_{xz}}\mathbf{D}_{\varepsilon_{yz}}\mathbf{D}_{\mathbf{S}_z}^{-1} \\ \tilde{\mathbf{D}}_{\mathbf{A}_{xy}}\mathbf{D}_{\varepsilon_{zx}}\mathbf{D}_{\mathbf{S}_x}^{-1} & \tilde{\mathbf{D}}_{\mathbf{A}_{xy}}\mathbf{D}_{\varepsilon_{zy}}\mathbf{D}_{\mathbf{S}_y}^{-1} & \tilde{\mathbf{D}}_{\mathbf{A}_{xy}}\mathbf{D}_{\varepsilon_{zz}}\mathbf{D}_{\mathbf{S}_z}^{-1} \end{pmatrix} \tag{7}$$

The off-diagonal terms of the material operator are coupling electric grid flux vectors in one direction to electric grid voltage vectors in another direction. To bring the vectors with different orientations to the same location on the grid, an interpolation process is necessary (Krüger, 2000) for which a local scheme is presented in Fig. 1.

In Fig. 1, the electric grid voltage $\hat{e}_x(i-1/2,\ j+1/2,\ k)$ is interpolated in two steps at the location of $\hat{\hat{d}}_y(i,\ j,\ k)$. In the first step, every pair of electric voltages having the same coordinate on the x-axis, are interpolated along the y-axis in the middle (each voltage contributing with a factor of a half) through an A-type interpolation, building the voltages $\hat{e}_x(i-1,\ j+1/2,\ k)$ and $\hat{e}_x(i,\ j+1/2,\ k)$. In the second step, these two voltages are interpolated along the x-axis in the middle (each contributing with a factor of a half) through a B-type interpolation, building the voltage $\hat{e}_x(i-1/2,\ j+1/2,\ k)$.

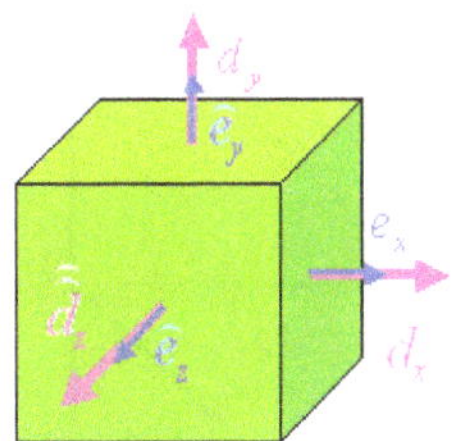
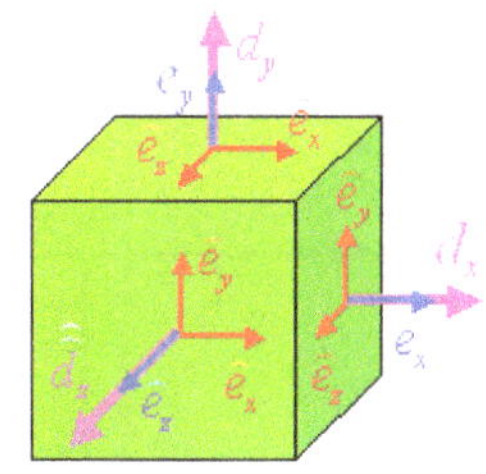

Fig. 2. Allocation of both electric grid voltages and electric grid fluxes relative to a dual FIT cell. Left: before the interpolation. Right: after the interpolation.

The local interpolation process is expressed by the following equation:

$$\hat{e}_x(i-1/2,\ j+1/2,\ k)=\frac{1}{4}\big[\hat{e}_x(i,\ j,\ k) \\ +\hat{e}_x(i,\ j+1,\ k)+\hat{e}_x(i-1,\ k,\ k)+\hat{e}_x(i-1,\ j+1,\ k)\big]. \tag{8}$$

Figure 2 shows the allocation of voltages relative to a dual cell after the interpolation (right-hand side) compared to the classical allocation (left-hand side).

To globally account the interpolation process, the following matrices are defined:

$$[Q_x]_{pq}=\begin{cases} \frac{1}{2} & p=q \text{ or } p=q+1 \\ 0 & \text{else} \end{cases} \tag{9}$$

$$\left[Q_y\right]_{pq}=\begin{cases} \frac{1}{2} & p=q \text{ or } p=q+I \\ 0 & \text{else} \end{cases} \tag{10}$$

$$\left[Q_z\right]_{pq}=\begin{cases} \frac{1}{2} & p=q \text{ or } p=q+IJ \\ 0 & \text{else} \end{cases}, \tag{11}$$

where I and J represent the maximum number of grid nodes in x- and y-directions, respectively (Krüger, 2000). Within the FIT algorithm, the interpolation matrices will be placed at the off-diagonal terms in the material operator.

In order to keep the symmetry of the physical permittivity tensor the off-diagonal terms $[\mathbf{M}_{\varepsilon}]_{\mathbf{u},\mathbf{v}}$ and $[\mathbf{M}_{\varepsilon}]_{\mathbf{v},\mathbf{u}}$ of the discrete material operator have to be equal. The symmetry of the material operator can be reached through the following steps:

- Assuming that primary and dual edges in the same direction have the same length, their cancellation leads to equal grid information in the above-mentioned terms of the material operator according to Eq. (12).

$$\frac{\tilde{A}_{vw}}{L_v}=\frac{\tilde{L}_v\tilde{L}_w}{L_v}=\frac{\tilde{A}_{uw}}{L_u}=\frac{\tilde{L}_u\tilde{L}_w}{L_u}\approx\tilde{L}_w. \tag{12}$$

The approximation introduced in Eq. (12) is given by $L_u=\tilde{L}_u + O(\Delta)$, where the first order error term vanishes if the grid is equidistant. This condition can be applied when meshing the human body model because it consists of cubic voxels.

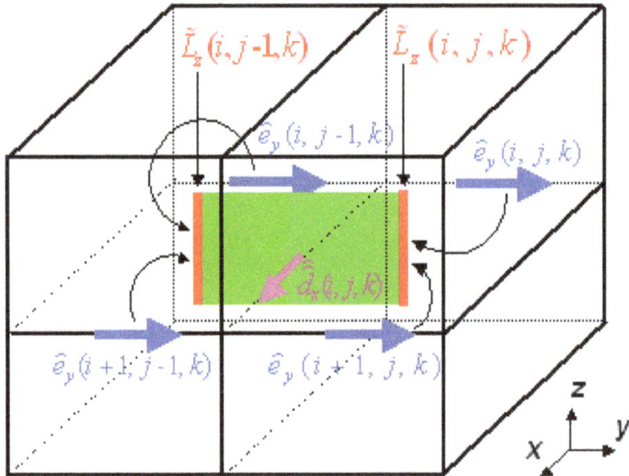

Fig. 3. Construction of electric voltages on the grid.

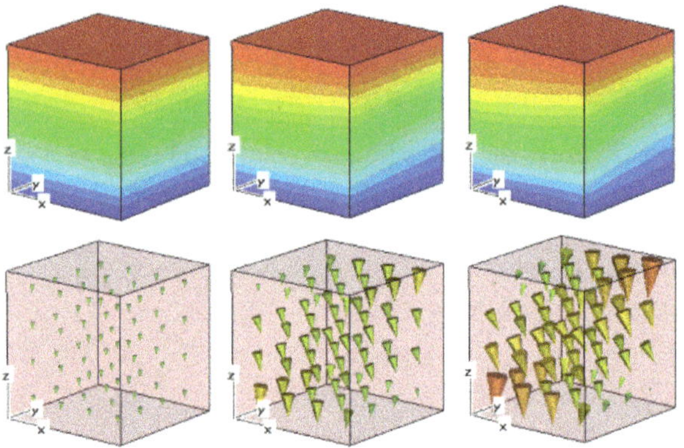

Fig. 4. Top row: electric potential distribtuion. Bottom row: electric flux density. Left column: $\varepsilon_l/\varepsilon_p$=1. Middle column: $\varepsilon_l/\varepsilon_p$=2. Right column: $\varepsilon_l/\varepsilon_p$=3

- Averaging the permittivity along the common dual edge $\tilde{L}_w$ resulting from the first step leads to equal material information in the considered terms. The permittivity average for each off-diagonal term of the material operator uses the approximation from Eq. (12):

$$\frac{\iint_{\tilde{A}_{uw}} \varepsilon \, dA + O(\Delta^{k+1})}{\int_{L_u} ds + O(\Delta^k)} = \frac{\bar{\varepsilon} \left|\tilde{L}_w\right| \int_{\tilde{L}_u} ds}{\int_{L_u} ds} + O(\Delta^k)$$
$$\approx \bar{\varepsilon} \left|\tilde{L}_w\right| + O(\Delta^{k^*}) \approx \bar{\varepsilon} \left|\tilde{L}_w\right|, \qquad (13)$$

  where $k^* \in [1, 3]$.

- Positioning the interpolation matrices within the material operator such as to ensure the symmetry regarding the interpolation process.

Table 1. Electric flux density (max value [C/m^2]) given in % relative to the max value in the case of $\varepsilon_l/\varepsilon_p$=3.

	$\varepsilon_l/\varepsilon_p$=1	$\varepsilon_l/\varepsilon_p$=2	$\varepsilon_l/\varepsilon_p$=3
$d_{\max}$ [C/m^2]	13.5%	60.8%	100%

Taking into account the above considerations, the material operator becomes:

$$\mathbf{M}_\varepsilon = \begin{pmatrix} \tilde{\mathbf{D}}_{A_{yz}} \mathbf{D}_{\varepsilon_{xx}} \mathbf{D}_{S_x}^{-1} & \mathbf{Q}_y^T \mathbf{D}_{\varepsilon_{xy}} \tilde{\mathbf{D}}_{S_z} \mathbf{Q}_x & \mathbf{Q}_z^T \mathbf{D}_{\varepsilon_{xz}} \tilde{\mathbf{D}}_{S_y} \mathbf{Q}_x \\ \mathbf{Q}_x^T \mathbf{D}_{\varepsilon_{yx}} \tilde{\mathbf{D}}_{S_z} \mathbf{Q}_y & \tilde{\mathbf{D}}_{A_{xz}} \mathbf{D}_{\varepsilon_{yy}} \mathbf{D}_{S_y}^{-1} & \mathbf{Q}_z^T \mathbf{D}_{\varepsilon_{yz}} \tilde{\mathbf{D}}_{S_x} \mathbf{Q}_y \\ \mathbf{Q}_x^T \mathbf{D}_{\varepsilon_{zx}} \tilde{\mathbf{D}}_{S_y} \mathbf{Q}_z & \mathbf{Q}_y^T \mathbf{D}_{\varepsilon_{zy}} \tilde{\mathbf{D}}_{S_x} \mathbf{Q}_z & \tilde{\mathbf{D}}_{A_{xy}} \mathbf{D}_{\varepsilon_{zz}} \mathbf{D}_{S_z}^{-1} \end{pmatrix}. \qquad (14)$$

This distribution of the interpolation matrices to the off-diagonal terms ensures the global symmetry of the material operator. Due to the permittivity average, when coupling two directions the A-type interpolation is weighted with the length of a dual edge on the third direction. To keep the symmetry, the B-type interpolation cannot be length-weighted. Figure 3 shows that the voltages $\hat{e}_y(i, j-1, k)$ and $\hat{e}_y(i+1, j-1, k)$ are built along the dual edge $\tilde{L}_z(i, j-1, k)$ and the voltages $\hat{e}_y(i, j, k)$, $\hat{e}_y(i+1, j, k)$ are built along the dual edge $\tilde{L}_z(i, j, k)$.

6 Simulations

The human body model (HUGO) offered by the simulation software package CST EMStudio™ (CST GmbH) was used for the import of a cubic volume of muscle tissues within a C++ code where the anisotropic FIT was implemented. The diagonal direction of muscle fibres is described by the angles ϕ=θ=45°. These muscles were placed in the electrostatic field determined by the imposed potential values of −10 V and +10 V, at the boundary planes Z= min and Z= max, through the Dirichlet boundary conditions. All the other boundaries, were treated with Neumann boundary conditions. Three simulations were performed using the CG solver from PETSc (Balay et al., 1997) corresponding to three different values for the ratio between longitudinal and perpendicular permittivity of muscle tissues with respect to their fibre direction, i.e. $\varepsilon_l/\varepsilon_p$=1; 2; 3.

7 Results and discussions

In Fig. 4, the top row presents three scalar plots of the electric potentials computed inside a muscle volume considered to be isotropic (left), anisotropic (having the ratio $\varepsilon_l/\varepsilon_p$=2 (middle)) and anisotropic (with $\varepsilon_l/\varepsilon_p$=3 (right)). Comparing the plots, the equipotential planes were aligned to the direction of muscle fibers (diagonal direction with regard to the Cartesian coordinate system) in the anisotropic cases. The magnitude of this alignment depends on the magnitude of the ratio between longitudinal and perpendicular permittivity ($\varepsilon_l/\varepsilon_p$).

The bottom row in Fig. 4 presents three vector plots of the electric flux density, each computed with the corresponding electric potentials in the same column and scaled to the same maximum value which was found in the anisotropic case of $\varepsilon_l/\varepsilon_p=3$. Relative to this value, the other maximum values are given in percentage in Table 1. Compared to the isotropic case where the electric flux is homogeneous in the entire muscle volume, in the anisotropic cases, the flux has higher values along the fiber's direction.

8 Conclusions

In this paper we presented the anisotropy of muscle tissues with regard to dielectric permittivity and its mathematical analytic model. This model was further discretized to conform with the numerical algorithm of the Finite Integration Technique and implemented in software code. With this code, more simulations were performed and the results were compared to the isotropic case.

Acknowledgements. The authors are grateful to M. Clemens (TU Darmstadt) for many useful suggestions as well as to F. Sachse (IBT Karlsruhe, now CVRTI Univ. of Utah) for providing us with the Orientation Data Set. V. Motrescu was supported by DFG (RI 814/12-1) and CST GmbH, Darmstadt.

References

Balay, S., Gropp, W. D., McInnes, L. C., and Smith, B. F.: PETSc – Efficient Management of Parallelism in Object Oriented Numerical Software Libraries, Modern Software Tools in Scientific Computing, 163–202, Birkhauser Press, 1997.

CST EMStudio™: CST GmbH, Bad Neuheimer Str. 19, D-64289, Darmstadt, Germany.

Durney, C. H., Massoudi, C. H., and Iskander, M. F.: Radiofrequency Radiation Dosimetry Handbook, Fourth Edition, Univ. of Utah, Salt Lake City, http://www.brooks.af.mil/AFRL/HED/hedr/reports/handbook/home.html, 1986.

Gabriel, S., Lau, R. W., and Gabriel, C.: The dielectric properties of biological tissues: II. Measurements in the frequency range 10 Hz to 20 GHz, Phys. Med. Biol., 41, 2251–2269, 1996.

Krüger, H.: Zur numerischen Berechnung transienter elektromagnetischer Felder in gyrotropen Materialien, Dissertation, Technische Universität, Darmstadt, 2000.

Sachse, F. B., Werner, C., Mueller, M., and Meyer-Waarden, K.: Preprocessing of the Visible Man dataset for the generation of macroscopic anatomical models, Proc. First Users Conference of the National Library of Medicine's Visible Human Project, 123–124, 1996a.

Sachse, F. B., Werner, C., Mueller, M., and Meyer-Waarden, K.: Segmentation and tissue-classification of the Visible Man dataset using the computer tomographic scans and the thin sections photos, Proc. First Users Conference of the National Library of Medicine's Visible Human Project, 125–126, 1996b.

Sachse, F. B., Werner, C., Meyer-Waarden, K., and Dössel, O.: Comparison of Solutions to the Forward Problem in Electrophysiology with Homogeneous, Heterogeneous and Anisotropic Impedance Models, Biomedizinische Technik, 42, 277–280, 1997.

Sachse, F. B., Wolf, M., Werner, C., and Meyer-Waarden, K.: Extension of Anatomical Models of the Human Body: Three-Dimensional Interpolation of Muscle Fiber Orientation Based on Restrictions, J. Computing and Information Technology, 6, 1, 95–101, 1998.

van Rienen, U., Flehr, J., Schreiber, U., and Motrescu, V.: Modeling and Simulation of Electro-Quasistatic Fields, Modeling, Simulation and Optimization of Integrated Circuits, International Series of Numerical Mathematics, Birkhäuser Verlag Basel/Switzerland, 146, 17–31, 2003.

van Rienen, U.: Numerical Methods in Computational Electrodynamics – Linear Systems in Practical Applications, Springer-LNCSE, Verlag Berlin Heidelberg, 12, 2001.

Weiland, T.: Eine Methode zur Lösung der Maxwellschen Gleichungen für sechskomponentige Felder auf discreter Basis, AEÜ, 31, 116–120, 1977.

17

Numerical quadrature for the approximation of singular oscillating integrals appearing in boundary integral equations

L. O. Fichte, S. Lange, and M. Clemens

Professur für Theoretische Elektrotechnik und Numerische Feldberechnung, Helmut-Schmidt-Universität Universität der Bundeswehr Hamburg, PO. Box 700 822, D-22 008 Hamburg, Germany

Abstract. Boundary Integral Equation formulations can be used to describe electromagnetic shielding problems. Yet, this approach frequently leads to integrals which contain a singularity and an oscillating part. Those integrals are difficult to handle when integrated naivly using standard integration techniques, and in some cases even a very high number of integration nodes will not lead to precise results.

We present a method for the numerical quadrature of an integral with a logarithmic singularity and a cosine oscillator: a modified Filon-Lobatto quadrature for the oscillating parts and an integral transformation based on the error function for the singularity. Since this integral can be solved analytically, we are in a position to verify the results of our investigations, with a focus on precision and computation time.

1 Introduction

We investigate the field properties of a rectangular bar of constant conductivity κ_{i} and permeability μ_{i}. The space outside the bar has the arbitrary constant permeability μ_{a} and conductivity $\kappa_{\mathrm{a}} = 0$. The vector $\boldsymbol{r}_q = x_q \boldsymbol{e}_x + y_q \boldsymbol{e}_y$ is pointing at the lower left corner of the bar, the dimensions of it being a and b, as shown in Fig. 1. We name the cross-section of the bar Ω and its contour $C = \partial\Omega$.

An exciting loop consists of two thin conductors, which carry the currents $i(t)$ and $-i(t)$, with $i(t) = \Re\{I e^{j\omega t}\}$, I being the phasor describing the complex current . They are located at

$$\boldsymbol{r}_{\mathrm{e}} = \pm x_e \boldsymbol{e}_x.$$

The angular frequency ω is considered to be low enough so that displacement currents can be neglected: $|\boldsymbol{J}| \gg |\frac{\partial}{\partial t}\boldsymbol{D}|$, where $\boldsymbol{J}$ is the current density inside the bar and $\boldsymbol{D}$ the electric flux.

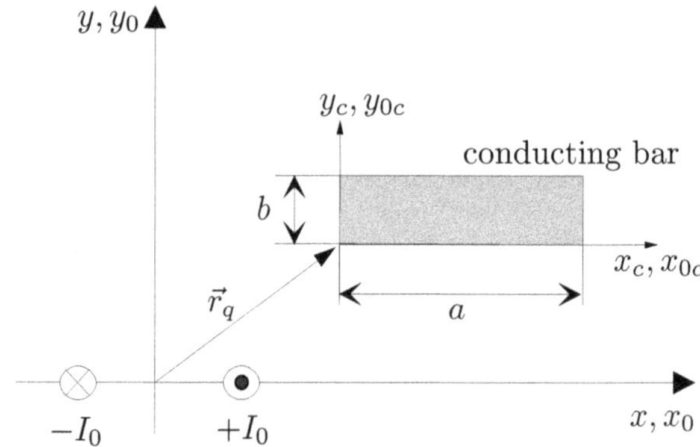

Fig. 1. Screening bar and exciting loop.

The z-directed dimensions of both bar and loop are supposed to be infinite.

2 Differential equations for the vector potential

As all exciting currents are oscillating at one single frequency, all fields will show the same time dependency and can be expressed by their complex phasors.

For a complex vector potential $\boldsymbol{A}$ as $\boldsymbol{B} = \mathrm{curl}\boldsymbol{A}$, with the magnetic flux $\boldsymbol{B}$, we find that for points inside the bar an additional term has to be added to the vector potential. The additional term can be expressed as the time domain integral of the gradient of a complex electrical scalar potential $\varphi_{\mathrm{i}}(z)$ by

$$\int (\mathrm{grad}\,\varphi_{\mathrm{i}})\mathrm{d}t = C\boldsymbol{e}_z,$$

with an unknown complex constant C. The vector potential $\boldsymbol{A}_{\mathrm{i}}$ inside the bar can be redefined by using the Buchholz convention

$$\boldsymbol{A}_{\mathrm{i}}^* = \boldsymbol{A}_{\mathrm{i}} + \int (\mathrm{grad}\,\varphi_{\mathrm{i}})\mathrm{d}t = \boldsymbol{A}_{\mathrm{i}} + C,$$

from which the fields can be computed with

$$\boldsymbol{B}_\mathrm{i} = \mathrm{curl}\,\boldsymbol{A}_\mathrm{i}^* \quad \text{and} \quad \boldsymbol{E}_\mathrm{i} = -j\omega \boldsymbol{A}_\mathrm{i}^*.$$

From Maxwell's equations we find that the modified vector potential inside the bar is governed by Helmholtz' equation, which for complex fields takes the form

$$\Delta \boldsymbol{A}_\mathrm{i}^* - j\omega\mu_\mathrm{i}\kappa_\mathrm{i}\boldsymbol{A}_\mathrm{i}^* = 0. \tag{1}$$

Outside the bar the equation

$$\Delta \boldsymbol{A}_\mathrm{a} = -\mu_\mathrm{a}\boldsymbol{J}_e, \tag{2}$$

is valid for the vector potential $\boldsymbol{A}_\mathrm{a}$. The exciting current density $\boldsymbol{J}_e$ can be expressed as

$$\boldsymbol{J}_e = \{\delta_1(x-x_e) - \delta_1(x+x_e)\}\delta_1(y)\cdot I\boldsymbol{e}_z.$$

The symbol $\delta_1(x-x_o)$ denotes the one-dimensional Dirac distrubution; accordingly, δ_2 and δ_3 are the two- and three-dimensional Dirac distribution.

Since we allow only z-directed exciting currents, all vector potentials will also be exclusively z-directed. They can be described by their z-components A_az and A_iz^*, respectively. The boundary conditions for A_az and A_iz^* are

$$A_\mathrm{az} = A_\mathrm{iz}^* - C \text{ and } \frac{1}{\mu_\mathrm{a}}\boldsymbol{n}_o\cdot \mathrm{grad}A_\mathrm{az} = \frac{1}{\mu_\mathrm{i}}\boldsymbol{n}_o\cdot \mathrm{grad}A_\mathrm{iz}^*,$$

where $\boldsymbol{n}_o$ is a unit vector normal to the boundary.

3 Derivation of boundary integral equation (BIE)

In Green's second theorem

$$\iiint_\tau (\Phi\,\Delta\Psi - \Psi\,\Delta\Phi)\mathrm{d}\tau = \oiint_{F=\partial\tau} (\Phi\,\mathrm{grad}\Psi - \Psi\,\mathrm{grad}\Phi)\cdot \boldsymbol{n}_o\,\mathrm{d}F,$$

we insert the single component of the vector potential A_a as Ψ and a kernel function K satisfying $\Delta K = -\delta_3$ as Φ. As a result, we get an integral representation valid for A_a:

$$A_\mathrm{a} = -\oiint_{F=\partial\tau} (A_\mathrm{a}\,\mathrm{grad}K - K\,\mathrm{grad}A_\mathrm{a})\cdot \boldsymbol{n}_o\,\mathrm{d}F + \mu_\mathrm{a}I\int_{-\infty}^{+\infty} K\Big|_{\substack{x_o=\pm x_e\\ y_o=0}} \mathrm{d}z_o.$$

The integrand of the surface integral at the right hand side of the above equation does not depend on z. Hence the z-directed integration is only affecting the kernel function K, for which the elementary Kernel function

$$K = \frac{1}{4\pi r},$$

is inserted. Thereby we can define a new kernel function G_{20} as

$$G_{20} = -\frac{1}{2\pi}\ln(\frac{\rho}{\rho_o}),$$

with $\rho = \sqrt{(x-x_o)^2 + (y-y_o)^2}$ and $\rho_o = \mathrm{const}$. While we are aware that the above integral is divergent, it can be solved by performing a suitable renormalization.

The function G_{20} satisfies $\Delta G_{20} = -\delta_2$, as is shown in Ehrich et al. (2000).

If we use the new kernel and define the influence of the exciting loop as

$$A_\mathrm{e} := \mu_\mathrm{a} I G_{20}\big|_{\substack{x_o=\pm x_e\\ y_o=0}},$$

we can write the integral representation for A_a as

$$A_\mathrm{a} = -\oint_{C=\partial\Omega} (A_\mathrm{a}\,\mathrm{grad}G_{20} - G_{20}\,\mathrm{grad}A_\mathrm{a})\cdot \boldsymbol{n}_o\,\mathrm{d}s + A_\mathrm{e}.$$

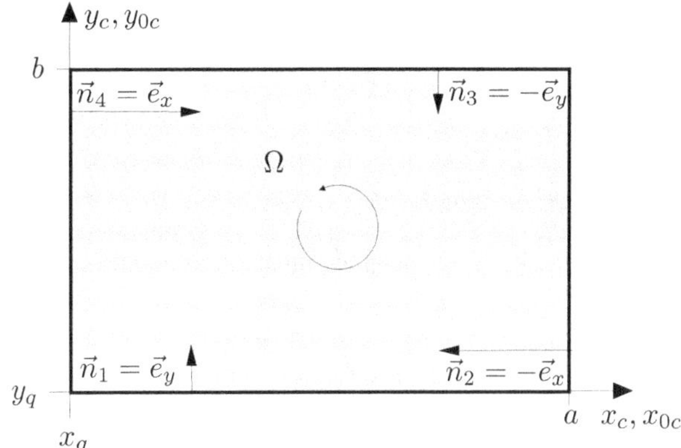

Fig. 2. Normals on boundary sections.

We use the cross-section of the bar as integration domain Ω with its facet normals defined as shown in Fig. 2.

By taking into account the boundary conditions discussed above, we get an integral representation for the vector potential outside the bar:

$$A_\mathrm{a} = -\oint_{C=\partial\Omega} [(A_\mathrm{i}^* - C)\,\mathrm{grad}\,G_{20} - G_{20}\frac{\mu_\mathrm{a}}{\mu_\mathrm{i}}\,\mathrm{grad}A_\mathrm{i}^*]\cdot \boldsymbol{n}_o\,\mathrm{d}s + A_\mathrm{e}. \tag{3}$$

It can be shown that

$$\oint_{C=\partial\Omega} (C\,\mathrm{grad}\,G_{20})\cdot \boldsymbol{n}_o\,\mathrm{d}s = 0,$$

so we can omit this term in Eq. (3).

In a last step we take the normal derivative of A_a on each of the four contour sections of Ω and move the observation point to the very same section, i.e.

$$(1)\ y = y_q, \quad (3)\ y = y_q + b,$$
$$(2)\ x = x_q + a, \quad (4)\ x = x_q.$$

As a result, we get four independent integral equations

$$\frac{\mu_a}{\mu_i} \boldsymbol{n}_o \cdot \mathrm{grad} A_i^* = \boldsymbol{n}_o \cdot \mathrm{grad}\{ \oint_{C=\partial\Omega} (A_i^* \,\mathrm{grad}\, G_{20} - G_{20} \frac{\mu_a}{\mu_i} \mathrm{grad} A_i^*) \,\boldsymbol{n}_o \,\mathrm{d}s + A_e\}\Big|_{\substack{x=x_q \vee x=x_q+a, y \\ x, y=y_q \vee y=y_q+b}} . \quad (4)$$

We find a solution to Eq. (1) by separation, using a product of functions depending only on one coordinate for the descrition of $\boldsymbol{A}_i^*$.

$$\begin{aligned} A_i^*(x_c, y_c) = \sum_{n=1}^{\infty} \big[& v_{1n} \cosh(\beta_n (b - y_c)) \\ & + v_{3n} \cosh(\beta_n y_c)\big] \cos(\alpha_n x_c) \\ & + \big[v_{2n} \cosh(\tilde{\beta}_n (a - x_c)) \\ & + v_{4n} \cosh(\tilde{\beta}_n x_c)\big] \cos(\tilde{\alpha}_n y_c), \end{aligned} \quad (5)$$

$$\alpha_n = \frac{n\pi}{a}, \quad \beta_n = \sqrt{\alpha_n^2 + j\omega\kappa\mu_i},$$
$$\tilde{\alpha}_n = \frac{n\pi}{b}, \quad \tilde{\beta}_n = \sqrt{\tilde{\alpha}_n^2 + j\omega\kappa\mu_i}.$$

Then, the unknown constants v_{in}, $i = 1, ..., 4$ will have to be determined.

We can now insert the result for A_i^* into Eq. (4). Using the orthogonality of the cosine functions

$$\int_0^{2\pi} \cos(n\xi)\cos(p\xi)\mathrm{d}\xi = \begin{cases} 0 & n \neq p \\ \pi & n = p \neq 0 , \\ 2\pi & n = p = 0 \end{cases}$$

we can isolate one coefficient v_{im} on the left hand side of Eq. (4). Expanding the right hand side into a series of cosine functions by multiplication with $\cos(\alpha_m x_c)$ (or $\cos(\tilde{\alpha}_m y_c)$, respectively) and computing its integral along one of the contour sections we get

$$\frac{\mu_a}{\mu_i} v_{im} d_{im} = \sum_{i=1}^{4} \sum_{n=1}^{\infty} [\gamma_{inm} v_{in}] + b_m, \quad m \int \mathbb{N}.$$

If we take only the first N series elements into account, this equation can be written as a matrix equation

$$\boldsymbol{\Gamma v} = \boldsymbol{b}, \quad (6)$$

with matrix $\boldsymbol{\Gamma}$ and vectors $\boldsymbol{v}$, $\boldsymbol{b}$ of finite dimensions. Equation (6) represents a system of linear equations from which the $4N$ unknown coefficients v_{in} can be computed.

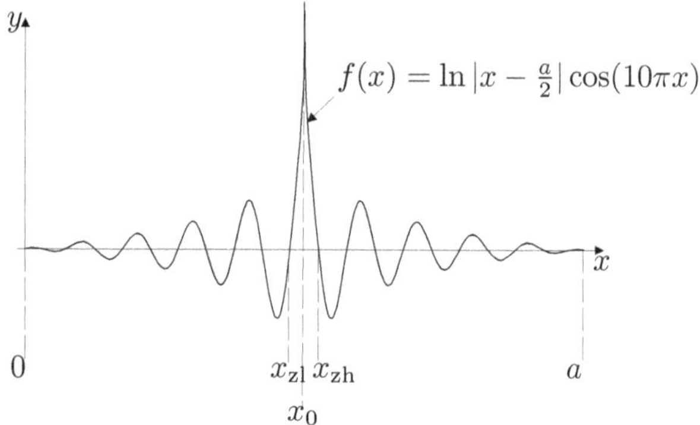

Fig. 3. Oscillating function with singularity.

The values of the coefficients γ_{inm}, d_{im} and b_m, needed for the solution of the matrix equation, can be obtained analytically (see Fichte et al., 2004). The calculations resulting from the analytical approach are discussed in detail in Fichte (@).

Once the values of these coefficients are known, they can be used to express the vector potential outside the shielding bar as a series of known functions $A_k(x, y)$, $k = 1, ..., 4$:

$$A_a(x, y) = \sum_{n=1}^{M} v_{1n} A_1(x, y) + v_{2n} A_2(x, y) + v_{3n} A_3(x, y) + v_{4n} A_4(x, y).$$

This method, using analytical solutions, has been used to calculate the shielding effects of a conducting bar (Fichte et al., 2005).

4 Numerical treatment of integrals

When we insert the series representation of A_i^* into the BIE (4), we can transform the resulting integrals into expressions like

$$\int_0^a \cos(n\xi) \cdot \ln(\xi)\,\mathrm{d}\xi \quad , n \int \mathbb{N}. \quad (7)$$

Standard schemes for numerical integration do not lead to appropriate results due to the oscillatory nature of the cosine function and the singularity of the logarithmic term.

An example of an integrand is displayed in Fig. 3, with $f(x) = \ln|x - \frac{a}{2}|\cos(10\pi x)$.

In a novel approach we use numerical integration methods which deliver highly precise results. First, a decomposition of the domain of the integral is necessary. It is divided into three subdomains: the singularity is in the non-oscillating subdomain $[x_{zl}, x_{zh}]$. The edge points of this domain are the highest zero of f below the singularity and the lowest zero of f above it.

The remaining two subdomains $[0, x_{zl}]$ and $[x_{zh}, a]$ represent the non-singular oscillating parts of the integral:

$$\int_0^a \cos(n\xi)\cdot\ln(\xi)\,d\xi = \underbrace{\int_0^{x_{zl}} \cos(n\xi)\cdot\ln(\xi)d\xi}_{I_{low}} + \underbrace{\int_{x_{zl}}^{x_{zh}} \cos(n\xi)\cdot\ln(\xi)d\xi}_{I_{sing}} + \underbrace{\int_{x_{zl}}^{a} \cos(n\xi)\cdot\ln(\xi)d\xi}_{I_{hi}}$$

Here, x_{zl} is the largest root of the integrand below the singularity and x_{zh} the smallest root above the singularity.

Note that by combination of the integral we only need three subdomains.

4.1 Treatment of the subdomain with oscillating integrand

We apply a Filon-Lobatto quadrature described in Iserles (2004) to the oscillating parts of the integral. As a first step the boundaries of the integral are mapped to the domain [0, 1]:

$$I_{low} = \int_0^{x_{zl}} \cos(n\xi)\cdot\ln(\xi)d\xi = x_{zl}\int_0^1 \cos(\underbrace{nx_{zl}}_{\eta}\xi)\cdot\ln(x_{zl}\xi)d\xi.$$

Then we approximate the logarithmic function in the integral using N Legendre polynomials:

$$\ln(x_{zl}\xi) \approx \sum_{k=1}^{N} P_k(\frac{\xi}{x_{zl}})\ln(c_k x_{zl}),$$

where P_k denotes the Legendre polynomial of order k.

For the first two abscissas we choose the start- and the endpoint of the integration domain. The remaining $n-2$ abscissas c_k, $k = 1, ..., N-1$, are the roots of $P'_{N-1}(x)$.

Thus, we can approximate the lower integral by:

$$I_{low} \approx x_{zl}\sum_{k=1}^{N} b_k(\eta)\ln(c_k h),$$

with known weight factors b_k:

$$b_k(\eta) = \int_0^1 P_k(\xi)\cos(\eta\xi)d\xi.$$

The integral I_{hi}, which covers the part of the integral for $\xi > x_{z(i+1)}$, is treated likewise.

The use of Filon-Lobatto quadrature results in a fairly low number of integration nodes neccessary for numerical quadrature of the highly oscillating integrals, since the accuracy of this method is improving with the number of zeros of the integrand.

4.2 Treatment of the subdomain containing the singularity

The subdomain of the original integral which contains the singularity is handled separately, acoording to a method presented in Ehrich et al. (1997).

First the integral is mapped to the domain [0, 1]. Then a Gauss quadrature is applied: the whole integrand is approximated by Legendre's polynomials and the zeros of Legendre's polynomial of the N-th order, c_k, are taken as abscissas. The corresponding weighting factors of the Gaussian integration are named w_k. With these expressions, I_{sing} is given by:

$$\int_0^1 \ln(x)\cos(\tilde{\eta}x)\,dx = \sum_{k=1}^{N/2} \gamma_k(\alpha_k(\ln(x)\cos(\tilde{\eta}x))|_{x=\alpha_k} + \beta_k(\ln(x)\cos(\tilde{\eta}x)|_{x=\beta_k}).$$

The constants α_k, β_k and γ_k are:

$$\alpha_k = \frac{1}{2}\left[1 + \frac{F\{qc_k\}}{F\{q\}}\cdot e^{q^2-(qc_k)^2}\right],$$

$$\beta_k = \frac{\int_{qc_k}^{q} e^{-t^2}\,dt}{2F(q)}e^{q^2},$$

$$\gamma_k = \frac{w_i\, q}{2F(q)}e^{q^2-(qc_k)^2},$$

By using the function $F(\xi)$,

$$F(\xi) = \xi\sum_{k=0}^{\infty}\frac{(2\xi^2)^k}{(2k+1)!!};$$

the error function, as defined in Abramowitz and Stegun (1970),

$$\mathrm{ERF}(\xi) := \frac{2}{\sqrt{\pi}}\int_0^{\xi} e^{-u^2}du,$$

can be expressed as

$$\mathrm{ERF}(\xi) = \frac{2}{\sqrt{\pi}}e^{-\xi^2}F(\xi).$$

The constant q is an arbitrary real constant which has to be chosen in advance. This method for dealing with the logarithmic singularity is discussed in depth in Ehrich et al. (1997) and a value of $q = 7$ has been established as an optimal value for precision and computation time.

Table 1. Integration nodes and computation time

Number	relative precision					
of	$10E{-}3$		$10E{-}7$		$10E{-}10$	
Zeros	k	t/s	k	t/s	k	t/s
10	14	$40E-3$	21	0.16	23	0.22
50	44	12.4	120	343	162	804
100	50	16.4	109	291	170	1001

While this method for computing singular integral has been used in the past, the combination with the Filon-Lobatto integration amounts to a new approach for the numerical quadrature of singular oscillating integrals.

5 Numerical results

The presented method has been used for the numerical approximation of an integral as appearing in Eq. 7. The results are displayed in Table 1. Here, n is the number of zeros of the integral, k is the number of polynomials used for the approximation and t is the computation time in seconds (on a 1 GHz Pentium system with 500 MB RAM).

6 Conclusion

A plane magneto-quasistatic eddy current problem has been described by a boundary integral equation. To obtain a solution to this integral equation, one has to solve a kind of integral which is highly oscillating and contains a singularity. A novel method for the approximation of those integrals has been developed. The results have been compared to analytical solutions.

References

Abramowitz, M. and Stegun, I. A.: Handbook of Mathematical Functions, New York 1970.

Ehrich, M., Fichte, L. O., and Lüer, M.: Contribution to Boundary Integrals by the Singularity of Kernels satisfying Helmholtz' Equation, CJMW'2000 China-Japan Joint Meeting on Microwaves, Nanjing, PR China, CD-ROM, 2000.

Ehrich, M., Kuhlmann, J., and Netzler, D.: High accuracy integration of boundary integral equations describing axisymmetric field problems, Asia-Pacific Microwave Conf., Hong Kong, Microwave Conf. Proc., CD-ROM, 1997.

Fichte, L. O.: Berechnung der Stromverteilung in einem System rechteckiger Massivleiter bei Wechselstrom mit Hilfe der Randintegralgleichungsmethode, PhD-Thesis, to be published.

Fichte, L. O., Ehrich, M., and Kurz, S.: An Analytical Solution to the Eddy Current Problem of a Conducting Bar, EMC 2004 Intern. Symposium on Electromagnetic Compatibility, Sendai Conf. Proc., CD-ROM, 2004.

Fichte, L. O., Lange, S., Steinmetz, T., Clemens, M.: Shielding Properties of a Conducting Bar calculated with a Boundary Integral Equation Method, Adv. in Radio Sci., 3, 119–123, 2005.

Hanson, G. W. and Yakovlev, A. B.: Operator Theory for Electromagnetics, Springer, New York, 2002.

Iserles, A.: On the numerical quadrature of highly-oscillating integrals, IMA J. of Numerical Anal., 24, 365–391, 2004.

18

Contactless vector network analysis using diversity calibration with capacitive and inductive coupled probes

T. Zelder, I. Rolfes, and H. Eul

Institut für Hochfrequenztechnik und Funksysteme, Universität Hannover, Appelstraße 9A, 30167 Hannover, Germany

Abstract. Contactless vector network analysis based on a diversity calibration is investigated for the measurement of embedded devices in planar circuits. Conventional contactless measurement systems based on two probes for each measurement port have the disadvantage that the signal-to-noise system dynamics strongly depends on the distance between the contactless probes.

In order to avoid a decrease in system dynamics a diversity based measurement system is presented. The measurement setup uses one inductive and two capacitive probes. As an inductive probe a half magnetic loop in combination with a broadband balun is introduced. In order to eliminate systematic errors from the measurement results a diversity calibration algorithm is presented. Simulation and measurement results for a one-port configuration are shown.

1 Introduction

The characterization of single subcircuits within complex planar microwave circuits can be realized by contactless measurement techniques. The measurement of the scattering parameters of each individual subcircuit is useful for the development as well as the quality control of complex microwave planar circuits. Different contactless measurement methods have already been investigated. For example, by Bridges (2004) the scanning probe microscopy, by Osofsky and Schwarz (1992), Gao and Wolff (1997) and Quardirhi and Laurin (2003) contactless electromagnetic probes and by Dudley et al. (1999) electro-optic probes are used to measure the forward and backward traveling waves in front of and behind of the device under test (DUT). Further on, by Hui and Weikle (2005) a non-contacting sampled-line reflectometer is implemented to determine the forward and backward traveling waves. An overview of different contactless measurement techniques is given by Sayil et al. (2005).

Another contactless measurement approach is based on the contactless vector network analysis. Thereby, the internal directional couplers of a conventional vector network analyzer (VNA) are replaced by contactless probes. For an accurate determination of the scattering parameters conventional calibration algorithms are used. De Groote et al. (2006) and Yhland and Stenarson (2006) applied electromagnetic loops as contactless coupling structures. In contrast to this approach, by Zelder et al. (2007) and Stenarson et al. (2001), pure inductive or capacitive probes are used. A principle setup of a contactless vector network analyzer system using two probes is shown in Fig. 1. The complex planar circuit consists, in this example, of three subcircuits and is fed by the signal source of a two-port vector network analyzer. By means of contactless probes placed in front of and behind of the DUT, a part of the complex signal power is coupled directly to the four receivers of the VNA. In this configuration the internal directional couplers of the VNA are not used. For the elimination of the systematic errors of the measurement setup conventional calibration algorithms can be used as e.g. Short-Open-Load-Thru (SOLT) Schiek (1999) or self-calibration methods like Thru-Reflect-Line (TRL) Engen and Hoer (1979). One disadvantage of this system is, that for certain probe distances a calibration is not possible which is shown by Zelder et al. (2007). This limits the measurement bandwidth. For the measurement system with two capacitive probes the critical distances have been calculated by Zelder and Eul (2006).

In this paper the analytical calculation of the critical probe distances is completed for a system with one inductive and one capacitive probe. Further on, to reach a broadband measurement bandwidth, a diversity calibration setup with more than two contactless probes for each measurement port is applied. Simulation and measurement results for a one-port contactless diversity network analyzer based on one

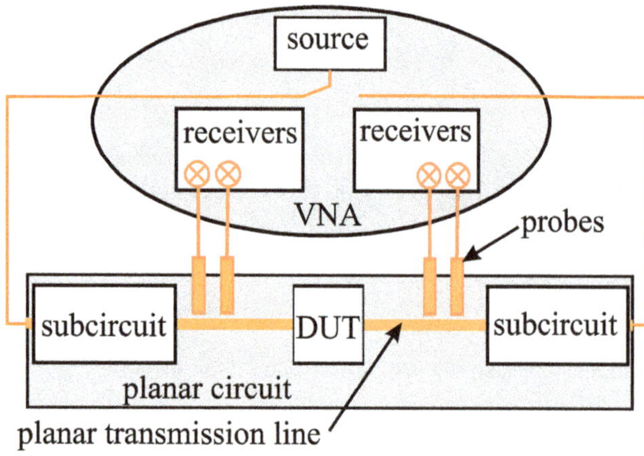

Fig. 1. Principle setup of a contactless vector network analyzer system using two probes for each measurement port.

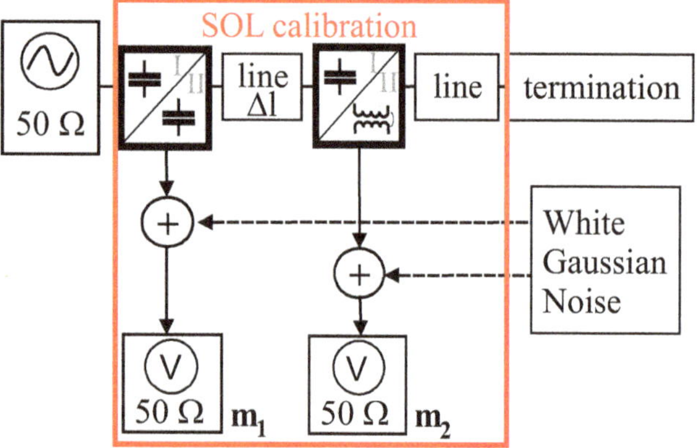

Fig. 2. Simulation model of a contactless measuring system with two electromagnetic probes.

inductive and two capacitive probes are presented. As an inductive probe a half magnetic loop in combination with a balun is used. An extension of the presented method to N-port measurements is possible. In this paper merely the principle function of a contactless diversity based system and its potentials are shown.

2 Problems using two contactless probes

The simulation model for a contactless measurement system with two electromagnetic probes is illustrated in Fig. 2. It consists of a signal source with a power level of 1dBm which feeds a lossless 50Ω transmission line. As a transmission line termination a calibration standard or the DUT can be used. By means of this model two configurations are considered. In the first configuration, CC, two capacitive probes are used. For the second configuration, LC, one inductive and one capacitive probe are applied. The capacitive coupling is modeled by an ideal capacitance and the inductive

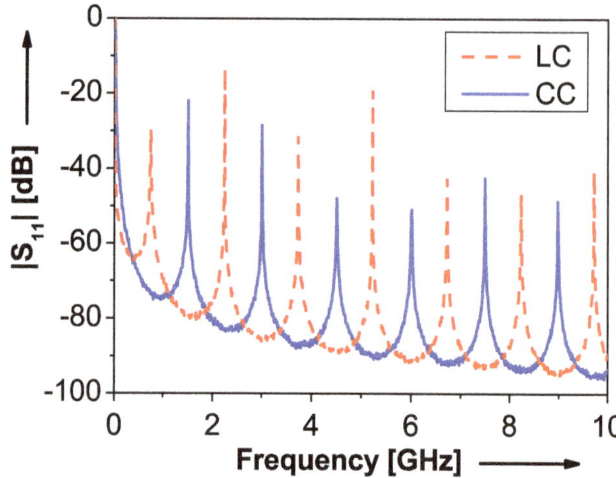

Fig. 3. System dynamics against the frequency for a probe distance of Δl=100 mm; DUT: 50Ω load – Data from Zelder and Eul (2006).

coupling by an ideal transformer. The distance between the two probes is Δl. The coupled voltages are determined in a 50Ω system at the receivers m_1 and m_2. To get a more realistic system, additive white Gaussian noise is applied at the receivers with a respective power level of -118dBm relative to a noise bandwidth of 10Hz, which is typical for modern vector network analyzers. For the accurate measurement of the scattering parameters of a DUT, the error terms of the system are determined using the conventional Short-Open-Load (SOL) calibration method Schiek (1999). For this purpose, the complex voltage ratios μ_x of the receivers m_1 and m_2 using the three calibration standards short, open and load are measured. In order to verify the calibration and to determine the signal-to-noise system dynamics, the reflection coefficient of a load standard is remeasured. In the following the signal-to-noise system dynamics is abbreviated as system dynamics.

A disadvantage of a contactless system, as illustrated in Fig. 2, is that the system dynamics depends on the distance between the two contactless probes. The system dynamics for two different probe pair combinations is presented in Fig. 3. To show the effect of probe distance, Δl is chosen to 100 mm which is equivalent to a frequency distance of 3 GHz. In Fig. 3 it is shown that at some critical frequencies, which are located in a distance of about 1.5 GHz, the dynamics decrease. The reason for these system failures is, that the three voltage ratios μ_x of the two probes measuring the calibration standards short, open and load become equal at these critical frequencies. By contrast, for an accurate calibration the three voltage ratios must be different. Depending on the probe distance, it is possible to determine the critical frequencies analytically, which is shown in detail, for the configuration with one inductive and one capacitive probe (LC), in the next section.

3 Analytical analysis of the dynamics decrease using two contactless probes

At first, the critical probe distances for a contactless one-port measurement system with one inductive and one capacitive probe will be calculated. For the mathematical analysis of the decrease in dynamics range the SOL correction algorithm Schiek (1999) described by Eq. (1) is used.

$$\frac{(\Gamma_D - \Gamma_L)\cdot(\Gamma_S - \Gamma_O)}{(\Gamma_D - \Gamma_O)\cdot(\Gamma_S - \Gamma_L)} = \frac{(\mu_D - \mu_L)\cdot(\mu_S - \mu_O)}{(\mu_D - \mu_O)\cdot(\mu_S - \mu_L)} \tag{1}$$

The reflection coefficients are represented by Γ_x for the calibration standards short (Γ_S), open (Γ_O), load (Γ_L) and the DUT (Γ_D). μ_x denotes the corresponding ratio of the voltages at the receivers m_1 and m_2:

$$\mu_x = \frac{U_{m1}(z_1)}{U_{m2}(z_2)} \tag{2}$$

where z_1 and z_2 represent the coupling positions along the transmission line. For the mathematical description of the system dynamics the calibration standards are assumed to be ideal. Thus, the reflection coefficients of the calibration standards are given by:

$$\Gamma_L = 0;\ \Gamma_S = -1;\ \Gamma_O = 1. \tag{3}$$

After inserting Eqs. (3) into Eq. (1) the reflection coefficient Γ_D of the DUT can be calculated:

$$\Gamma_D = \frac{1}{1 - 2\frac{(\mu_D - \mu_O)\cdot(\mu_S - \mu_L)}{(\mu_D - \mu_L)\cdot(\mu_S - \mu_O)}}. \tag{4}$$

For the mathematical description of the voltage ratios μ_x, the voltage $U_{m,V}(z_1)$ obtained by the capacitive probe and the voltage $U_{m,I}(z_2)$ obtained by the inductive probe are considered first. The expressions of the voltages are given by:

$$\begin{aligned} U_{m,V}(z_1) &= K_V \cdot U_h(l)(e^{j\varphi(z_1)} \\ &\quad + \Gamma_n(l)\cdot e^{-j\varphi(z_1)}) + u_{r1} \end{aligned} \tag{5}$$

$$\begin{aligned} U_{m,I}(z_2) &= K_I \cdot U_h(l)(e^{j\varphi(z_2)} \\ &\quad - \Gamma_n(l)\cdot e^{-j\varphi(z_2)}) + u_{r2} \end{aligned} \tag{6}$$

$U_h(l)$ represents the voltage amplitude of the forward traveling wave at the position $z{=}l$ of the planar transmission line. The position $z{=}l$ indicates the reference plane of the system, $\Gamma_n(l)$ represents the reflection coefficient at the reference plane, K_V is the voltage coupling coefficient of the capacitive probe and K_I is the coupling coefficient of the inductive probe. The terms u_{r1} and u_{r2} represent complex noise values. The phase angles of the forward and backward traveling waves are $\pm\varphi$:

$$\varphi = \frac{2\cdot\pi\cdot f}{v}\cdot(l - z) \tag{7}$$

where f represents the frequency and v the phase velocity.

The voltage ratios can be expressed by inserting Eqs. (5) and (6) into Eq. (2):

$$\mu_x = \frac{K_V\cdot U_h(l)(e^{j\varphi(z_1)} + \Gamma_x(l)\cdot e^{-j\varphi(z_1)}) + u_{r1}}{K_I\cdot U_h(l)(e^{j\varphi(z_2)} - \Gamma_x(l)\cdot e^{-j\varphi(z_2)}) + u_{r2}}. \tag{8}$$

To set up an equation for the voltage ratio μ_L, the load reflection coefficient Γ_L of Eq. (3) is inserted into Eq. (8). With $\varphi(z_1){=}\varphi_1$ and $\varphi(z_2){=}\varphi_2$ it results in:

$$\mu_{L,LC} = \frac{K_V\cdot U_h(l)\cdot e^{j\varphi_1} + u_{r1}}{K_I\cdot U_h(l)\cdot e^{j\varphi_2} + u_{r2}}. \tag{9}$$

In order to reduce the arithmetical complexity, the noise contributions for the voltage ratios μ_S and μ_O of the short and open standard are neglected. This results in:

$$\mu_S = \frac{K_V\cdot U_h(l)(e^{j\varphi_1} - 1\cdot e^{-j\varphi_1})}{K_I\cdot U_h(l)(e^{j\varphi_2} + 1\cdot e^{-j\varphi_2})} = jK_x\frac{\sin(\varphi_1)}{\cos(\varphi_2)} \tag{10}$$

and

$$\mu_O = \frac{K_V\cdot U_h(l)(e^{j\varphi_1} + 1\cdot e^{-j\varphi_1})}{K_I\cdot U_h(l)(e^{j\varphi_2} - 1\cdot e^{-j\varphi_2})} = -jK_x\frac{\cos(\varphi_1)}{\sin(\varphi_2)} \tag{11}$$

with

$$K_x = \frac{K_V}{K_I}. \tag{12}$$

For the determination of the system dynamics the load standard is used as a DUT. By use of Eq. (9) we have

$$\mu_{D,LC} = \frac{K_V\cdot U_h(l)\cdot e^{j\varphi_1} + u_{r3}}{K_I\cdot U_h(l)\cdot e^{j\varphi_2} + u_{r4}} \tag{13}$$

with changed noise values u_{r3} and u_{r4}. Now, the raw data from Eqs. (9) to (13) are inserted into Eq. (4) and it can be transformed to:

$$\begin{aligned} \Gamma_{D,LC} &= j\cdot K_x\cdot(\mu_{D,LC} - \mu_{L,LC})\cos(\Delta\varphi) \\ &\quad \cdot [\mu_{L,LC}\cdot\mu_{D,LC}\cdot\sin(2\varphi_2) \\ &\quad + K_x^2\cdot\sin(2\Delta\varphi + 2\varphi_2) \\ &\quad + j\cdot K_x\cdot(\mu_{L,LC} + \mu_{D,LC})\cos(\Delta\varphi + 2\varphi_2)]^{-1} \\ &= N_{LC}(\Delta\varphi)\cdot D_{LC}(\Delta\varphi)^{-1}. \end{aligned} \tag{14}$$

The phase difference $\Delta\varphi$ between the two capacitive probes is given by:

$$\Delta\varphi = \varphi_1 - \varphi_2. \tag{15}$$

A similar analytical calculation can be done using two identical capacitive probes (CC) Zelder and Eul (2006). The dynamics for a contactless measurement system with two capacitive probes results in:

$$\begin{aligned} \Gamma_{D,CC} &= (\mu_{D,CC} - \mu_{L,CC})\sin(\Delta\varphi) \\ &\quad \cdot [\mu_{L,CC}\cdot\mu_{D,CC}\cdot\sin(2\varphi_2) \\ &\quad + \sin(2\Delta\varphi + 2\varphi_2) \\ &\quad - (\mu_{L,CC} + \mu_{D,CC})\sin(\Delta\varphi + 2\varphi_2)]^{-1} \\ &= N_{CC}(\Delta\varphi)\cdot D_{CC}(\Delta\varphi)^{-1}. \end{aligned} \tag{16}$$

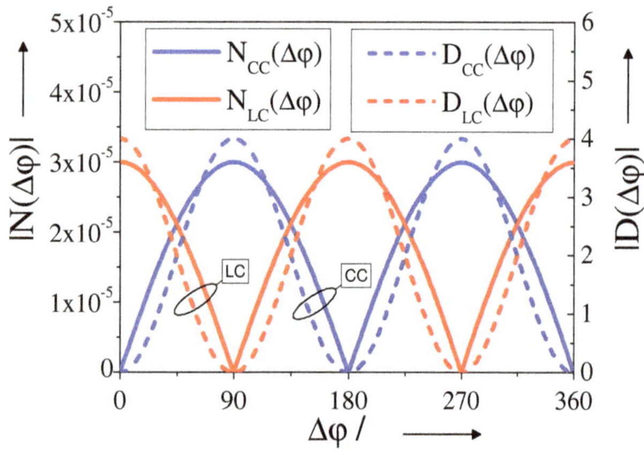

Fig. 4. Calculated results of a contactless measuring system using two probes.

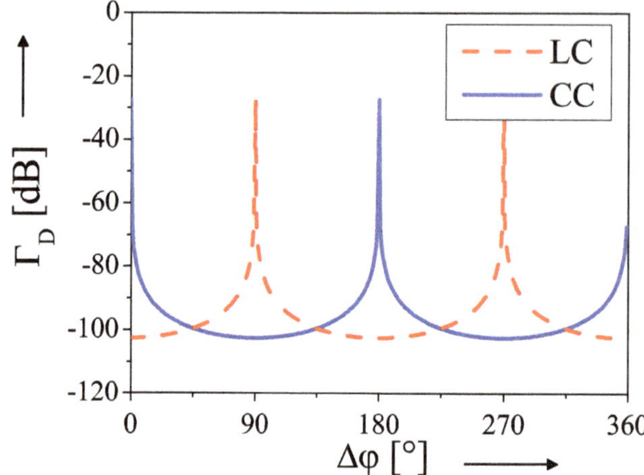

Fig. 5. Calculated reflection coefficients $\Gamma_{D,CC}$ and $\Gamma_{D,LC}$ for both probe pair combinations; DUT: 50Ω load.

For the determination of the critical probe distances the zeros of the numerator and denominator of Eqs. (14) and (16) must be quantified. Therefore the following simplifications are introduced for the system with two identical capacitive probes:

$$(\mu_{D,CC} - \mu_{L,CC}) \approx \Delta u_r \quad (17)$$

$$(\mu_{D,CC} + \mu_{L,CC}) \approx 2 \cdot e^{j \cdot \Delta\varphi} \quad (18)$$

$$(\mu_{D,CC} \cdot \mu_{L,CC}) \approx e^{j \cdot 2 \cdot \Delta\varphi}. \quad (19)$$

Further on, for the measurement system using one inductive and one capacitive probe (LC) the simplifications are as follows:

$$(\mu_{D,LC} - \mu_{L,LC}) \approx K_x \cdot \Delta u_r \quad (20)$$

$$(\mu_{D,LC} + \mu_{L,LC}) \approx 2 \cdot K_x \cdot e^{j \cdot \Delta\varphi} \quad (21)$$

$$(\mu_{D,LC} \cdot \mu_{L,LC}) \approx K_x^2 \cdot e^{j \cdot 2 \cdot \Delta\varphi}. \quad (22)$$

Inserting Eqs. (17) to (19) into Eq. (16) and the Eqs. (20) to (22) into Eq. (14), the dynamics of the contactless measurement system with two contactless probes can be calculated:

$$\Gamma_{D,CC} = \frac{2 \cdot j \cdot \Delta u_r \cdot \sin(\Delta\varphi) \cdot e^{j \cdot 2 \cdot (\Delta\varphi + \varphi_2)}}{-1 + 2 \cdot e^{j \cdot 2 \cdot \Delta\varphi} - e^{j \cdot 4 \cdot \Delta\varphi}} = \frac{N_{CC}}{D_{CC}} \quad (23)$$

$$\Gamma_{D,LC} = \frac{2 \cdot \Delta u_r \cdot \cos(\Delta\varphi) \cdot e^{j \cdot 2 \cdot (\Delta\varphi + \varphi_2)}}{1 + 2 \cdot e^{j \cdot 2 \cdot \Delta\varphi} + e^{j \cdot 4 \cdot \Delta\varphi}} = \frac{N_{LC}}{D_{LC}}. \quad (24)$$

In Fig. 4 the numerators N_{CC}, N_{LC} and the denominators D_{CC}, D_{LC} of Eqs. (23) and (24) are shown as a function of $\Delta\varphi$. The value of the noise difference Δu_r is obtained from measurements with a VNA model PNA E8361A. Thereby, the raw data of the reflection coefficient of the 50Ω-load is measured twice in a frequency range between 10 MHz and 20 GHz, respectively. The mean value of the difference between the twice measured reflection coefficients is approximately $1.5 \cdot 10^{-5}$, which is used for the analytical investigation. Figure 4 shows distances where the zeros of the numerator and denominator occur are identical for both probe pair configurations. At these probe distances a measurement is not possible. The critical phase differences can be determined by calculating the zeros of the numerators N_{CC}, N_{LC} and the denominators D_{CC}, D_{LC} of Eqs. (23) and (24):

$$\Delta\varphi_{CC,0} = m \cdot \pi \quad \text{with } m = 0, 1, 2, 3, ... \quad (25)$$

$$\Delta\varphi_{LC,0} = n \cdot \frac{\pi}{2} \quad \text{with } n = 1, 3, 5, 7, ... \quad (26)$$

For these phase differences the system dynamics decreases. The characteristics of the system dynamics is illustrated in Fig. 5. At the critical phase differences $\Delta\varphi$, peaks in the system dynamics occur, so that a measurement is not possible. For two capacitive probes a calibration is not possible, if the distance is in the range of integer multiples of $\lambda/2$. In case of one inductive and one capacitive probe the critical distances between the probes are odd-numbered multiples of $\lambda/4$. For a given probe distance Δl and for an effective permittivity $\epsilon_{r,\text{eff}}$ of the planar transmission line we have the critical probe frequencies:

$$f_{\text{crit},CC} = m \cdot \frac{c_0}{2 \cdot \sqrt{\epsilon_{r,\text{eff}}} \cdot \Delta l} \quad (27)$$

$$f_{\text{crit},LC} = n \cdot \frac{c_0}{4 \cdot \sqrt{\epsilon_{r,\text{eff}}} \cdot \Delta l} \quad (28)$$

4 Simulation model of a contactless diversity measurement system using different probe types

For contactless measurements using two probes for each port of the DUT, the measurement bandwidth is limited due to the probe distance, as shown in Sects. 2 and 3. By means of a diversity based measurement system a large measurement bandwidth can be achieved. For a contactless diversity measurement setup, more than two contactless probes for each port of the DUT are used. One-port measurement results using such a diversity based system are given by Zelder and Eul

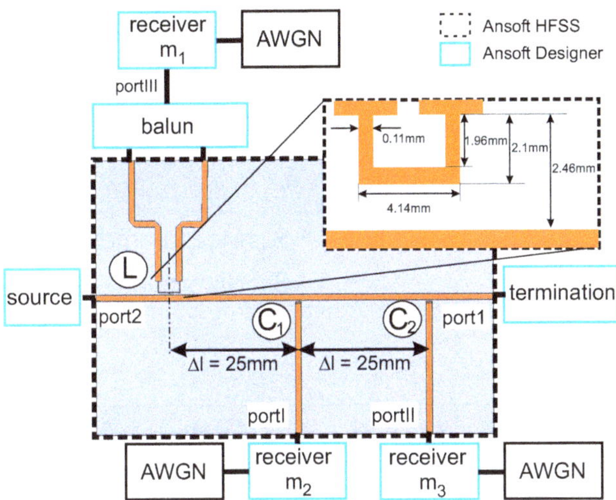

Fig. 6. Simulation model for the basic analysis of the contactless diversity system.

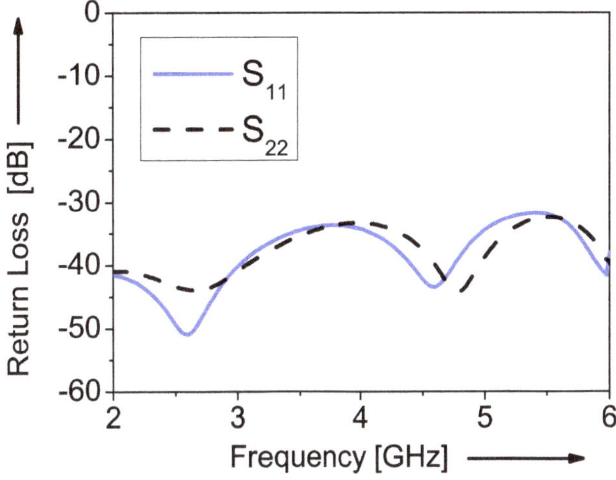

Fig. 7. Return Loss of the main microstrip line for port 1 and port 2.

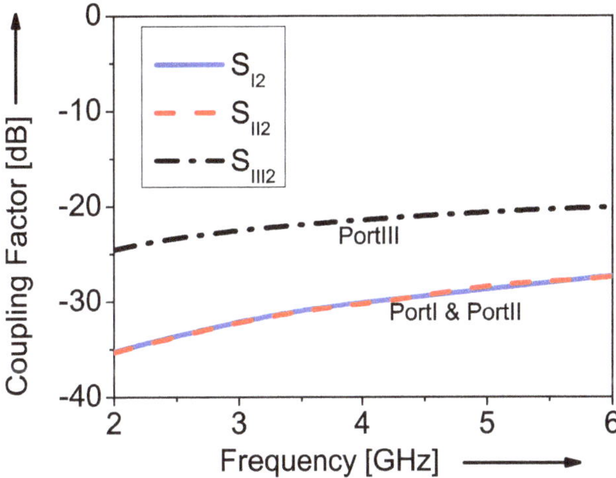

Fig. 8. Coupling factor between port 2 of the first microstrip line and port I, port II and port III.

(2006). There, three identical contactless capacitive probes made from very small semi-rigid coaxial lines with an outer diameter of 86 μm, are applied.

In this paper, a diversity based measurement setup using two identical capacitive (C_1, C_2) and one inductive probe (L) is examined. At first, a simulation of the system is performed. The simulation model is given in Fig. 6. A continuous wave of a power level of 1 dBm is led by a signal source to a microstrip line at port 2. During the calibration and measurement the signal is swept over the frequency. The width of the first microstrip line is 1.12 mm equivalent to a line impedance of 50Ω. As terminations of the planar transmission line, the calibration standards short, open, load as well as a DUT are used. Without loss of generality, the three contactless probes are realized on the same substrate (Rogers 4003) as the microstrip line. Of course, it is also possible to use separated probes. To illustrate the principal diversity method and its potentials, this setup is sufficient. For the setup all distances between the probes are chosen to Δl=25 mm. The size of the substrate is 76.2 mm×50.8 mm.

In the setup the two capacitive probes are modeled by means of a second and third microstrip line, which are not connected to the main line. The distance between the capacitive probes and the main microstrip line is 300 μm. The inductive probe is realized by a half magnetic loop. The geometry of the loop is also given in Fig. 6. The loop is directly connected to two additional microstrip lines. One characteristic of a half loop is that the energy is coupled through the loop by the electric and magnetic field at the main microstrip line. To get rid of the capacitive coupled voltage, a balun is used at the end of the loop. The ports of the two capacitive probes and the inductive probe are denoted by port I, port II and port III. The return losses of the main microstrip line are given in Fig. 7 and the coupling factors between the first microstrip line and the probes are illustrated in Fig. 8. For the simulation of this setup two different programs were considered. To determine the scattering parameters of the planar six-port circuit, HFSS™ of Ansoft was used. The other elements like the balun, the receivers, the source and the terminations were simulated using ANSOFT DESIGNER. Introducing white Gaussian noise, the system was calibrated by using the diversity SOL (DSOL) calibration algorithm which was implemented in MATLAB. A description of the DSOL calibration is given in the next section.

5 Diversity SOL calibration

For a better understanding of the diversity calibration, a general block diagram of the probe pair selection is given in

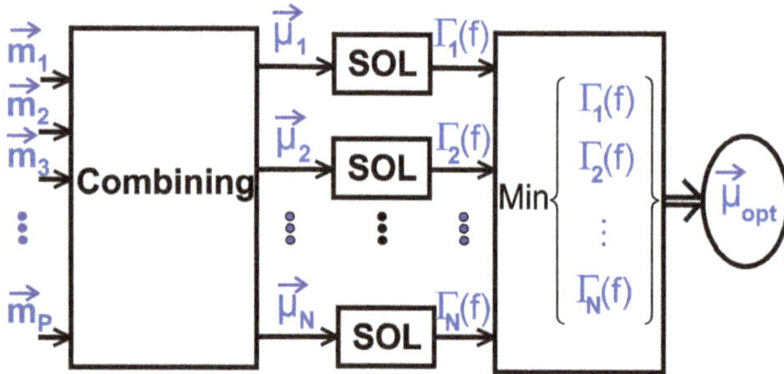

Fig. 9. General block diagram of the probe pair selection.

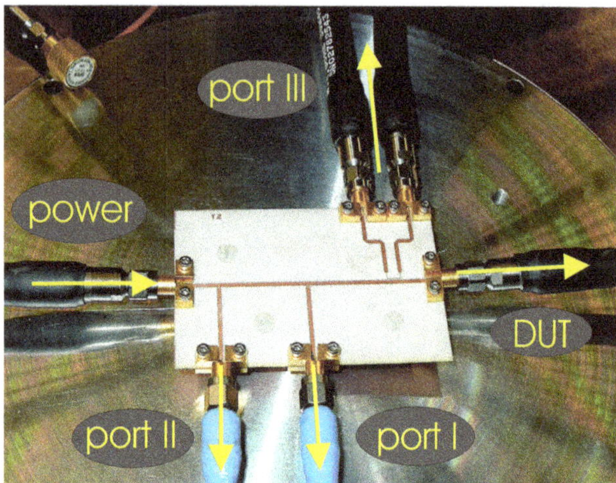

Fig. 10. Photograph of the contactless measurement setup with three probes.

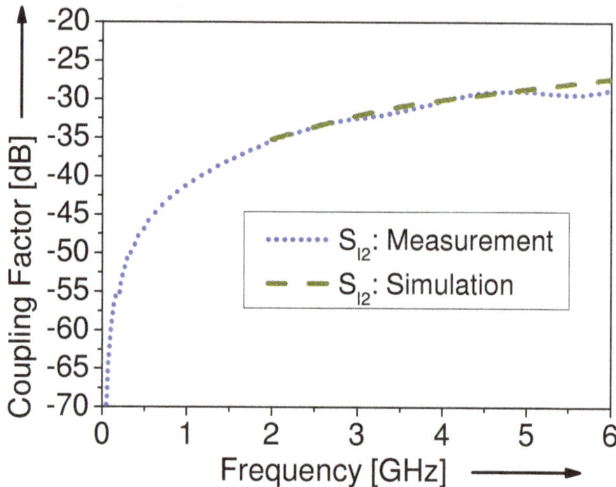

Fig. 11. Comparison of the coupling factors of the simulation and the measurement results.

Fig. 9. The input quantities of the algorithm are the frequency dependent, complex voltage vectors $\boldsymbol{m}_1$ to $\boldsymbol{m}_P$ of the P contactless probes. Each voltage vector $\boldsymbol{m}_p$ consists of four voltage values resulting from the measurement of the calibration standards short, open and load with $U_{m_p,S}$, $U_{m_p,O}$, $U_{m_p,L}$ and the DUT with $U_{m_p,D}$ for the probe p:

$$\boldsymbol{m}_p = [U_{m_p,S}(f); U_{m_p,O}(f); U_{m_p,L}(f); U_{m_p,D}(f)]. \quad (29)$$

In order to determine the system dynamics, the load standard is employed also as the DUT.

The voltage ratios μ_n for each possible probe pair combination are calculated using all measured probe voltages. With P contactless probes, N different probe pair combinations are possible. N is given by the binomial coefficient:

$$N = \binom{P}{2} = \frac{P!}{2 \cdot (P-2)!}. \quad (30)$$

One example of the voltage ratio μ_n for the nth probe pair combination using the probes p_1 and p_2 is given by:

$$\mu_n = [\frac{U_{m_{p_1},S}}{U_{m_{p_2},S}}; \frac{U_{m_{p_1},O}}{U_{m_{p_2},O}}; \frac{U_{m_{p_1},L}}{U_{m_{p_2},L}}; \frac{U_{m_{p_1},D}}{U_{m_{p_2},D}}] \quad (31)$$
$$= [\mu_{n,S}; \mu_{n,O}; \mu_{n,L}; \mu_{n,D}]$$

With the knowledge of all voltage ratios μ_1 to μ_N the reflection coefficients Γ_1 to Γ_N for all possible probe pair combinations are determined using the conventional SOL correction algorithm of Eq. (4). As the load standard is used as DUT, the determined reflection coefficients Γ_1 to Γ_N describe the system dynamics. In order to get, for each frequency, the probe pair combinations showing the highest dynamics the minima of the reflection coefficients have to be found. With these minima, the best probe pair combinations are known for which the largest system dynamics can be achieved. These probe pairs are used for further measurements of an unknown DUT. The described diversity combining can also be implemented with other calibration methods like SOLT, TRL, etc. For a practical test, a real diversity measurement setup is discussed in the next section.

6 Measurement setup

The one-port measurement system is realized according to the simulation setup of Fig. 6. The setup is shown in Fig. 10. The line is fed by the signal source of the VNA of a power level of −14 dBm. The forward and backward traveling signals propagating on the main line are partly coupled to three coherent receivers of the VNA, by two capacitive probes and one inductive probe. The geometries of the inductive and capacitive coupling structures are described in Sect. 4. The measured coupling factors of the capacitive probes are approximately −35 dB at 2 GHz and −25 dB at 10 GHz and fit very well to the simulated values. As an example, the simulated and measured coupling factors between port I and port 2 are compared in Fig. 11. The difference of measurement and simulation can be explained by influence of microstrip to coaxial transitions used in the measurement. These were connected to coaxial cables used to feed the

Table 1. Critical frequencies.

f_{LC1} [GHz]	f_{LC2} [GHz]	f_{C1C2} [GHz]
1.85	0.92	0.00
5.54	2.77	3.69
9.23	4.62	7.39
12.92	6.46	11.08
⋮	⋮	⋮

device and to connect the DUT and precise coaxial calibration standards at a well-defined reference plane. The obtained simulation and measurement results are given in Sect. 7.

7 Results

7.1 50Ω standard

A comparison of the system dynamics for the simulation model and the measurement system using the 50Ω calibration load as DUT is shown in Figs. 12 and 13. The results show a good accordance. The remaining differences are due to fabrication tolerances of the coupling substrate and to the constant noise level used in the simulation. According to a 10 Hz intermediate frequency bandwidth of the PNA E8361A, reference AWGN was applied in the simulation with a respective power level of −118 dBm. For example, Fig. 12 shows the system dynamics using the probe pair combination L and C_2. The system was calibrated with a conventional SOL algorithm. Here peaks occur showing at which frequencies the calibration fails and consequently where accurate measurements are not possible. By knowledge of the probe distances and of the mean value of the effective permittivity of the main microstrip line, which is $\epsilon_{r,\text{eff}}$=2.64 at 6 GHz, the critical frequencies can be calculated using Eqs. (27) and (28). The results are given in Table 1. The calculated critical frequencies f for the three possible probe pair combinations are equal to the frequencies obtained by simulation and measurement. In comparison to the system using two contactless probes, the diversity system achieves a broad measurement bandwidth. Critical frequencies do not exist any more. Figure 13 shows the achieved measurement bandwidth of the diversity based system. The system dynamics is approximately −70 dB. Compared to the LC_2 combination (Fig. 12), a gain of about 30 dB is achieved.

7.2 Open standard

Here a high reflective DUT is considered, realized by the open calibration standard. In Fig. 14, the measured reflection coefficients are given for all three possible probe pair combinations LC_1, LC_2 and C_1C_2. A SOL calibration was applied. We observe, that at the critical frequencies given in Table 1, peaks occur with an amplitude of up to 0.3 dB. An improvement can be achieved using the diversity measurement system. The measurement setup of Fig. 10 is calibrated using the DSOL calibration, and then the open calibration load is remeasured. The results are shown in Fig. 15. Compared with the results given in Fig. 14, a significant improvement can be achieved: Only small peaks with a maximum amplitude of 0.04 dB exist.

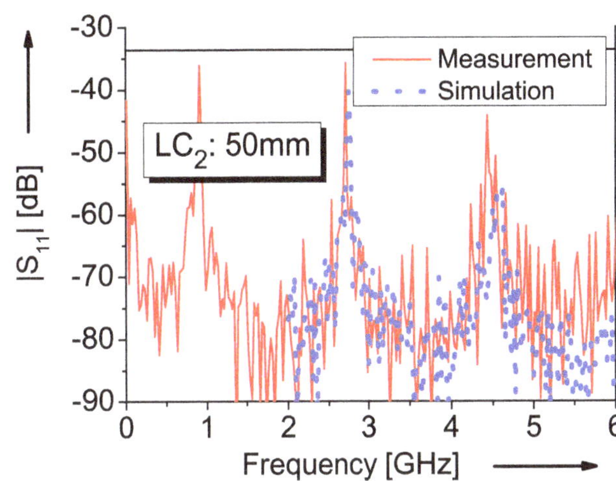

Fig. 12. Comparison of simulated and measured system dynamics for combination LC_2.

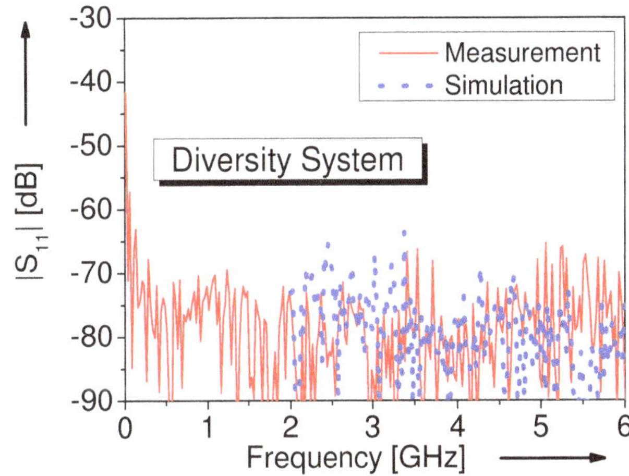

Fig. 13. Comparison between the simulated and measured system dynamics for the diversity system.

7.3 Comparison of contactless and conventional vector network analysis using an offset-load

The contactless system using two probes and the diversity based system was compared here using a conventional

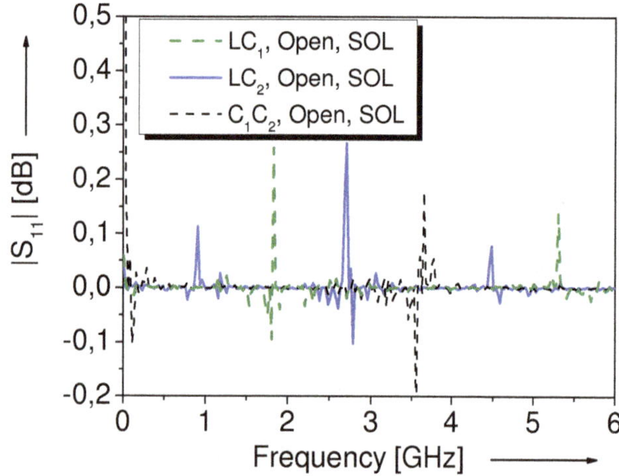

Fig. 14. Measurements results of the open standard as a DUT, after a SOL calibration, for three probe pair combinations.

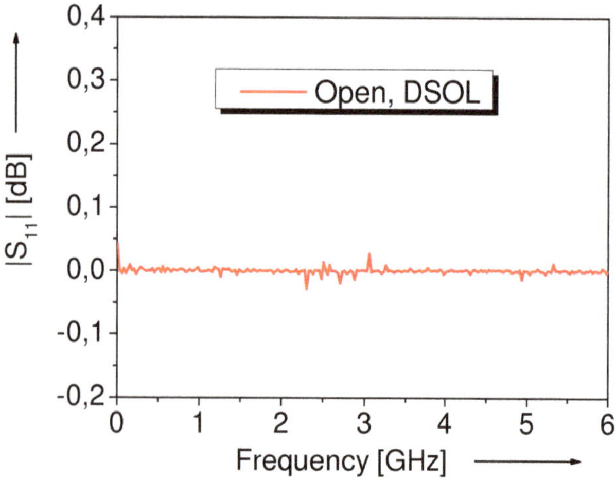

Fig. 15. Measurements results of the open standard as a DUT, after a DSOL calibration, for the diversity system.

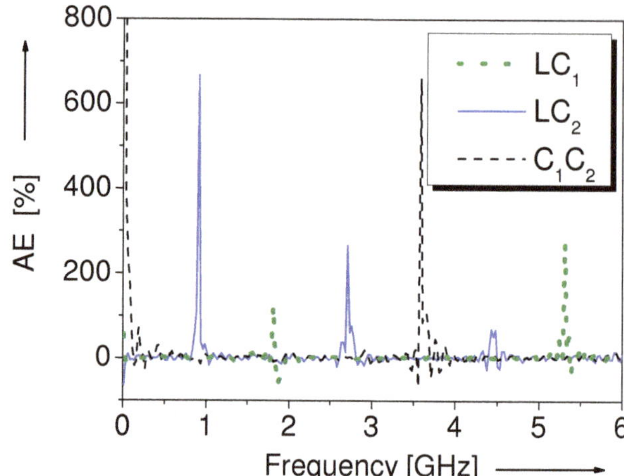

Fig. 16. Deviations between the contactless system with two probes and the measurement results using the conventional VNA, DUT: Offset-50Ω-load.

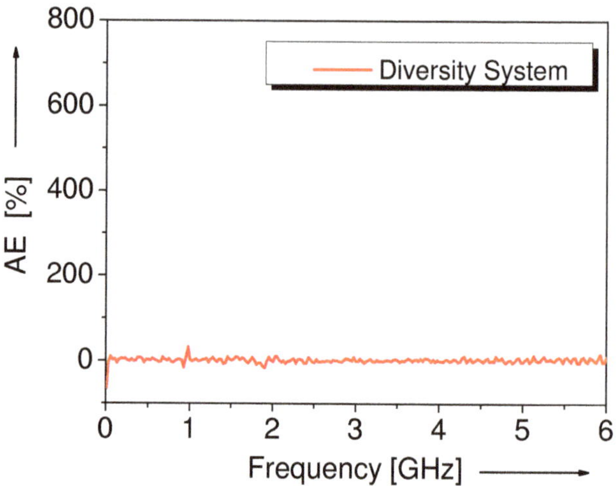

Fig. 17. Deviations between the contactless diversity based system with three probes and the measurement results using the conventional VNA, DUT: Offset-50Ω-load.

vector network analyzer. For this purpose an offset-50Ω-load, which is not used for the calibration, is applied as a DUT. The results are shown in Figs. 16 to 19. Thereby, the measurement data from a conventional vector network analysis was used as a reference.

In particular Fig. 16 shows the percentage deviations of the absolute values of the measured reflection coefficients using the system with the probe pair combinations LC_1, LC_2 and C_1C_2. Thereby, the amplitude errors (AE) of the contactless systems were calculated by using Eqs. (32):

$$AE = 100\% \cdot \frac{|\Gamma_{\text{contactless}}| - |\Gamma_{\text{conventional}}|}{|\Gamma_{\text{conventional}}|}. \tag{32}$$

At the critical frequencies large errors in the characteristics of the measured reflection coefficients occur. These result in measurement errors of up to 670% between 35 MHz and 6 GHz. In contrast to these results, the errors are minimized using the diversity based system. The described diversity based system shows a significant improvement of the contactless measurement system, as can be seen in Fig. 17. The results of the diversity based system show a maximal error of 32%. The remaining maximal error seems to be quite high, but it has to be annotated that for the calculation of the percentage deviation a very small reference value of about −35 dB of the offset load is used.

Considering the phase, for the contactless system of two probes, phase errors of up to 178° occur at the critical frequencies. By means of the diversity based system a maximal error of 22° arises in the frequency range between 35 MHz and 6 GHz, as shown in Figs. 18 and 19. Thus, a clear

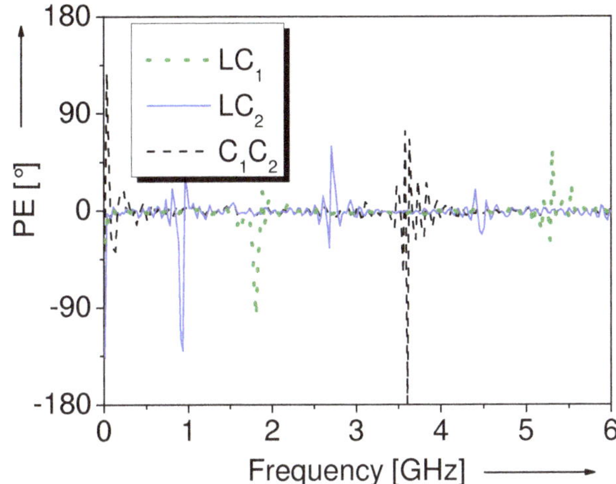

Fig. 18. Deviations between the contactless diversity based system with three probes and the measurement results using the conventional VNA, DUT: Offset-50Ω-load.

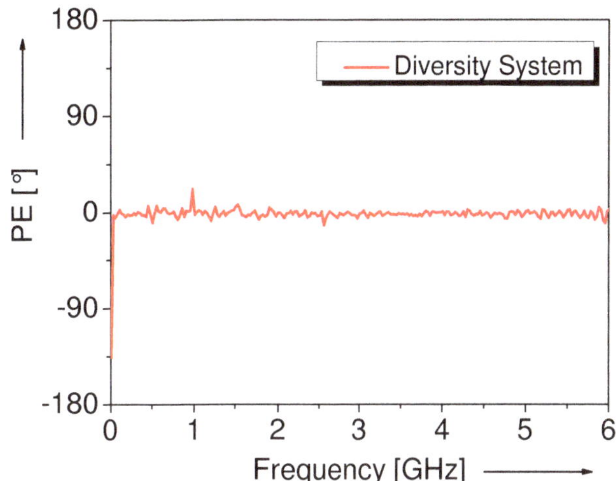

Fig. 19. Deviations between the contactless diversity based system with three probes and the measurement results using the conventional VNA, DUT: Offset-50Ω-load.

reduction of the measurement errors can be achieved. The phase errors (PE) were calculated by using Eqs. (33):

$$PE = arg(\Gamma_{\text{contactless}}) - arg(\Gamma_{\text{conventional}}). \qquad (33)$$

A part of the remaining errors of the diversity system are caused by non-perfect reproducibility of the system and by cross-coupling of signals in the planar structure. An improvement of the measurement accuracy can be achieved by using shielded probes, which is shown by Zelder et al. (2007).

8 Conclusions

In the first part of this paper the contactless vector network analysis using two probes is presented. It is shown that dependent on the distance between the two probes, a calibration is not possible at certain frequencies. For the two configurations using either two capacitive probes or one inductive and one capacitive probe, these critical frequencies are obtained analytically. An enhanced, diversity based, contactless measurement system is presented which eliminates these system failures. In a one-port setup two capacitive and one inductive probe are integrated in a system. A diversity calibration algorithm is applied to determine probe pair combinations, for which the largest system dynamics can be achieved. The most convenient probe pair combinations are then used for the measurement of the reflection coefficient of a DUT. Simulation and measurement results show a significant improvement of the system dynamics of up to 30 dB. It is also illustrated that when using two probes, peaks of up to 0.3 dB in the return loss occur in case of a high-reflective DUT, which are avoided using the contactless diversity based system. Further on, the reflection of an offset-load was determined using a conventional vector network analyzer. The results are compared with reflection measured by the contactless system using two probes and by the contactless diversity based system. And in this case, the measurement errors can be reduced significantly by applying the diversity based system.

References

Bridges, G. E.: Non-contact probing of integrated circuits and packages, IEEE MTT-S International Microwave Symposium Digest, vol. 3, pp. 1805–1808, June, 2004.

De Groote, F., Verspecht, J., Tsironis, C., Barataud, D., and Teyssier, J.-P.: An improved coupling method for time domain load-pull measurements, European Microwave Conference, vol. 1, October, 2005.

Dudley, R. A., Roddie, A. G., Bannister, D. J., Gifford, A. D., Krems, T., and Facon, P.: Electro-optic S-parameter and electric-field profiling measurement of microwave integrated circuits, IEE Proceedings Science, Measurement and Technology, vol. 146, no. 3, pp. 117–122, May, 1999.

Engen, G. F. and Hoer, C. A.: Thru-Reflect-Line: An improved technique for calibrating the dual six-port automatic network analyzer, IEEE Transaction on Microwave Theory and Techniques, vol. 12, pp. 987–993, December, 1979.

Gao, Y. and Wolff, I.: Measurements of field distributions and scattering parameters in multiconductor structures using an electric field probe, IEEE MTT-S International Microwave Symposium Digest, vol. 3, pp. 1741–1744, June, 1997.

Hui, D. and Weikle, R. M.: A non-contacting sampled-line reflectometer for microwave scattering parameter measurements, 64th ARFTG Microwave Measurements Conference, Fall 2004, pp. 131–137, Dec., 2004.

Osofsky, S. S. and Schwarz, S. E.: Design and performance of a noncontacting probe for measurements on high-frequency planar circuits, IEEE Transactions on Microwave Theory and Techniques, vol. 40, no. 8, pp. 1701–1708, Aug., 1992.

Quardirhi, Z. and Laurin, J. J.: Méthode de measures des paramétres s sans contact, CCECE 2003-CCGEI 2003, Montréal, May, 2003.

Sayil, S., Kerns, D. V., and Kerns, S. E.: A survey contactless measurement and testing technique potentials, IEEE Potentials, vol. 24, no. 1, pp. 25–28, February–March, 2005.

Schiek, B.: Grundlagen der Systemfehlerkorrektur von Netzwerkanalysatoren, Grundlagen der Hochfrequenz-Messtechnik, 1th ed., Berlin Heidelberg, Germany, Springer-Verlag, chapter 4, pp. 141–174, 1999.

Stenarson, J., Yhland, K., and Wingqvist, C.: An in-circuit noncontacting measurement method for S-parameters and power in planar circuits, IEEE Transactions on Microwave Theory and Techniques, vol. 49, no. 12, pp. 2567–2572, December, 2001.

Yhland, K. and Stenarson, J.: Noncontacting measurement of power in microstrip circuits, in 65th ARFTG, pp. 201–205, June, 2006.

Zelder, T., Rabe, H., and Eul, H.: Contactless electromagnetic measuring system using conventional calibration algorithms to determine scattering parameters, Adv. Radio Sci., 5, 2007.

Zelder, T. and Eul, H.: Contactless network analysis with improved dynamic range using diversity calibration, Proceedings of the 36th European Microwave Conference, Manchester, UK, pp. 478–481, September, 2006.

Usage of the contactless vector network analysis with varying transmission line geometries

T. Zelder, B. Geck, I. Rolfes, and H. Eul

Institut für Hochfrequenztechnik und Funksysteme, Leibniz Universität Hannover, Appelstr. 9A, 30167 Hannover, Germany

Abstract. The scattering parameters of embedded devices can be measured by means of contactless vector network analysis. To achieve accurate measurement results, the contactless measurement setup has to be calibrated. However, if the substrate material or the planar transmission lines on the substrate changes, a new calibration is necessary. In this paper a method will be examined, which reduces the number of calibration cycles by using a database. Analytical results show that by using this database method, errors occur which depend on the coupling coefficients and on the load impedances of the contactless probes. However, the measurement results show deviations smaller than 7% in comparison to the conventional vector network analysis, which is sufficient for the most pratical applications.

1 Introduction

The scattering parameters of embedded devices can be determined by contactless measurement methods described by Sayil et al. (2005), Zelder and Eul (2006), Dudley et al. (1999). A promising method is the contactless vector network analysis proposed by Stenarson et al. (2001), Zelder et al. (2007a). A simplified two-port setup of this contactless measurement technique is illustrated in Fig. 1. The two-port setup consists of a conventional vector network analyzer (VNA) with a source, a switch and four vectorial receivers m_1 to m_4. In Fig. 1 the output ports of the VNA are connected to a planar circuit board. In contrast to the conventional setup, contactless probes are connected directly to the receivers. Furthermore, the probes are positioned above the planar transmission lines of the device under test (DUT). In this paper a setup using capacitive probes is described. Zelder and Eul (2006) also used inductive probes, whereas Zelder et al. (2007a) and Yhland and Stenarson (2005) applied loop couplers as contactless probes. After a conventional calibration such as Thru-Reflect-Line (TRL) given by Engen and Hoer (1979), a DUT can be characterized. By moving the probes from one planar DUT to another, the scattering parameters of all DUTs which have equal planar connection lines, can be determined. If the geometry of the planar connection lines or the substrate changes, a new calibration has to be carried out because the coupling coefficients between the contactless probes and the planar transmission lines change. This leads to variations of the calibration error terms. Thus, for each substrate material and each planar transmission line geometry to be measured, a separate calibration substrate must be available. With the introduction of the database method the calibration complexity can be reduced, because only one reference calibration substrate and a database are necessary.

2 The database method

In the database, differences of the error terms are stored using on the one hand a planar transmission line of a reference substrate and on the other hand any other planar transmission line. Without loss of generality in this paper, the database method is described using the conventional Short-Open-Load (SOL) calibration algorithm. However, the method can also be applied to other calibration methods like TRL.

At first the well-known one-port error box model of a reflectometer used by Rehnmark (1974) is shown as a flow graph in Fig. 2. In the graph a_1, b_1, a_{DUT} and b_{DUT} represent the measured waves and the real traveling waves to and from the DUT, respectively. The other parameters illustrate

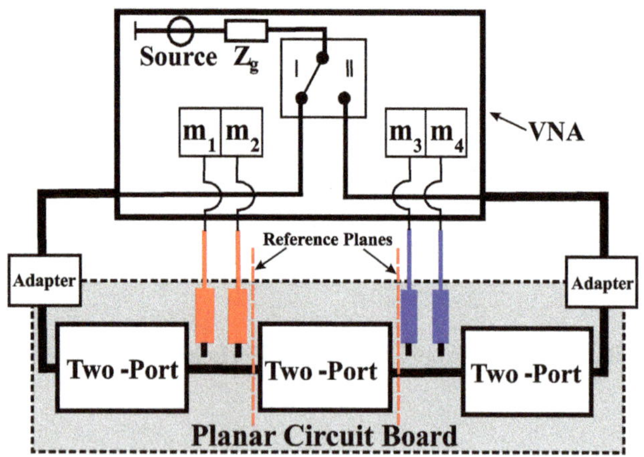

Fig. 1. Contactless vector network analysis setup shown in Zelder et al. (2007b).

the error terms, which can be expressed by a matrix

$$[\mathbf{R}] = \begin{pmatrix} R_{11} & R_{12}R_{21} \\ 1 & R_{22} \end{pmatrix}. \tag{1}$$

The error terms R_{11}, $R_{12}R_{21}$ and R_{22} can be determined using three known calibration standards like short, open and load. The ratio μ_x of the measured traveling waves

$$\mu_x = \frac{b_1}{a_1} \tag{2}$$

for a reflection coefficient Γ_x in the reference plane can be expressed in dependence of the error terms shown in Eq. (3).

$$\mu_x = R_{11} + \frac{R_{12} \cdot R_{21} \cdot \Gamma_x}{1 - R_{22} \cdot \Gamma_x} \tag{3}$$

Applying Eq. (3) for three calibration standards, the error terms can be calculated, whereby an unknown reflection coefficient Γ_{DUT} can be determined

$$\Gamma_{\text{DUT}} = \frac{\mu_{\text{DUT}} - R_{11}}{R_{22} \cdot \mu_{\text{DUT}} + R_{12} \cdot R_{21} - R_{22} \cdot R_{22}}. \tag{4}$$

For the database method, the error term matrices $[\mathbf{R}_{\text{sub}_r}]$, $[\mathbf{R}_{\text{sub}_1}]$, $[\mathbf{R}_{\text{sub}_2}]$, ..., $[\mathbf{R}_{\text{sub}_n}]$, ... $[\mathbf{R}_{\text{sub}_N}]$ according to Eq. (1) using (N+1) different calibration substrates with different transmission line geometries or substrate permittivities are determined by means of the SOL algorithm and a contactless one-port setup. Thereby, a constant and repeatable probe positioning relative to the DUT and to the transmission line must be warranted. One of the applied substrates can be defined arbitrarily as the reference substrate (index sub_r). For the database method the following basic approach is arranged

$$[\mathbf{T}_{\text{sub}_n}] = [\mathbf{T}_{\text{sub}_r}] \cdot [\Delta\mathbf{T}_{\text{sub}_{\mathbf{rn}}}]. \tag{5}$$

In Eq. (5) the cascade matrix $[\mathbf{T}_{\text{sub}_r}]$ corresponds to the reference error term matrix $[\mathbf{R}_{\text{sub}_r}]$ by

$$[\mathbf{T}_{\text{sub}_r}] = \begin{pmatrix} R_{12}R_{21} - R_{11}R_{22} & R_{11} \\ -R_{22} & 1 \end{pmatrix}. \tag{6}$$

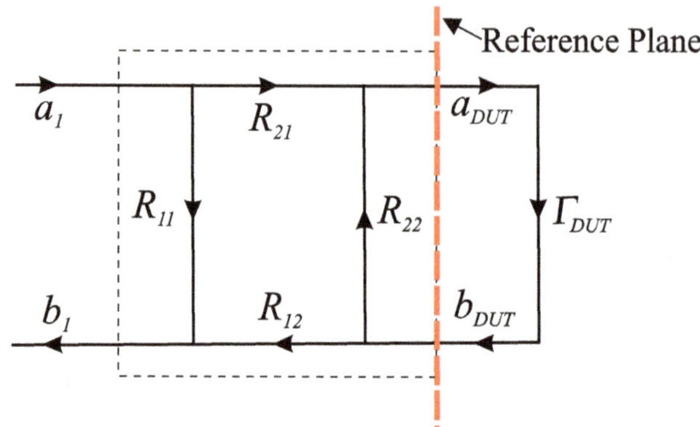

Fig. 2. One-port error box model.

Further more, it is assumed that the cascade matrix $[\mathbf{T}_{\text{sub}_r}]$ multiplied by a matrix $[\Delta\mathbf{T}_{\text{sub}_{rn}}]$ equals the error term cascade matrix $[\mathbf{T}_{\text{sub}_n}]$ of the substrate sub_n. In this case the matrix $[\Delta\mathbf{T}_{\text{sub}_{rn}}]$ represents the deviations of the error terms between the setup using the reference substrate sub_r and the substrate sub_n. The deviation matrices $[\Delta\mathbf{T}_{\text{sub}_{rn}}]$ can be calculated relatively to the reference transmission line by

$$[\Delta\mathbf{T}_{\text{sub}_{rn}}] = [\mathbf{T}_{\text{sub}_{ref}}]^{-1} \cdot [\Delta\mathbf{T}_{\text{sub}_{rn}}] \tag{7}$$

where $[\Delta\mathbf{T}_{\text{sub}_{rn}}]$ corresponds to

$$[\Delta\mathbf{T}_{\text{sub}_{rn}}] = \begin{pmatrix} \Delta\mathbf{T}_{11,\text{sub}_{rn}} & \Delta\mathbf{T}_{12,\text{sub}_{rn}} \\ \Delta\mathbf{T}_{21,\text{sub}_{rn}} & \Delta\mathbf{T}_{22,\text{sub}_{rn}} \end{pmatrix}. \tag{8}$$

These deviation matrices $[\Delta\mathbf{T}_{\text{sub}_{rn}}]$ are stored in the database for all given substrates and transmission lines, respectively.

An alternative way to determine the deviation matrices $[\Delta\mathbf{T}_{\text{sub}_{rn}}]$ is the usage of a two-tiered calibration. Thereby, the contactless measurement setup is calibrated, for example with a SOL algorithm using the reference substrate. After that, the reference transmission line is replaced by another transmission line, which has a different geometry and/or another substrate material. Then, a second calibration is performed with the changed transmission line and the resulting error term matrix of the second tier is equal to the deviation matrix $[\Delta\mathbf{T}_{\text{sub}_{rn}}]$.

The database method can be used for example in the quality control of printed circuit boards (PCB), where different transmission line geometries often exist on one board or devices on different substrate materials have to be characterized. Usually, for each change of the transmission line geometry a new calibration has to be carried out. However, by using the database method the contactless measurement setup has to be calibrated only once using the reference substrate. If during the measurement the planar transmission line geometry changes, the error terms can be corrected easily by a software using the database and thus Eq. (5).

Usually the VNA, which is used for generating the database, is different to the VNA which is used during the

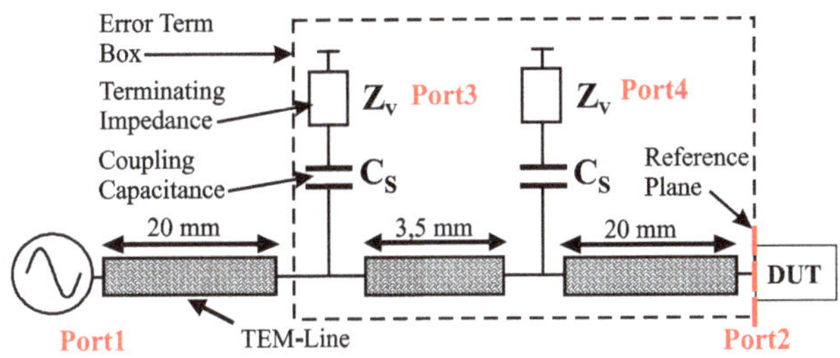

Fig. 3. Analytical one-port model of a contactless measurement setup.

measurement; also the connecting cables between the receivers of the VNA and the contactless probes are different. These facts result in a changed load impedance of the contactless probes. If the database methods are used with a changed load impedance an error occurs. An error estimation using an analytical model will be given in the next section.

3 Simulation

3.1 Analytical model

In this section the influence of different load impedances Z_V of the contactless probes and different coupling coefficients are examined. Therefore, an analytical model is used as illustrated in Fig. 3. In the model a source with a power level of 0 dBm is connected to a coaxial line having conductive loss of 1 dB/m. Two capacitively coupled probes are used in the one-port scenario which are modeled by two capacitances C_S. In the simulation results the capacitances for the reference substrate are named C_{Sr} and for the measurement substrate C_{S1}. Each capacitance is terminated with the impedance Z_V, which represents the receiver impedance of the VNA and the interferences of the transmission lines between receiver and probe. The load impedances which are used while the database is set up are named Z_{Vr} and the impedances which exist during the measurement of a DUT are named Z_{V1}. The distance between the probes is chosen to be 3.5 mm and the distance between the last probe and the reference plane is 20 mm. Furthermore, two different substrates are examined. Therefore, two different permittivities $\epsilon_{r,\mathrm{subr}}=2.0$, $\epsilon_{r,\mathrm{sub1}}=2.8$ for the coaxial line between source and DUT are used in the simulation. These values correspond roughly to the effective permittivities of a 50Ω microstrip line on a Rogers RT5870 and RO4003 substrate. In this paper the coaxial line with the permittivity $\epsilon_{r,\mathrm{subr}}$ is chosen as the reference substrate and the line with the $\epsilon_{r,\mathrm{sub1}}$ represents the measurement substrate.

3.2 Analytical results

Comparing both database methods, exactly the same scattering parameters are obtained, thus, the following results are valid for both methods. In the first analysis the capacitances are chosen to be C_{Sr}=15 fF and C_{S1}=10 fF which results in a deviation of $\Delta|S_{31}|$=3.4 dB and $\Delta\angle(S_{31})$=61° at 10 GHz. The values of capacitances C_S and the coupling factors at 10 GHz are given in Table 1. In Fig. 4, two scenarios are compared, whereas on the one hand the load impedances are chosen to be Z_{Vr}=40Ω and Z_{V1}=60Ω and on the other hand the impedances are exchanged. The deviations regarding an ideal DUT with an input impedance of Z_L=200Ω and a constant reflection coefficient of S_{11}=0.6 are smaller than ±0.5% up to 20 GHz. The phase deviations which are not illustrated are smaller than ±0.12°.

Table 1. Capacitances and coupling coefficients at 10 GHz.

C_S [fF]	10	15	22.5	34
$\|S_{31}\|$ [dB]	−30.2	−26.8	−23.4	−20.0

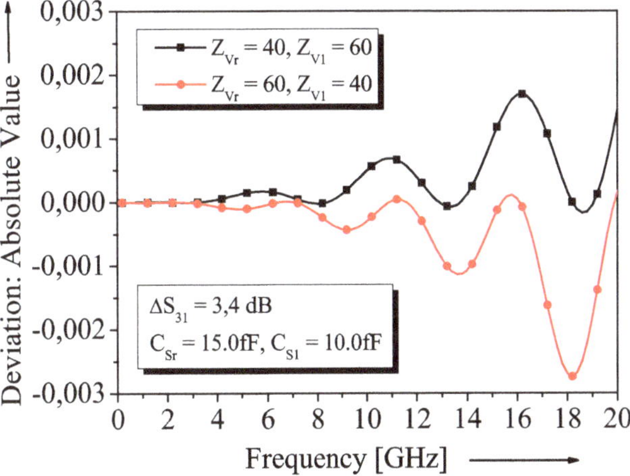

Fig. 4. Deviations of the reflection coefficient in dependency of the the port impedances Z_V and the frequency. DUT: Z_L=200Ω.

The increase of the errors in dependence of the difference of the load impedances $\Delta Z=Z_{V1}-Z_{Vr}$ is shown in Fig. 5. For this examination the capacitances are chosen to be C_{Sr}=1 fF and C_{S1}=10 fF. In Fig. 5 it is illustrated that the error is zero if ΔZ=0. With an increasing ΔZ the errors also increase. In Fig. 6 the influence of the parameter S_{31} is examined. Using different substrates the complex coupling coefficients depend on the transmission line geometry, its permittivity and the vertical, and horizontal probe positions. Three scenarios are examined, where for all scenarios the same deviation of the coupling coefficient of about $\Delta|S_{31}|$=3.4 dB exists. Figure 6 makes it obvious that the errors increase if the coupling coefficients $|S_{31}|$ increase. The reason for that is the increasing interaction between the probes and the main transmission line. Thus, there must be a tradeoff, because low coupling coefficients result

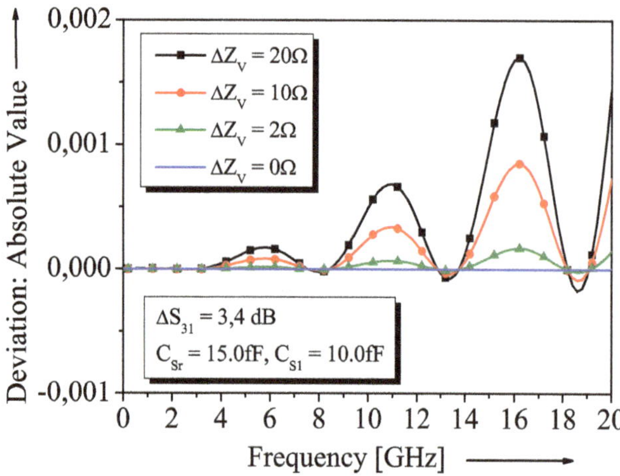

Fig. 5. Deviations of the reflection coefficient in dependency of the differences of the port impedances ΔZ and the frequency. DUT: Z_L=200Ω.

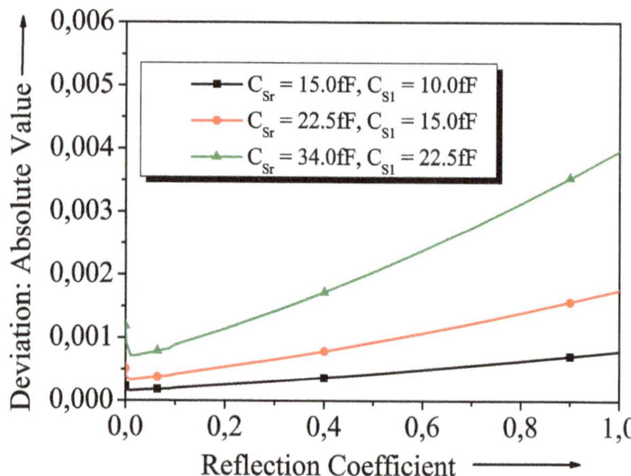

Fig. 7. Deviations of the reflection coefficient in dependency of the coupling capacitances at 10 GHz.

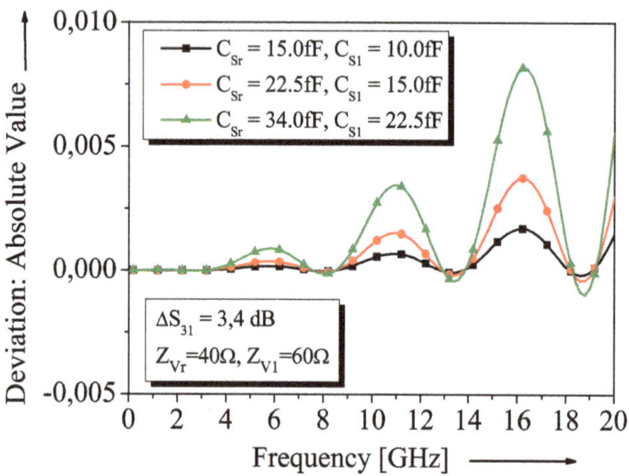

Fig. 6. Deviations of the reflection coefficient in dependency of the coupling capacitances and the frequency. DUT: Z_L=200Ω.

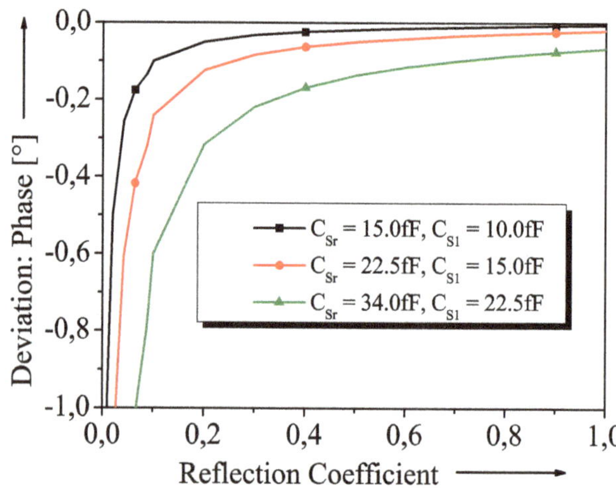

Fig. 8. Deviations of the reflection coefficient in dependency of the coupling coefficient S_{31} at 10 GHz.

in small errors, however, they also reduce the dynamic range of the contactless measurement. Furthermore, in Figs. 7 and 8 the deviations are separated into absolute error values and phase errors and are shown in dependence of the reflection coefficient. It is obvious that the deviation increases for high reflection coefficients. For the phase deviations the error increases with decreasing reflection coefficients such as the absolute error if it is depicted in percent. In Fig. 9 the deviations depending on the differences $\Delta|S_{31}|$ of the coupling coefficients using either the reference or the measurement substrate are shown. By adjusting the capacitances C_{Sr} and C_{S1}, the differences of the coupling coefficients $\Delta|S_{31}|$ are changed between 3.4 dB and 10.2 dB. As shown in Fig. 9, the errors increase with an increasing $\Delta|S_{31}|$. But even if a strong coupling of about −20 dB (C_{Sr}=34 fF) and a difference of about $\Delta|S_{31}|$=10.2 dB exist, the measurement error of a 200Ω load is smaller than ±2,5% and ±1°. Figure 10 shows that a high coupling coefficient for the reference substrate as well as for the measurement substrate is even worse. For the chosen coupling coefficients it is insignificant that the probes above the reference substrate or above the measurement substrate have a higher coupling. It can be summarized that using the database method a small error occurs. The errors decrease with decreasing differences of the load impedances ΔZ of the contactless probes, with decreasing the coupling coefficients S_{31} and S_{41} as well as with decreasing differences of the coupling coefficients ΔS_{31} and ΔS_{41} of the reference substrate in comparison to the measurement substrate.

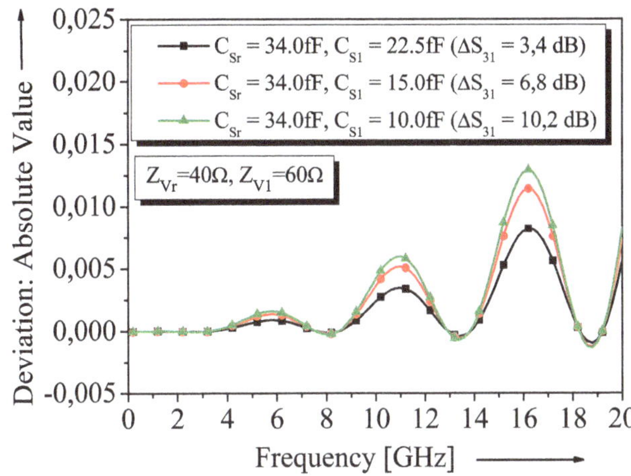

Fig. 9. Deviations of the reflection coefficient in dependency of the differences $\Delta|S_{31}|$ of the coupling coefficient. DUT: Z_L=200Ω.

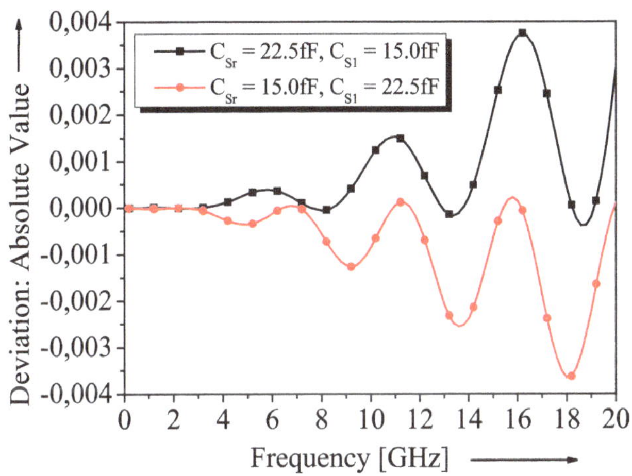

Fig. 10. Deviations of the reflection coefficient in dependency of the order of the capacitances C_{Sr} and C_{S1} and the frequency. DUT: Z_L=200Ω.

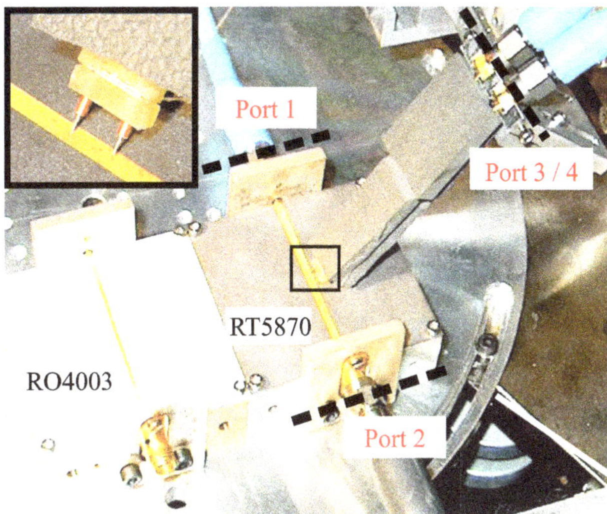

Fig. 11. Contactless measurement setup with capacitive probes.

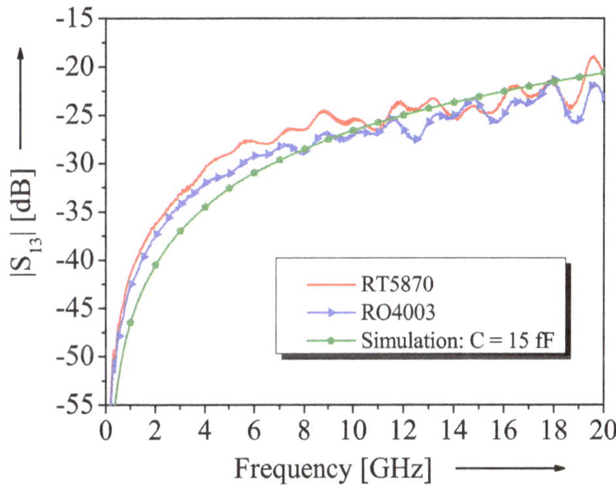

Fig. 12. Comparison of the measured coupling coefficients with a model using a single capacitance.

4 Measurement

4.1 Measurement setup

The database method is tested using the measurement setup shown in Fig. 11. For the measurement two small semi-rigid coaxial lines are used as capacitive probes. The inner conductors of the coaxial lines have a diameter of 200 μm. The probe ends are positioned in a distance of 3.5 mm from each other and about 50 μm above the microstrip lines. Two microstrip lines on Rogers RO4003 (ϵ_r=3.55) and on RT5870 (ϵ_r=2.33) are used. The width of the microstrip lines are 1.1 mm (RO4003) and 1.49 mm (RT5870) which results in characteristic impedances of about 50Ω. Setting up the reference database, the calibration substrate RT5870 and a VNA (Agilent PNA E8361A) are used. The database method is utilized on the measurement substrate RO4003 using a VNA (Rhode & Schwarz ZVA24).

4.2 Measurement results

At first the characteristics of both the contactless probes and the substrates have been examined. Therefore, a four-port measurement setup is used which is calibrated with a four-port TOSM procedure regarding the coaxial reference planes at port 1 to port 4. The absolute values of the coupling coefficients using both substrates are plotted over the frequency in Fig. 12. For the RO4003 substrate, the coupling coefficient is about $|S_{31}|$=−26.8 dB which is comparable with a coupling capacitance of 15fF as illustrated in Fig. 12. The differences of the coupling coefficients S_{13} and S_{14} using the

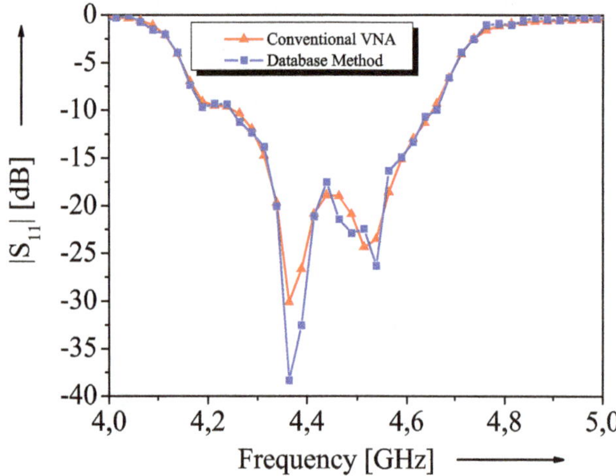

Fig. 13. Measured reflection coefficient of a filter using on the one hand a conventional VNA and on the other hand the contactless database method.

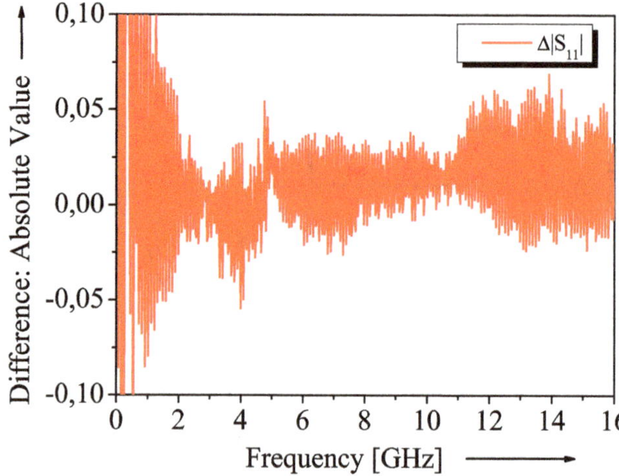

Fig. 14. Deviations of the absolute values of the reflection coefficients of the filter using on the one hand a conventional VNA and on the other hand the contactless database method.

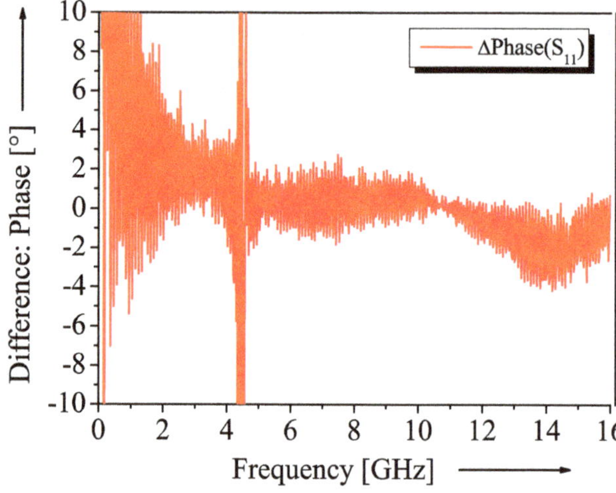

Fig. 15. Deviations of the phase of the reflection coefficients of the filter using on the one hand a conventional VNA and on the other hand the contactless database method.

RT5870 and the RO4003 substrate is smaller than ±3.5 dB in magnitude and ±25° in phase up to 20 GHz. Furthermore, the load impedances of the probes are compared using on the one hand the PNA and on the other hand the ZVA. Over the whole frequency range, the load impedances including the coaxial connecting lines and the receiver impedances vary by ±13Ω between 1 GHz and 20 GHz. The match of the load impedances with respect to a reference impedance of 50Ω is smaller than −10 dB. The reflection coefficient of a passband filter is shown in Fig. 13. The measurement results using a conventional VNA and the contactless setup in combination with the database method are in close concordance. The deviations between both measurement results are shown in Figs. 14 and 15. The deviations of the absolute values of the reflection coefficient for high reflective DUTs are smaller than ±7% and for the phase smaller than ±6° between 2 GHz and 16 GHz. The deviations can be reduced if the measurement results are averaged which has not been done for the results shown in Figs. 13 to 15. With the measurement the deviations are higher than with the simulations, as in the simulations noise, positioning errors, and surrounding influences are neglected.

5 Conclusions

Using the contactless vector network analysis it is necessary to recalibrate the setup for any changes of the planar transmission lines which are connected to a DUT. With the database method only one calibration on a reference substrate is necessary to use different transmission line geometries during the measurement. Simulation and measurement results show that the errors of the database method can be neglected in comparison to the errors caused by noise, the surroundings, and insufficient reproducibility of the probe positioning.

Acknowledgements. The authors are indebted to Rosenberger Hochfrequenztechnik GmbH & Co. KG for financial support and to Rhode & Schwarz GmbH & Co. KG for sponsoring of the vector network analyzer.

References

Dudley, R., Roddie, A., Bannister, D., Gifford, A., Krems, T., and Facon, P.: Electro-optic S-parameter and electric-field profiling measurement of microwave integrated circuits, IEE Proceedings – Science, Measurement and Technology, 146, 117–122, 1999.

Engen, G. and Hoer, C.: Thru-Reflect-Line: An Improved Technique for Calibrating the Dual Six-Port Automatic Network Analyzer, IEEE Transactions on Microwave Theory and Techniques, 27, 987–993, 1979.

Rehnmark, S.: On the calibration process of automatic network analyzer systems, IEEE Transactions on Microwave Theory and Techniques, 22, 457–458, 1974.

Sayil, S., Kerns, D. V. J., and Kerns, S.: Comparison of contactless measurement and testing techniques to a all-silicon optical test and characterization method, IEEE Transactions on Instrumentation and Measurement, 54, 2082–2089, 2005.

Stenarson, J., Yhland, K., and Wingqvist, C.: An in-circuit noncontacting measurement method for S-parameters and power in planar circuits, IEEE Transactions on Microwave Theory and Techniques, 49, 2567–2572, 2001.

Yhland, K. and Stenarson, J.: Noncontacting measurement of power in microstrip circuits, 65th ARFTG Conference Digest, 201–205, 2005.

Zelder, T. and Eul, H.: Contactless network analysis with improved dynamic range using diversity calibration, Proceedings of the 36th European Microwave Conference, 478–481, 2006.

Zelder, T., Geck, B., Wollitzer, M., Rolfes, I., and Eul, H.: Contactless Network Analysis System for the Calibrated Measurement of the Scattering Parameters of Planar Two-Port Devices, Proceedings of the 37th European Microwave Conference, 246–249, 2007a.

Zelder, T., Rabe, H., and Eul, H.: Contactless electromagnetic measuring system using conventional calibration algorithms to determine scattering parameters, Adv. Radio Sci., 5, 427–434, 2007b, http://www.adv-radio-sci.net/5/427/2007/.

Novel algorithms for the characterization of n-port networks by using a two-port network analyzer

B. Will[1], **I. Rolfes**[2], **and B. Schiek**[1]

[1]Ruhr-Universität Bochum, Institut für Hochfrequenztechnik, Universitätsstrasse 150, 44801 Bochum, Germany
[2]Leibniz-Universität Hannover, Institut für Hochfrequenztechnik und Funksysteme, Appelstrasse 9A, 30167 Hannover, Germany

Abstract. The measurement of the scattering matrices of n-port networks is an important task. For this purpose two ports of the n-port network are connected with the network analyzer and the remaining ports are connected to reflecting terminations. In order to specify the scattering matrix of a n-port network with the multi-port method (Rolfes et al., 2005), n reflecting terminations are required from which at least one reflection factor needs to be known.

There are some cases, in which the multi-port method shows weak convergence properties. For example, a T-junction cannot be identified if the reflecting terminations used are short circuits and if the line length is equivalent to a multiple of a half wavelength. This is due to the fact that the two ports connected to the network analyzer become isolated.

Two new algorithms, named the sub-determinant method and the wave-identification method, respectively, which employ a second set of reflection terminations that have to differ from the first set, allow to identify every n-port network without the necessity to distinguish different cases. Both methods are based on least square algorithms and allow to determine all scattering parameters of a n-port-network directly and uniquely.

1 Introduction

The measurement of multi-ports with more than two ports gets more and more important. Modern circuits are often complex. Additionally the number of ports may be higher than the number of ports of the network analyzer. Therefore, methods for the measurement of multi-ports with a two-port network analyzer are needed, independent of the number of ports. The principle of these methods is to combine all possible two-port measurements of the multi-port so that the scattering parameters of the multi-port can be identified. An easy method is to connect all ports of the multi-port which are not connected with the network analyzer to a match. In this case the scattering parameters can directly be determined by the two-port measurements. But in fact, it is often not possible to connect all ports to a match, for example if non-contacting measurements are performed. Furthermore the available matches might not be accurate enough, especially in the range of higher frequencies. Thus, another method which is independent of the external reflections has to be developed. The following article describes different methods to evaluate a multi-port by using a two-port network analyzer and unknown external reflections. Finally a very general method to evaluate any multi-port is described, which so far has proven to be successful in all cases considered.

2 Multi-port methods

An already known method (Tippet et al., 1982) offers the possibility to characterize almost every multi-port by connecting each port to one external reflection Γ_i, but does not allow reflection coefficients of ± 1 or values in the vicinity of ± 1. An improvement was made with the multi-port method (Rolfes et al., 2005), which allows to use external reflections independent of their value and additionally offers the possibility to obtain the value of all reflection coefficients as unknowns, except for one. But there are some multi-ports, which cannot be characterized with only one external reflection on every port, because this reflection might e. g. isolate the measurement ports. One example for such a problem is a tee junction as shown in Fig. 1, connected to short circuits or opens as external reflections. If one port of this 3-port is connected to an open circuit and the length l of the lines is an odd multiple of a quarter of a wavelength, the two other ports which are connected to the network analyzer become

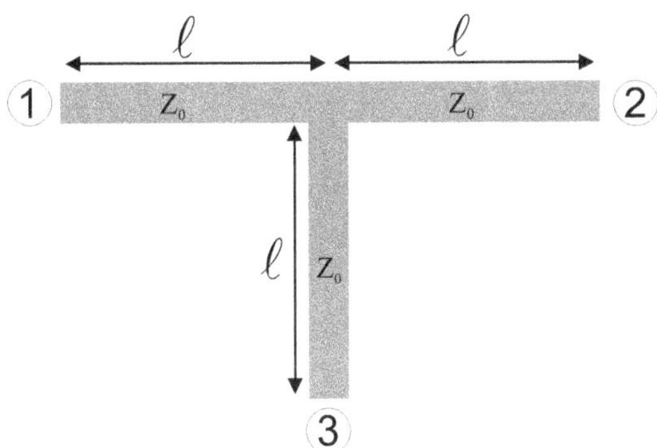

Fig. 1. Setup of a tee junction.

isolated. Then, it is not possible to characterize this 3-port with the multi-port method, if at least two of the external reflections are opens.

A solution of this problem is to use a second set of external reflections which have to differ from the first set. Thus, the different ports of the tee junction are coupled at every frequency point at least once because an isolation only occurs for one of the two external reflections.

The following methods are all described for the identification of a 3-port. In fact, the methods can be used for the identification of multi-ports with any number of ports because every multi-port with n ports can be subdivided into n 3-ports. This subdivision (Rolfes et al., 2005) offers the possibility to characterize all multi-ports with a method which is defined for 3-ports. Although it is possible to extend the following method also for the calculation of a 4-port, this yields a very complex system of equations and furthermore a subdivision remains necessary for all multi-ports with more than four ports. Thus, the following methods are described for the calculation of a 3-port only because it is the simplest structure.

3 The wave-identification method

In general a multi-port can be described by its scattering parameters in the following way

$$\boldsymbol{b} = \mathbf{S} \cdot \boldsymbol{a} \ , \tag{1}$$

where $\boldsymbol{a}$, $\boldsymbol{b}$ are the vectors of incident and reflected waves, respectively, and $\mathbf{S}$ is the scattering matrix. If the port i is connected to the external reflection Γ_i it holds

$$a_i = \Gamma_i \cdot b_i \ . \tag{2}$$

Thus, it follows for a 3-port, if port 2 and port 3 are connected to the network analyzer and port 1 is connected to the external reflection Γ_i,

$$\begin{pmatrix} b_1 \\ m_2 \\ m_3 \end{pmatrix} = \begin{pmatrix} S_{11} & S_{12} & S_{13} \\ S_{21} & S_{22} & S_{23} \\ S_{31} & S_{32} & S_{33} \end{pmatrix} \begin{pmatrix} \Gamma_1 \cdot b_1 \\ a_2 \\ a_3 \end{pmatrix} \ . \tag{3}$$

Here m_2 and m_3 are the waves measured by the network analyzer and b_1 is the unknown wave at port 3. The measurement between port 2 and port 3 yields a two-port scattering matrix. With the help of this matrix it is possible to choose the incident waves and calculate the corresponding reflected waves m_2 and m_3. Thus, the incident waves a_2 and a_3 can be chosen as two linearly independent vectors

$$\begin{pmatrix} 0 \\ a_2 \\ a_3 \end{pmatrix} = \begin{pmatrix} 0 \\ 1 \\ 0 \end{pmatrix} \quad \text{and} \quad \begin{pmatrix} 0 \\ \tilde{a}_2 \\ \tilde{a}_3 \end{pmatrix} = \begin{pmatrix} 0 \\ 0 \\ 1 \end{pmatrix} \ . \tag{4}$$

With these two different excitations, one two-port measurement yields six equations

$$\begin{aligned} b_1^{(1)} &= S_{11}\Gamma_1 b_1^{(1)} + S_{12} & \tilde{b}_1^{(1)} &= S_{11}\Gamma_1 \tilde{b}_1^{(1)} + S_{13} \\ m_2^{(1)} &= S_{21}\Gamma_1 b_1^{(1)} + S_{22} & \tilde{m}_2^{(1)} &= S_{21}\Gamma_1 \tilde{b}_1^{(1)} + S_{23} \\ m_3^{(1)} &= S_{31}\Gamma_1 b_1^{(1)} + S_{32} & \tilde{m}_3^{(1)} &= S_{31}\Gamma_1 \tilde{b}_1^{(1)} + S_{33} \ . \end{aligned} \tag{5}$$

The unknown variables b_1 and $\tilde{b}_1$ can be eliminated and one gets four non-linear equations for the identification of the scattering parameters S_{ij} of the 3-port

$$\begin{aligned} m_2^{(1)} &= S_{22} + \frac{\Gamma_1 S_{21} S_{12}}{1-\Gamma_1 S_{11}} & \tilde{m}_2^{(1)} &= S_{23} + \frac{\Gamma_1 S_{21} S_{13}}{1-\Gamma_1 S_{11}} \\ m_3^{(1)} &= S_{32} + \frac{\Gamma_1 S_{31} S_{12}}{1-\Gamma_1 S_{11}} & \tilde{m}_3^{(1)} &= S_{33} + \frac{\Gamma_1 S_{31} S_{13}}{1-\Gamma_1 S_{11}} \ . \end{aligned} \tag{6}$$

Thus, the three necessary two-port measurements yield twelve non-linear equations for the identification of a 3-port

$$\begin{aligned} S_{22} &= m_2^{(1)} - \frac{\Gamma_1 S_{21} S_{12}}{1-\Gamma_1 S_{11}} & S_{23} &= \tilde{m}_2^{(1)} - \frac{\Gamma_1 S_{13} S_{21}}{1-\Gamma_1 S_{11}} \\ S_{32} &= m_3^{(1)} - \frac{\Gamma_1 S_{31} S_{12}}{1-\Gamma_1 S_{11}} & S_{33} &= \tilde{m}_3^{(1)} - \frac{\Gamma_1 S_{31} S_{13}}{1-\Gamma_1 S_{11}} \\ S_{11} &= m_1^{(2)} - \frac{\Gamma_2 S_{21} S_{12}}{1-\Gamma_2 S_{22}} & S_{13} &= \tilde{m}_1^{(2)} - \frac{\Gamma_2 S_{12} S_{23}}{1-\Gamma_2 S_{22}} \\ S_{31} &= m_3^{(2)} - \frac{\Gamma_2 S_{32} S_{21}}{1-\Gamma_2 S_{22}} & S_{33} &= \tilde{m}_3^{(2)} - \frac{\Gamma_2 S_{32} S_{23}}{1-\Gamma_2 S_{22}} \\ S_{11} &= m_1^{(3)} - \frac{\Gamma_3 S_{31} S_{13}}{1-\Gamma_3 S_{33}} & S_{12} &= \tilde{m}_1^{(3)} - \frac{\Gamma_3 S_{13} S_{32}}{1-\Gamma_3 S_{33}} \\ S_{21} &= m_2^{(3)} - \frac{\Gamma_3 S_{31} S_{23}}{1-\Gamma_3 S_{33}} & S_{22} &= \tilde{m}_2^{(3)} - \frac{\Gamma_3 S_{32} S_{23}}{1-\Gamma_3 S_{33}} \ , \end{aligned} \tag{7}$$

where the upper index (i) indicates an external reflection Γ_i at port i. These equations can either be solved with numerical methods as the least squares method or they can be linearized for small variations of the parameters, i.e. applying Newton's method.

Therefore, the scattering parameters are defined as a constant $\hat{S}_{ij}$ and a variation ε_{ij}:

$$S_{ij} = \hat{S}_{ij} + \varepsilon_{ij} \ . \tag{8}$$

The products of the variables $\varepsilon_{ij} \cdot \varepsilon_{nm}$ are neglected. These assumptions yield a linear system of equations for the nine new variables ε_{ij}, if the scattering parameters in the system of Eqs. (7) are replaced by the new expressions defined in Eq. (8). The new system of equations with the coefficient matrix $\mathbf{A}_\varepsilon$ and the solution vector of the right hand side $\boldsymbol{m}_\varepsilon$

$$\mathbf{A}_\varepsilon \cdot \begin{pmatrix} \varepsilon_{11} \\ \varepsilon_{12} \\ \varepsilon_{13} \\ \varepsilon_{21} \\ \varepsilon_{22} \\ \varepsilon_{23} \\ \varepsilon_{31} \\ \varepsilon_{32} \\ \varepsilon_{33} \end{pmatrix} = \boldsymbol{m}_\varepsilon \tag{9}$$

is linear concerning the variables ε_{ij} and can easily be extended for a second set of external reflections. This system of equations is over-determined and can be solved by using e.g. a linear regression.

For a good convergence the initial value $\hat{S}_{ij}$ should be close to the solution vector. Good starting values can e.g. be obtained by the sub-determinant method, which is described below. In fact, several simulations and measurements have shown, that there is a very good convergence with only a few iterations, even if the initial vector is taken as the zero vector. Thus, the wave-identification method is a very robust method for the identification of the scattering parameters of multi-ports.

4 The sub-determinant method

The sub-determinant method is another method to identify a multi-port, if a second set of external reflections is used. A transformation of each Eq. (7) yields the following system of equations

$$\begin{aligned}
m_2^{(1)} &= m_2^{(1)}\Gamma_1 S_{11} + S_{22} + \Gamma_1(S_{12}S_{21} - S_{11}S_{22}) \\
\tilde{m}_2^{(1)} &= \tilde{m}_2^{(1)}\Gamma_1 S_{11} + S_{23} + \Gamma_1(S_{13}S_{21} - S_{11}S_{23}) \\
m_3^{(1)} &= m_3^{(1)}\Gamma_1 S_{11} + S_{32} + \Gamma_1(S_{31}S_{12} - S_{11}S_{32}) \\
\tilde{m}_3^{(1)} &= \tilde{m}_3^{(1)}\Gamma_1 S_{11} + S_{33} + \Gamma_1(S_{13}S_{31} - S_{11}S_{33}) \\
m_1^{(2)} &= m_1^{(2)}\Gamma_2 S_{22} + S_{11} + \Gamma_2(S_{12}S_{21} - S_{11}S_{22}) \\
\tilde{m}_1^{(2)} &= \tilde{m}_1^{(2)}\Gamma_2 S_{22} + S_{13} + \Gamma_2(S_{12}S_{23} - S_{13}S_{22}) \\
m_3^{(2)} &= m_3^{(2)}\Gamma_2 S_{22} + S_{31} + \Gamma_2(S_{21}S_{32} - S_{31}S_{22}) \\
\tilde{m}_3^{(2)} &= \tilde{m}_3^{(2)}\Gamma_2 S_{22} + S_{33} + \Gamma_2(S_{23}S_{32} - S_{22}S_{33}) \\
m_1^{(3)} &= m_1^{(3)}\Gamma_3 S_{33} + S_{11} + \Gamma_3(S_{13}S_{31} - S_{11}S_{33}) \\
\tilde{m}_1^{(3)} &= \tilde{m}_1^{(3)}\Gamma_3 S_{33} + S_{12} + \Gamma_3(S_{13}S_{32} - S_{12}S_{33}) \\
m_2^{(3)} &= m_2^{(3)}\Gamma_3 S_{33} + S_{21} + \Gamma_3(S_{31}S_{23} - S_{21}S_{33}) \\
\tilde{m}_2^{(3)} &= \tilde{m}_2^{(3)}\Gamma_3 S_{33} + S_{22} + \Gamma_3(S_{23}S_{32} - S_{22}S_{33})
\end{aligned} \tag{10}$$

for the first set of external reflections and similarly for the second set. A closer look to the equations shows that all terms which are non-linear in S_{ij} are sub-determinants of the scattering matrix $\mathbf{S}$ of a 3-port. This system of equations becomes linear if these sub-determinants are assumed as further variables Δ_k. Thus, one obtains 24 linear equations with 18 variables. Twelve equations result from the measurements with the first set of external reflections

$$\begin{aligned}
m_2^{(1)} &= m_2^{(1)}\Gamma_1 S_{11} + S_{22} + \Gamma_1\Delta_1 \\
\tilde{m}_2^{(1)} &= \tilde{m}_2^{(1)}\Gamma_1 S_{11} + S_{23} + \Gamma_1\Delta_2 \\
m_3^{(1)} &= m_3^{(1)}\Gamma_1 S_{11} + S_{32} + \Gamma_1\Delta_3 \\
\tilde{m}_3^{(1)} &= \tilde{m}_3^{(1)}\Gamma_1 S_{11} + S_{33} + \Gamma_1\Delta_4 \\
m_1^{(2)} &= m_1^{(2)}\Gamma_2 S_{22} + S_{11} + \Gamma_2\Delta_1 \\
\tilde{m}_1^{(2)} &= \tilde{m}_1^{(2)}\Gamma_2 S_{22} + S_{13} + \Gamma_2\Delta_5 \\
m_3^{(2)} &= m_3^{(2)}\Gamma_2 S_{22} + S_{31} + \Gamma_2\Delta_6 \\
\tilde{m}_3^{(2)} &= \tilde{m}_3^{(2)}\Gamma_2 S_{22} + S_{33} + \Gamma_2\Delta_7 \\
m_1^{(3)} &= m_1^{(3)}\Gamma_3 S_{33} + S_{11} + \Gamma_3\Delta_4 \\
\tilde{m}_1^{(3)} &= \tilde{m}_1^{(3)}\Gamma_3 S_{33} + S_{12} + \Gamma_3\Delta_8 \\
m_2^{(3)} &= m_2^{(3)}\Gamma_3 S_{33} + S_{21} + \Gamma_3\Delta_9 \\
\tilde{m}_2^{(3)} &= \tilde{m}_2^{(3)}\Gamma_3 S_{33} + S_{22} + \Gamma_3\Delta_7 \ ,
\end{aligned} \tag{11}$$

and twelve further equations result from the measurements with the second set of external reflections $\tilde{\Gamma}_i$ and waves $n_i^{(j)}$ measured by the VNA

$$\begin{aligned}
n_2^{(1)} &= n_2^{(1)}\tilde{\Gamma}_1 S_{11} + S_{22} + \tilde{\Gamma}_1\Delta_1 \\
\tilde{n}_2^{(1)} &= \tilde{n}_2^{(1)}\tilde{\Gamma}_1 S_{11} + S_{23} + \tilde{\Gamma}_1\Delta_2 \\
n_3^{(1)} &= n_3^{(1)}\tilde{\Gamma}_1 S_{11} + S_{32} + \tilde{\Gamma}_1\Delta_3 \\
\tilde{n}_3^{(1)} &= \tilde{n}_3^{(1)}\tilde{\Gamma}_1 S_{11} + S_{33} + \tilde{\Gamma}_1\Delta_4 \\
n_1^{(2)} &= n_1^{(2)}\tilde{\Gamma}_2 S_{22} + S_{11} + \tilde{\Gamma}_2\Delta_1 \\
\tilde{n}_1^{(2)} &= \tilde{n}_1^{(2)}\tilde{\Gamma}_2 S_{22} + S_{13} + \tilde{\Gamma}_2\Delta_5 \\
n_3^{(2)} &= n_3^{(2)}\tilde{\Gamma}_2 S_{22} + S_{31} + \tilde{\Gamma}_2\Delta_6 \\
\tilde{n}_3^{(2)} &= \tilde{n}_3^{(2)}\tilde{\Gamma}_2 S_{22} + S_{33} + \tilde{\Gamma}_2\Delta_7 \\
n_1^{(3)} &= n_1^{(3)}\tilde{\Gamma}_3 S_{33} + S_{11} + \tilde{\Gamma}_3\Delta_4 \\
\tilde{n}_1^{(3)} &= \tilde{n}_1^{(3)}\tilde{\Gamma}_3 S_{33} + S_{12} + \tilde{\Gamma}_3\Delta_8 \\
n_2^{(3)} &= n_2^{(3)}\tilde{\Gamma}_3 S_{33} + S_{21} + \tilde{\Gamma}_3\Delta_9 \\
\tilde{n}_2^{(3)} &= \tilde{n}_2^{(3)}\tilde{\Gamma}_3 S_{33} + S_{22} + \tilde{\Gamma}_3\Delta_7 \ .
\end{aligned} \tag{12}$$

Thus, a linear system of equations for the calculation of the 3-port scattering parameters is obtained, which can be written in a compact matrix notation with the coefficient matrix $\mathbf{A}_\Delta$, which includes the measured waves $m_i^{(j)}$, $n_i^{(j)}$ and the reflections Γ_i, and the solution vector of the right hand side $\boldsymbol{m}_\Delta$, which includes the measured waves $m_i^{(j)}$ and $n_i^{(j)}$, as

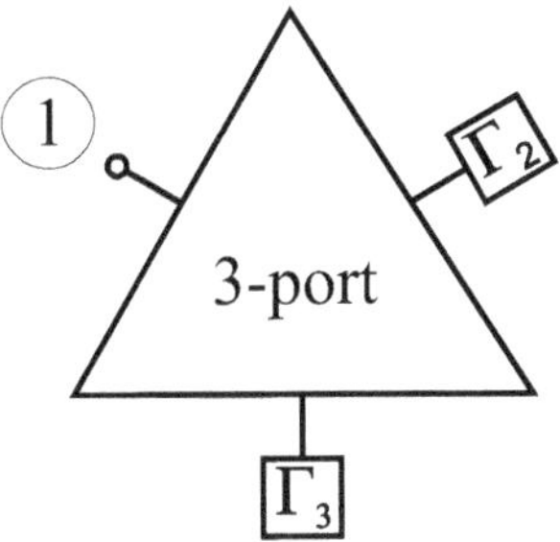

Fig. 2. One-port measurement.

follows:

$$\mathbf{A}_{\Delta} \cdot \mathbf{\Delta}_S = \boldsymbol{m}_{\mathbf{\Delta}} \quad . \tag{13}$$

Here the vector $\mathbf{\Delta}_S$ consists of the nine unknown scattering parameters $S_{11}, S_{12}, \cdots, S_{33}$ and additionally the nine unknown sub-determinants $\Delta_1, \Delta_2, \cdots, \Delta_9$. This system of equations is over-determined and can be solved with the help of a linear regression. Numerical experiments have shown that the sub-determinant method is less robust against disturbances or measurement errors as compared to the wave-identification method. This can be explained by the number of variables. While the wave-identification method yields a solution concernig nine variables the sub-determinant method is based on 18 variables, although the number of equations is the same in both cases. It is therefore a very successful strategy to use the sub-determinant method to create good starting values for the wave-identification method in order to reduce the number of necessary iterations, although the wave-identification method shows a very good convergence for each starting vector, for example the null vector.

5 The one-port method

For the described methods the assumption was made that the external reflections are known. In fact both methods presented above offer the possibility to obtain the value of all external reflections except for one which has to be known. Alternatively, the one-port method can be used for the characterization of the external reflections. For this purpose, additionally to the necessary two-port measurements two one-port measurements yield the values of all external reflections. This so-called one-port method is performed by connecting one port of the multi-port to the network analyzer and all other ports to their external reflections. The resulting input reflection can be linked with the corresponding two-port measurement for the characterization of the external reflections, also known as de-embedding (Bauer et al., 1974). Figure 2 shows the one-port method applied to port 1. The input reflection ϱ_{11} given by the one-port measurement can be inserted into the equation of the scattering matrix $\hat{\mathbf{S}}$ of the two-port between port 1 and port 2 in the following way

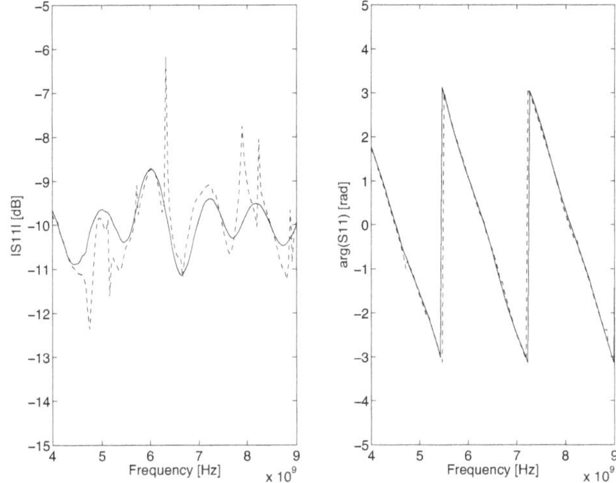

Fig. 3. Measurement results with a single set of external reflections with opens (−−) and with matches (—).

$$\begin{pmatrix} \varrho_{11} \\ \hat{b}_2 \end{pmatrix} = \begin{pmatrix} \hat{S}_{11} & \hat{S}_{12} \\ \hat{S}_{21} & \hat{S}_{22} \end{pmatrix} \cdot \begin{pmatrix} 1 \\ \Gamma_2 \hat{b}_2 \end{pmatrix} \quad . \tag{14}$$

Thus, it holds for the unknown external reflection Γ_2 after the elimination of $\hat{b}_2$

$$\Gamma_2 = \frac{\varrho_{11} - \hat{S}_{11}}{\varrho_{11}\hat{S}_{22} + (\hat{S}_{12}\hat{S}_{21} - \hat{S}_{11}\hat{S}_{22})} \quad . \tag{15}$$

Furthermore this one-port measurement can be linked to the two-port measurement between port 1 and port 3 which offers the possibility to define the unknown external reflection Γ_3. Thus, at least two one-port measurements are necessary to identify all external reflections independently of the number of ports.

6 Results

The measured device which was used to verify the different methods is a 3-port signal divider, consisting of three lines connected in the form of a T-junction, with open circuits as external reflections. The results produced with the multi-port method show some singularities. This can be explained by the isolation of the measurement ports if the length of the third line connected to an open circuit is equal to an odd multiple of a quarter wavelength. Figure 3 shows the results for the scattering parameter S_{11} of the multi-port method and additionally the results achieved with matched terminations as external reflections. It can clearly be seen that it is not possible to characterize this 3-port with a single set of unknown external reflections, because the multi-port method shows measurement singularities in contrast to the results produced with matched terminations as external reflections which have a smooth behaviour versus frequency.

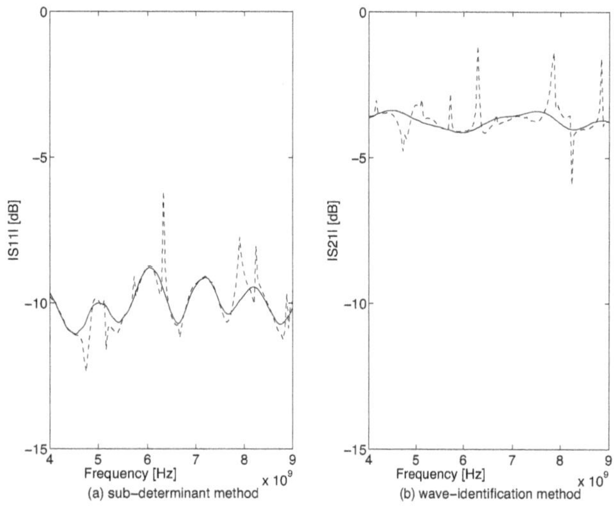

Fig. 4. Measurement results with a single set (– –) and a double set (—) of external reflections.

A further measurement performed with two sets of external reflections, namely three open circuits and three short circuits, is shown in Fig. 4.

In Fig. 4a the wave-identification method with a double set of external reflections is compared to the results achieved with a single set of external reflections, in this case three open circuits. While a single set of external reflections yields several measurement singularities as expected the wave-identification method offers a very smooth behaviour. Part (b) of Fig. 4 shows a comparison of the sub-determinant method and with the results for a single set of external reflections. Similarly to the wave-identification method the sub-determinant method shows a very smooth behaviour in contrast to the singularities of the measurements with a single set of external reflections.

7 Conclusions

In conclusion it can be stated that both described methods offer the possibility to characterize every multi-port, if a double set of external reflections is used. In fact the use of a second set of external reflections doubles the number of necessary measurements, but it yields a solution for the identification of every multi-port independent of its structure and of the values of the external reflections. Thus, no distinction of different cases is needed even if several ports become isolated. Furthermore the one-port method allows to deal with all external reflections as unknowns with only two further measurements. A very robust method is to combine both presented methods in such a manner that the sub-determinant method provides the starting values for the wave-identification method. This approach allows to apply the wave-identification method with just one iteration.

References

Bauer, R. F. and Penfield, P.: De-Embedding and Unterminating, in: IEEE Trans. Microw. Theory Tech., vol. 22, pp. 282–288, March, 1974.

Engen, G. F. and Hoer, C. A.: Thru-Reflect-Line: An improved technique for calibrating the dual six port automatic network analyzer, in: IEEE Trans. Microw. Theory Tech., vol. 27, pp. 987–993, December, 1979.

Eul, H.-J. and Schiek, B.: A Generalized Theory and New Calibration Procedures for Network Analyzer Self-Calibration, in: IEEE Trans. Microw. Theory Tech., vol. 39, pp. 724–731, April, 1991.

Lu, H.-C. and Chu, T.-H.: Multiport Scattering Matrix Measurement Using a Reduced-Port Network Analyzer, in: IEEE Trans. Microw. Theory Tech., vol. MTT-51, pp. 1525–1533, May, 2003.

Rolfes, I. and Schiek, B.: Multiport Method for the Measurement of the Scattering Parameters of N-Ports, in: IEEE Trans. Microw. Theory Tech., vol. 53, pp. 1990–1996, June, 2005.

Tippet, J. C. and Speciale, R. A.: A Rigorous Technique for Measuring the Scattering Matrix of a Multiport Device with a 2-Port Network Analyzer, in: IEEE Trans. Microw. Theory Tech., vol. MTT-30, pp. 661–666, May, 1982.

21

Implementation and evaluation of coherent synthetic aperture radar processing for level measurements of bulk goods with an FMCW-system

M. Vogt[1], M. Gerding[2], and T. Musch[2]

[1]Forschungsgruppe Hochfrequenztechnik, Ruhr-Universität Bochum, 44780 Bochum, Germany
[2]Lehrstuhl für Elektronische Schaltungstechnik, Ruhr-Universität Bochum, 44780 Bochum, Germany

Abstract. In industrial process measurement instrumentation, radar systems are well established for the measurement of filling levels of liquids in tanks. Level measurements of bulk goods in silos, on the other hand, are more challenging because the material is heaped up and its surface has typically a relatively complex shape. In this paper, the application of synthetic aperture radar (SAR) reconstruction with a frequency modulated continuous wave (FMCW) radar system for level measurements of bulk goods is evaluated. In the proposed monostatic setup, echo signals are acquired at discrete antenna positions on top of the silo. Spatially resolved information about the surface contour of a bulk good heap is reconstructed by coherent 'delay and sum' processing. The concept has been experimentally evaluated with a 24 to 26 GHz FMCW radar system mounted on a linear stepping motor positioning unit. Measurements on a thin metal wire at different range and on a curved test-object with a diffusely scattering surface have been performed to analyze the system's point spread function (PSF) and performance. Constant range and azimuth resolutions ($-6\,\mathrm{dB}$) of 15 cm and 8 cm, respectively, have been obtained up to a range of 6 m, and results of further evaluations show that the proposed concept allows more accurate and reliable level reconstructions of surface profiles compared to the conventional approach with measurements at a single antenna position.

1 Introduction

The conventional approach in radar and time-domain reflectometry (TDR) level measurement is to perform echo measurements along a fixed radiation beam with an antenna at a fixed position on top of a tank or silo (Reindl et al., 2001; Gerding et al., 2003, 2006; Rabe et al., 2009). In the case of liquids, this technique usually delivers adequate and sufficiently good results, because the surface of the liquid is planar and the surface level is the only relevant parameter. Level measurements on bulk goods are more challenging because the material is heaped up and its surface profile is aimed with the measurement.

Figure 1 shows the scenario given with radar level measurements of *liquids* in tanks in comparison to *bulk goods* in silos. In Fig. 1a, microwaves are emitted from an antenna at a *fixed* position on top of the tank and are *reflected* at the planar boundary between the air and the liquid. From the time of flight (TOF) of propagating waves, the distance between the antenna and the boundary is measured, and the filling level H of the liquid can be calculated by taking the overall height of the tank into account. Different from ultrasound based systems, radar level measurements are largely independent from the composition of the air, fog, temperature, pressure, etc., and high precision measurements can be performed. A single-point measurement delivers sufficient information in the case of level measurements of liquids, because only a *single* parameter, the filling level H, is relevant.

Level measurements on *bulk goods*, on the other hand, are much more challenging because the material is heaped up and its surface typically builds a complicated shape, as is shown in Fig. 1b. Microwaves are *diffusely scattered* at the boundary between the air and the bulk good, and, in general, penetrate into the bulk good and are diffusely scattered inside the heap. Typically, the amplitude of echo signals from bulk goods is much smaller compared to echoes from strong reflecting surfaces like boundaries between air and water. With respect to the measurement problem, which is to assess the bulk good volume inside the silo, a single-point measurement is insufficiently in many cases because of the more complicated geometry, see Fig. 1b.

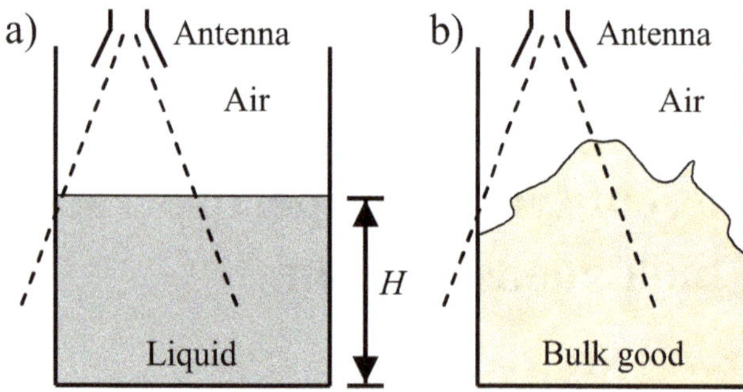

Fig. 1. Radar level measurement: **(a)** Liquid in tank, **(b)** Bulk good in silo.

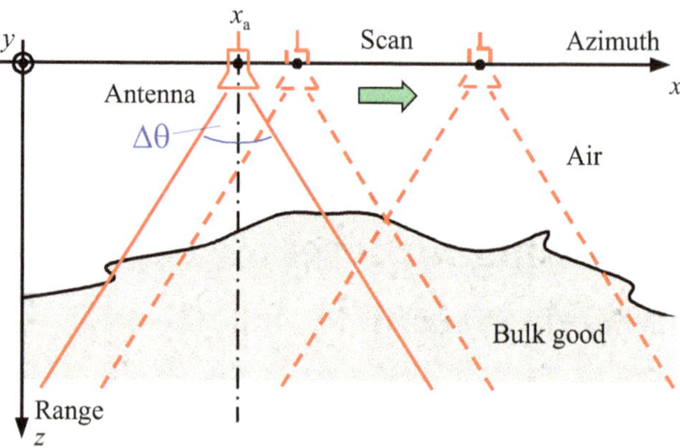

Fig. 2. Scanning radar system in monostatic system configuration.

In this paper, the utilization of synthetic aperture radar (SAR) processing for the imaging and assessment of surface profiles of bulk goods with a frequency modulated continuous wave (FMCW) radar system is evaluated. In the SAR approach, echo signals are acquired along a relatively long azimuth scanning path with a single antenna, and signals are superimposed after correction for the TOFs the waves need to travel on the round-trip between the antenna and each individual image point to be reconstructed. Coherent 'delay and sum' processing in the time-domain is applied for high-resolution reconstruction of information about reflection and scattering properties of objects. Calculations and measurements have been performed for the specific test-setup presented below.

2 Scanning radar system

Echo measurements at *multiple* antenna positions on top of a silo are proposed as a concept to cope with the aforementioned problems. SAR processing is applied to reconstruct backscattering and reflection images of the bulk goods from acquired echo signal data (Brenner and Roessing, 2008; Meta et al., 2007; Charvat et al., 2006; Ermert and Karg, 1979).

2.1 System setup

In the monostatic radar system configuration in Fig. 2, microwaves are emitted from a horn antenna, and backscattered and reflected waves are received by the same antenna.

Echo signals at discrete antenna positions x_a are sampled, digitized and stored for SAR processing and the reconstruction of two-dimensional (2-D) images of backscattering and reflecting structures in the 2-D x/z image plane.

2.2 System's point spread function (PSF)

Below, time-domain receive signals over TOF t at aperture positions x_a for a single point-object, i.e. a point-like radar target, which is small against the wavelength, at a position (x',z') are denoted as the system's point spread function (PSF) $h(t,x_a,x',z')$ (Vogt et al., 2010). For point-objects in the far field of the antenna, the PSF can be described as follows, see Fig. 2:

$$\begin{aligned} &h(t,x_a,x',z') = s(t-2/c\cdot r(x_a,x',z')) \\ &\text{with}: r(x_a,x',z') = \sqrt{z'^2+(x_a-x'^2)} \end{aligned} \quad (1)$$

In Eq. (1), $r(x_a,x',z')$ is the radial distance between the antenna and the point-object, c the speed of light, and the pulsed signal $s(t)$ is the impulse response of the band-pass radar system. Both, the radial distance $r(x_a,x',z')$ as well as the PSF $h(t,x_a,x',z')$, are *invariant* against translations of the point-object along the azimuth direction x (Vogt et al., 2010):

$$\begin{aligned} r(x_a,x',z') &= r'(x_a-x',z') \\ h(t,x_a,x',z') &= h'(t,x_a-x',z') \end{aligned} \quad (2)$$

Echo signals from any backscattering or reflecting radar target can be described based on the composition of objects by discrete point-objects and the corresponding superposition of PSFs, as long as multiple reflections are negligible and the radar system is linear (Opretzka et al., 2010).

3 SAR reconstruction

In the implemented radar system, SAR processing is performed by 'delay and sum' operation in the time-domain. 2-D backscattering and reflection images $b(x,z)$ are reconstructed from the acquired receive signals $s_R(t,x_a)$ at aperture positions x_a as follows (Vogt et al., 2010):

$$\begin{aligned} b(x,z) = |&\sum_{n=n_1}^{n_2} w(n-n_1) \\ &\cdot s_{R+}(t+\tau(x_{a,n},x,z),x_{a,n})|_{t=2/c\cdot z'} \\ \text{with}: \quad &\tau(x_{a,n},x,z,x_{a,n}) = 2/c\cdot(r(x_{a,n},x,z)-z') \end{aligned} \quad (3)$$

In Eq. (3), $s_{R+}(t)$ is the *analytical* receive signal obtained from the receive signal $s_R(t)$ by means of the Hilbert Transform. SAR reconstruction is performed by delaying the acquired receive signals at antenna positions $x_{a,n}$ by $\tau(x_{a,n}, x, z)$ and *coherent summation* of delayed signals after apodization with a weighting function $w(n)$. The backscattering and reflection image $b(x,z)$ is calculated as the magnitude of the *analytical* focused receive signal obtained after delay and sum operation. The summation in Eq. (3) is extended over receive signals from within the antenna's *main lobe* (aperture angle $\Delta\Theta$, see Fig. 2), which corresponds to a summation along a synthetic aperture of length $L_{SAR}(z) \approx z' \cdot \Delta\Theta$ along the azimuth direction x.

4 Evaluation of proposed imaging concept

The proposed imaging concept has been evaluated by measurements with an FMCW radar system, which was mounted on a linear stepping motor positioning unit.

4.1 FMCW radar system

In Fig. 3a, the block diagram of the utilized FMCW radar system (25 GHz center frequency, 2 GHz bandwidth, 8 ms sweep time, 1 MHz intermediate signal sampling frequency) is shown (Musch, 2003).

With this system, the transfer function $H(j\omega)$ of the radar path is assessed by the intermediate frequency (IF) echo signal $s_{IF}(t)$. The latter is obtained by mixing the frequency modulated transmit signal $s_T(t)$ with linearly swept instantaneous frequency $\omega(t)$ with the echo signal $s_E(t)$:

$$\begin{aligned} s_T(t) &= a_0 \cdot \cos(\varphi_T(t)),\ \omega(t) = 2\pi \cdot (f_1 + B/T \cdot t) \\ \varphi_T(t) &= 2\pi \cdot (f_1 \cdot t + b/(2 \cdot T) \cdot t^2) + \varphi_0 \\ s_E(t) &\approx a_0 \cdot |H(j\omega(t))| \cdot \cos(\varphi_T(t) + \arg(H(j\omega(t)))) \\ s_{IF}(t) &\approx k \cdot a_0 \cdot |H(j\omega(t))| \cdot \cos(\arg(H(j\omega(t)))) \\ s_{IF}(t) &\sim \mathrm{Re}\{H(j\omega(t))\} \end{aligned} \tag{4}$$

It can be seen in Fig. 3b that the echo signal $s_E(t)$ also shows a linearly swept instantaneous frequency $\omega_E(t)$, but with a constant delay τ for a *single* radar target. The receive signal $s_R(t)$ containing the impulse response $h(t)$ of the radar path is obtained from the Fourier Transform of $s_{IF}(t)$ as follows, see Eq. (4):

$$\begin{aligned} S_R(\omega) &= s_{IF}(t)_{t \to \omega(t)} = \mathrm{Re}\{H(j\omega(t))\} \\ S_R(\omega) &= \frac{1}{2} \cdot \left(H(j\omega) + H^*(j\omega)\right) \\ s_R(t) &= \frac{1}{2} \cdot \left(h(t) + h^*(-t)\right) \end{aligned} \tag{5}$$

4.2 Measurement setup

The FMCW radar system was equipped with a conical horn antenna with 75 mm aperture diameter and mounted on a linear stepping motor positioning unit. Echo measurements have been performed at 256 equidistant antenna positions with a distance of 6.6 mm from each other (55% of wavelength at 25 GHz center frequency), covering a scanning path with a length of 1.7 m. A thin metal wire (3 mm diameter) has been imaged at different range in order to assess the system's PSF. Furthermore, a test-object with a curved profile along the azimuth scanning direction and a diffusely scattering surface has been utilized to evaluate the proposed concept with respect to applications in bulk good level measurements.

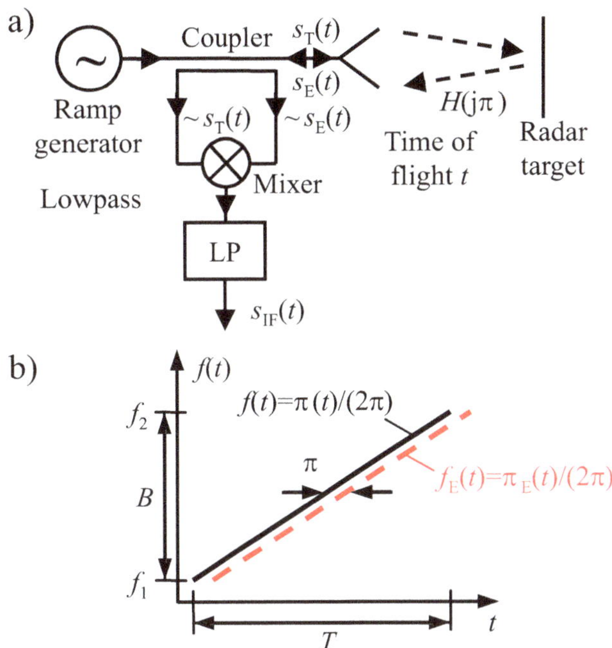

Fig. 3. FMCW radar system: **(a)** Block diagram, **(b)** Instantaneous frequencies $f(t)$ and $f_E(t)$ of transmit and echo signal, respectively, for a single radar target.

5 Measurement results

In Fig. 4a, a schematic of the laboratory environment for the PSF measurements with concrete walls at the sides of the laboratory room are shown.

The metal wire was positioned along the elevational direction y at a chosen distance z in range direction mounted on a tripod with an arm, and scanned along the azimuth direction x. Under these conditions, the metal wire is a point-like object in the x/z azimuth/range image plane. Consequently, the system's PSF can be analyzed by echo measurements from the wire positioned at different range z. In the *unfocused* 2-D backscattering and reflection image $b(x,z)$ in Fig. 4b, the envelope of the receive signals *without* delay and sum focusing is shown over a dynamic range of 30 dB. Echoes from the metal wire at a range of 5.5 m are spread along the azimuth

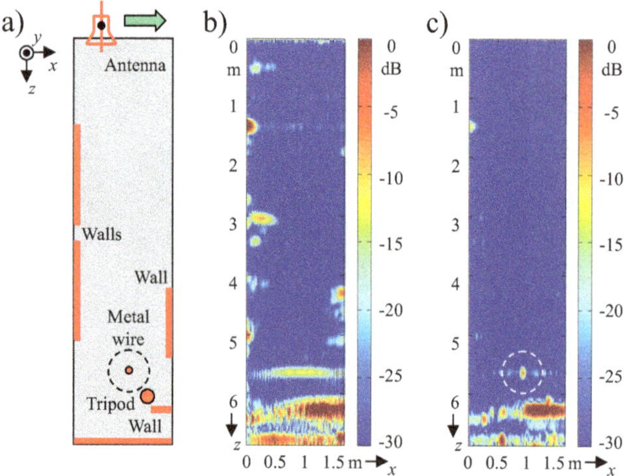

Fig. 4. Point spread function (PSF) measurement: **(a)** Schematic of measurement environment (laboratory), **(b)** Unfocused radar image, **(c)** Focused radar image.

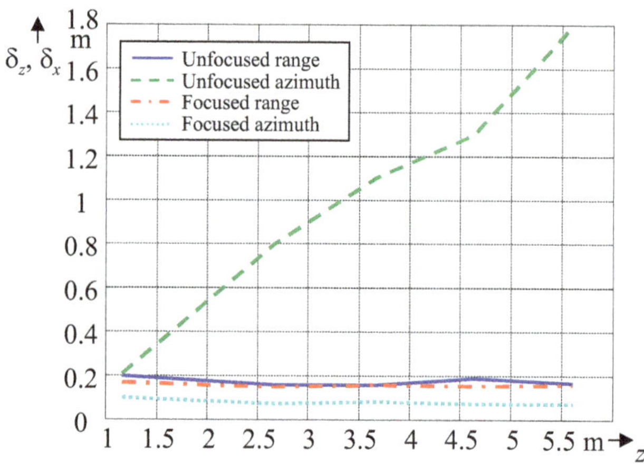

Fig. 5. Results of point spread function (PSF) measurements: Range resolution δ_z and azimuth resolution δ_x (FWHM) in unfocused and focued radar images over range z.

coordinate x due to the diverging beam pattern of the antenna in the far field. Furthermore, echoes from the corners of the side wall are visible, as well as strong reflections from the back wall. In Fig. 4c, the *focused* backscattering and reflection image $b(x,z)$ of the same scenario obtained by SAR processing is shown. Now, the result is a point-like, focused image of the metal wire in the x/z azimuth/range image plane, and echoes from the corners of the side walls are suppressed.

The -6 dB-widths of the echo from the wire along the azimuth and range coordinates x and z, respectively, have been assessed in cross-sectional images at different range z of the wire in order to analyze the system's range and azimuth resolutions δ_z and δ_x, respectively (Opretzka et al., 2010). Results are shown in Fig. 5.

Almost constant range resolutions (-6 dB) of 15 cm are obtained in the unfocused and focused images up to a range of 6 m (solid line and dotted/dashed line). This is in good agreement with the expectation that the range resolution only depends on the pulse-width of the system's impulse response, which is inversely proportional to the constant bandwidth of the system of 2 GHz. The azimuth resolution in unfocused images (dashed line), on the other hand, is proportional to depth, because of the diverging beam pattern (constant aperture angle $\Delta\Theta$) in the far field of the antenna. As indicated by the example in Fig. 4c, a much smaller and almost constant azimuth resolution (-6 dB) of 8 cm is obtained in focused images up to a range of 6 m with the implemented system (dotted line).

The setup shown in Fig. 6a has been used to evaluate the proposed concept in the context of level measurements on bulk goods. A test-object with a curved profile along the azimuth scanning direction and a diffusely scattering surface has been created by gluing a crunched aluminum foil on a paperboard. This structure has been bended along the azimuth coordinate x, as is shown in Fig. 6a, and was positioned at a mean range of about 5.3 m.

The *unfocused* 2-D backscattering and reflection image $b(x,z)$ in Fig. 6b again shows the envelope of receive signals *without* delay and sum focusing. It can be seen that the image contains information about the curved surface profile of the test-object. As a result of the superposition of diffusely backscattered coherent waves, 'speckle' patterns appear in the image. Furthermore, because of the relatively bad azimuth resolution, echoes from the surface of the test object along the azimuth coordinate, i.e. echoes from different depths, are superimposed in the unfocused image. Consequently, a precise detection of the object's surface is difficult. In the *focused* image in Fig. 6c of the same scenario, the speckle and the surface profile are less spread over range and azimuth directions compared to the unfocused image in Fig. 6b. Consequently, the surface of the object can be more easily and more accurately detected from the focused image.

6 Summary and conclusions

In this paper, the application of SAR processing has been proposed and evaluated to cope with the problem of radar level measurements of bulk goods in silos, which is challenging compared to level measurements of liquids in tanks. Echo measurements with a single antenna in a monostatic configuration along a 1-D scanning path have been discussed as a concept for the reconstruction of 2-D backscattering and reflection images of bulk goods. SAR processing in the time-domain by 'delay and sum' operation has been suggested for this purpose. The concept has been evaluated by measurements with a 24 to 26 GHz frequency range FMCW radar system, which was equipped with a conical

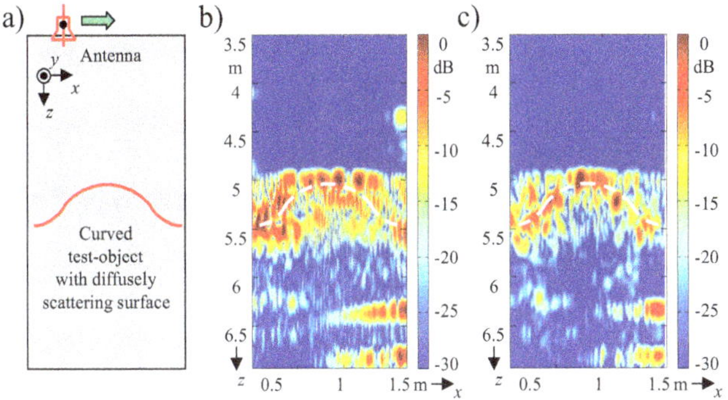

Fig. 6. Measurement on curved test-object: **(a)** Schmeatic of measurement environment (laboratory), **(b)** Unfocused radar image, **(c)** Focused radar image.

horn antenna with 75 mm aperture diameter. In the implemented system, the antenna is scanned along the azimuth coordinate by means of a linear stepping motor positioning unit, and 2-D backscattering and reflection images inside the azimuth/range image plane are reconstructed. Based on echo measurements on a metal wire, the system's PSF has been evaluated, showing that constant range and azimuth resolutions (-6 dB) of 15 cm and 8 cm, respectively, are achieved up to a range of 6 m. Results of measurements on a curved test-object with diffusely scattering surface show that the implemented system along with the proposed SAR concept allows to accurately assess curved profiles from reconstructed backscatter and reflection images. The goal of our future work is to extend the concept towards a 2-D scanning and 3-D SAR reconstruction and to further evaluate the proposed concept.

References

Brenner, A. R. and Roessing, L.: Radar imaging of urban areas by means of very high-resolution SAR and interferometric SAR, IEEE Trans. Geosc. Remote Sensing, 46(10), 2971–2982, 2008.

Charvat, G. L. and Kempel, L. C.: Synthetic aperture radar imaging using a unique approach to frequency-modulated continuous-wave radar design, IEEE Ant. Prop. Magazine, 48(1), 171–177, 2006.

Ermert, H. and Karg, R.: Multifrequency acoustical holography, IEEE Trans. Sonics, Ultrasonics, SU-26, 4, 279–286, 1979.

Gerding, M., Musch, T., and Schiek, B.: Precision level measurement based on time-domain reflectometry, Adv. Radio Sci., 1, 27–31, 2003, http://www.adv-radio-sci.net/1/27/2003/.

Gerding, M., Musch, T., and Schiek, B.: A novel approach for a high-precision multitarget-level measurement system based on time-domain reflectometry, IEEE Trans. Instr. Meas., 54(6), 2768–2773, 2006.

Meta, A., Hoogeboom, P., and Ligthart, L. P.: Signal processing for FMCW SAR, IEEE Trans. Geosc. Remote Sensing, 45(11), 3519–3532, 2007.

Musch, T.: A high precision 24-GHz FMCW radar based on a fractional-N ramp-PLL, IEEE Trans. Instr. Meas., 52(2), 324–327, 2003.

Opretzka, J., Vogt, M., and Ermert, H.: A model-based synthetic aperture imaging reconstruction technique for high-frequency ultrasound, IEEE 2009 Ultrason. Symp. Proc., 377–380, 2010.

Rabe, H., Denicke, E., Armbrecht, G., Musch, T., and Rolfes, I.: Considerations on radar localization in multi-target environments, Adv. Radio Sci., 7, 5–10, 2009, http://www.adv-radio-sci.net/7/5/2009/.

Reindl, L., Ruppel, C. W., Berek, S., Knauer, U., Vossiek, M., Heide, P., and Oréans, L.: Design, fabrication, and application of precise SAW delay lines used in a FMCW radar system, IEEE Trans. Microw. Theory Tech., 49(4), 787–794, 2001.

Vogt, M., Opretzka, J., and Ermert, H.: Synthetic aperture focusing technique for high-resolution imaging of surface structures with high-frequency ultrasound, IEEE 2009 Ultrason. Symp. Proc., 1514–1517, 2010.

22

Numerical dosimetric calculations for in vitro field expositions in the THz frequency range

C. Jastrow, T. Kleine-Ostmann, and T. Schrader

Physikalisch-Technische Bundesanstalt, Braunschweig and Berlin, Germany

Abstract. Field exposition experiments have been initiated by the German Federal Office for Radiation Protection (Bundesamt für Strahlenschutz – BfS) to examine genotoxic effects of THz radiation in vitro. Two different human skin cell types are exposed to continuous-wave radiation at six distinct frequencies between 100 GHz and 2.52 THz originating from different sources of THz radiation under defined environmental conditions. The cell containers are irradiated with free space power flux densities between 0.1 mW/cm^2 and 2 mW/cm^2 measured traceable to the SI units. For meaningful results, dosimetric calculations using the finite differences time-domain method have been performed in order to access the fields and consequently the specific absorption rate (SAR) in the cell layer.

1 Introduction

With the increasing amount of applications utilizing THz radiation appearing on the market (e.g. communication links, Piesiewicz et al., 2007, spectroscopy and quality control inspection systems, and security screening systems, Siegel, 2002) the question of health protection in non-ionizing electromagnetic fields arises for sub-mm radiation, also (Kleine-Ostmann et al., 2006). Like in the frequency range below, it is still not clear whether non-ionizing RF radiation below the thermal damage threshold could cause detrimental effects in living organisms. The International Commission for Non-Ionizing Radiation Protection (ICNIRP) limits the power flux density for general public exposure for the frequency range between 2 GHz and 300 GHz to 1 mW/cm^2 (ICNIRP, 1998, Physical Agents Directive 2004/40/EC, 2004, NiSG, 2009). The limit is based on the thermal damage threshold which has been examined extensively in the microwave frequency range, only. Above 300 GHz no such limits for public exposure exist and safety limits concern laser radiation, only. Depending on the specific laser source, the safety limits are in the range between 1 mW/cm^2 and 100 mW/cm^2 (ANSI,

(a)

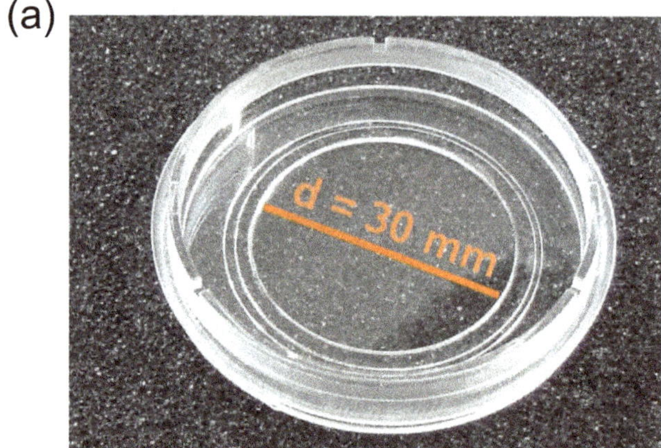

(b)

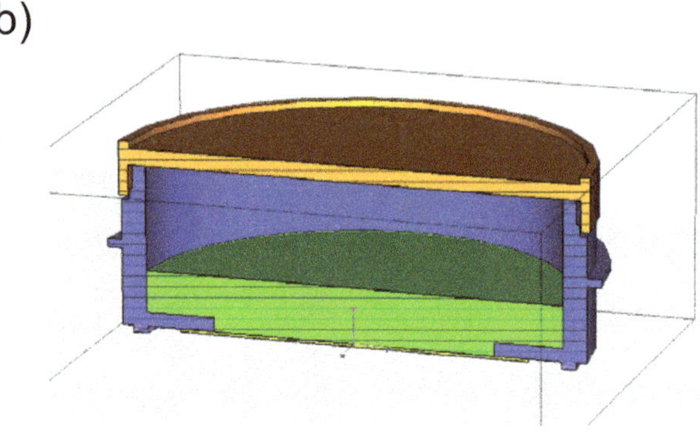

Fig. 1. **(a)** Sample container for field exposition experiments with skin cell lines (lid removed). **(b)** Computer model (CST Microwave Studio) for SAR calculation in a cell container filled with culture medium.

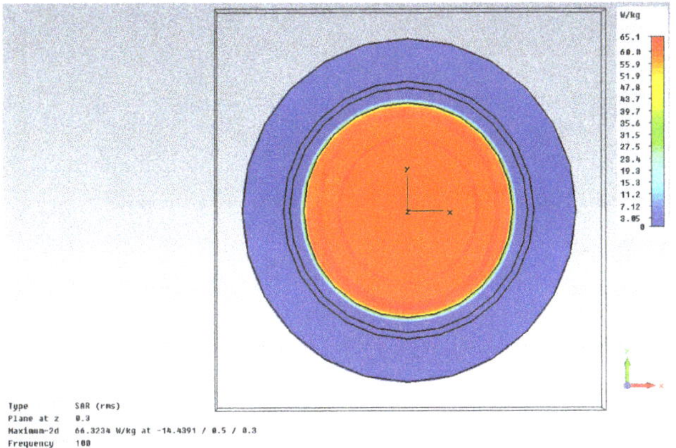

Fig. 2. Results of the SAR calculation (0.001 g RMS) at 100 GHz in the cell layer for a plane wave with 10 mW/cm^2 travelling in the z-direction. The cut plane is located 100 μm above the bottom. Results are valid both for isothermic and open boundary conditions.

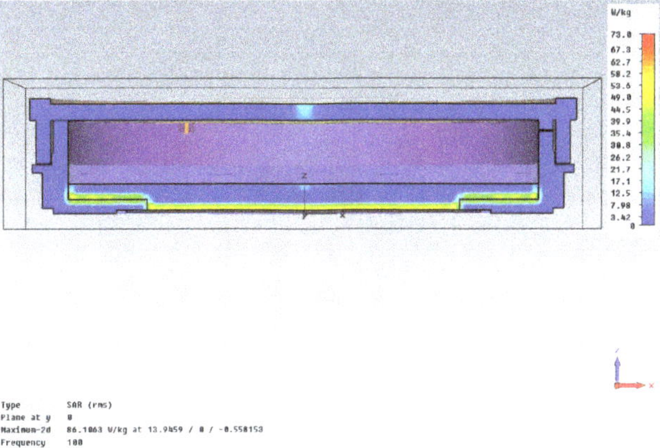

Fig. 3. Results of the SAR calculation (0.001 g RMS) at 100 GHz in the cell container for a plane wave with 10 mW/cm^2 travelling in the z-direction. The cut plane is a cross section through the center. Results are valid both for isothermic and open boundary conditions.

2007). An early study by Berry et al. (2002) shows that human exposure due to existing THz imaging systems is usually well below 1 mW/cm^2. However, the non-thermal effects of non-ionizing radiation are still discussed controversially. In the THz frequency range, only one comprehensive study exists so far (THz-Bridge, 2004). Korenstein-Ilan et al. (2008) found increased genomic instability in human lymphocytes for exposure to 100 GHz radiation well below the safety limit (0.031 mW/cm^2 in culture medium). Initiated by the BfS field exposition experiments to examine genotoxic effects of THz radiation in vitro are underway (Kleine-Ostmann et al., 2009). Under defined environmental conditions, two different human skin cell types are exposed to continuous-wave radiation at distinct frequencies between 100 GHz and 2.52 THz originating from different sources of THz radiation. The cell containers are irradiated with free space power densities between 0.1 mW/cm^2 and 2 mW/cm^2 measured traceable to the SI units (Kleine-Ostmann et al., 2008). In order to quantify and reliably evaluate the stress imposed on the living cells, it is necessary to determine the SAR within a monolayer of cells covered by culture medium in a cell container. In this contribution we show first dosimetric calculations to quantify the power introduction into the cell layer. The numerical calculations have been performed using the finite-differences time-domain method as implemented in the program CST Microwave Studio (2009). A detailed geometric model of the sample container and realistic dielectric properties of the materials are used to obtain reliable SAR values at 100 GHz. The results indicate that heating effects probably cannot be neglected at higher exposure levels when evaluating the outcome of the experiments. The calculated heat introduction is in good compliance with the temperature profile measured during field exposition using a fibre-coupled thermometer.

2 Field exposition experiments

In the project, two different human skin cell types, i.e. the Human adult low Calcium Temperature (HaCaT) keratinocyte cell line and primary dermal fibroblasts, are exposed to THz radiation, since the penetration depth into the human body above 100 GHz is well below 1 mm (THz-Bridge, 2004). The cells are cultivated in culture medium (Dulbecco's Modified Eagle Medium – DMEM) in a custom built sample container (ibidi GmbH) as shown in Fig. 1. The bottom of this container consists of a 200 μm thick foil which is transparent in the frequency range of interest (due to its low dielectric loss angle) on which the cells are fixed as a monolayer in a spot of 12 mm diameter in the center. The cells are irradiated from below in a modified incubator at six distinct frequencies of 100 GHz, 130 GHz, 385 GHz, 604 GHz, 1.63 THz and 2.52 THz to verify the results from the previous study and to cover a larger part of the THz frequency range. Four different sources of THz radiation are used to irradiate the exposition zone with a 2 cm diameter Gaussian beam (full width half maximum) from below, respectively: a frequency multiplier cascade at 100 GHz, a Gunn oscillator at 130 GHz, a backward-wave oscillator at 385 GHz and a far-infrared gas laser at 604 GHz, 1.63 THz and 2.52 THz. For each source, three power flux densities below (0.1 mW/cm^2), at (1 mW/cm^2) and above (2 mW/cm^2) the limit are chosen for the exposure. All power densities will be measured traceable to the SI units in order to allow for a precise and reliable assessment of the exposure dose. The traceable measurements were initially performed using a bolometer calibrated in front of a black body radiation source (Kleine-Ostmann et al., 2008). Later, to improve the reliability, a gas pressure membrane sensor traceable to an AC voltage (Thomas Keating Instruments, 2009) and a pyro-electric detector calibrated with a cryogenic substitution radiometer (Werner et al., 2009) were used.

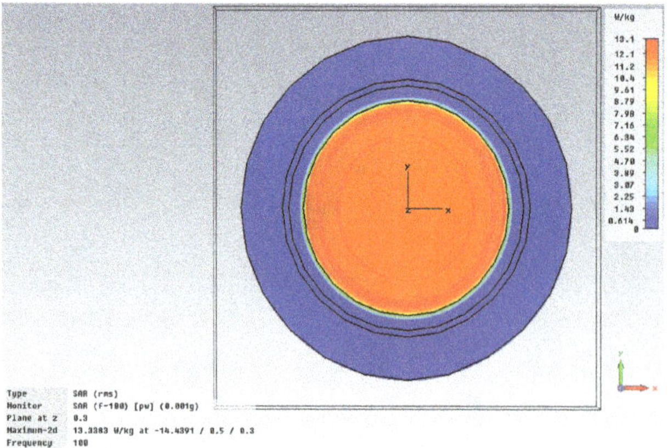

Fig. 4. Results of the SAR calculation (0.001 g RMS) at 100 GHz in the cell layer for a plane wave with 2 mW/cm^2 travelling in the z-direction. The cut plane is located 100 μm above the bottom. Results are valid both for isothermic and open boundary conditions.

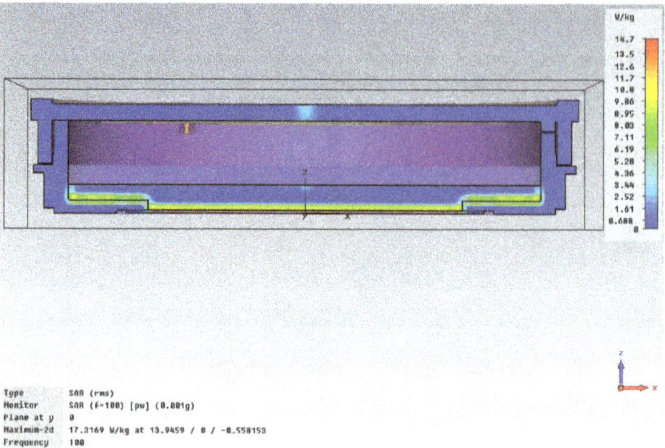

Fig. 5. Results of the SAR calculation (0.001 g RMS) at 100 GHz in the cell container for a plane wave with 2 mW/cm^2 travelling in the z-direction. The cut plane is a cross section through the center. Results are valid both for isothermic and open boundary conditions.

During exposition, the sample containers are kept at defined environmental conditions of approximately 37 °C and 5% CO_2 content of the atmosphere in a modified incubator as required by the cells. The temperature of the incubator is reduced to 36 °C (1 K below the targeted value) to account for a certain heating of the cells during illumination with RF radiation. All experimental parameters are permanently monitored during the exposition experiments. Two different exposure times of 2 h and 8 h are chosen. The comet assay is used as biological end point for strand breaks (Singh, 1988), whereas the micronucleus test is chosen to search for possible aneugenic and clastogenic effects (Matter and Schmid, 1971). The validity of the results is guaranteed by positive controls, sham exposition experiments and a blinded experimental procedure. In order to obtain statistically significant results, four independent experimental series are conducted. Three different independent exposition series will be performed to validate the results from an initial run. From each exposed cell container, two different samples will be taken for the evaluation of the end points.

3 Dosimetric calculations

First dosimetric calculations to quantify the power introduction into the cell layer have been performed using the finite-differences time-domain method as implemented in the program CST Microwave Studio (2009). Due to the small vacuum wavelength of the THz radiation compared to the size of the cell container ranging from 3 mm at 100 GHz to 0.119 mm at 2.52 THz, the calculations are strongly limited by computational resources. With a double quad-core 3.2 GHz processor machine and 64 GByte RAM the calculations were restricted to approximately 140 million voxels. SAR calculations with our model using 35 million voxels (restricted to one core) took about two weeks, each. Therefore, initial calculations had to concentrate on the lowest frequency of 100 GHz used for the field exposition experiments. Due to program restrictions the excitation had to be simplified assuming a planar wave of given power flux density applied from below. The three dimensional model of the sample container is shown in Fig. 1b). Both, cell layer and DMEM culture medium are assumed to have similar properties and have been modelled with ε_r=5.98 and tanδ=1.83 using the dielectric properties measured for the culture medium using THz time-domain spectroscopy (Pupeza et al., 2007). Dish, bottom foil and lid of the sample container are modelled using the dielectric properties specified by the manufacturer for 100 GHz (dish: ε_r=2.28 and tanδ=0.0001, foil: ε_r=2.34 and tanδ=0.005, lid: ε_r=3 and tanδ=0.01). For the lateral surfaces, open boundary conditions have been chosen, always, since the sample container is surrounded by a heat-insulating Rohacell™ foam. In contrast to this, top and bottom of the sample container are modelled with open or isothermic boundary conditions corresponding to a good thermal insulation or good heat transfer by convection of the surrounding air, respectively.

The calculated SAR distributions turn out to be identical for both open and isothermic boundary conditions which is consistent with the SAR value representing the power absorption, only. Figures 2 and 3 show the SAR distributions at 100 GHz in two perpendicular cut planes after irradiation of the sample container from below with a homogeneous wave having a power flux density of 10 mW/cm^2. The SAR distributions represent RMS averages over very small sample volume masses of 0.001 g. For this excitation, the maximum SAR value in the layer of the cells reaches values up to 66.3 W/kg. Since this is much above a value of 2 W/kg considered to cause negligible heating during local body exposition at lower frequencies (ICNIRP, 1998) we

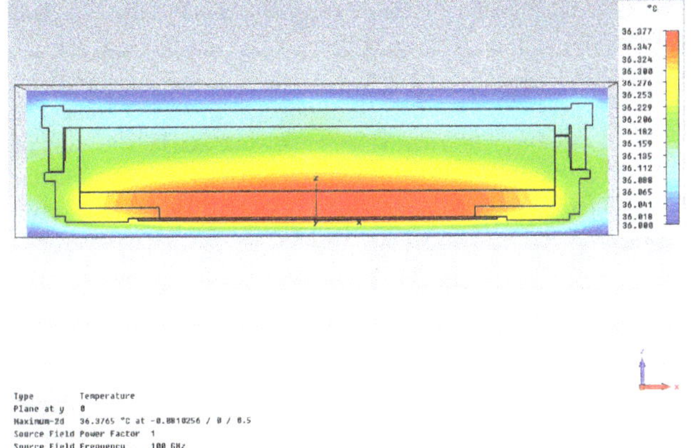

Fig. 6. Results of the heat distribution calculation at 100 GHz in the cell container for a plane wave with 2 mW/cm^2 travelling in the z-direction. The cut plane is a cross section through the center. Isothermic boundary conditions have been chosen above and underneath the sample container.

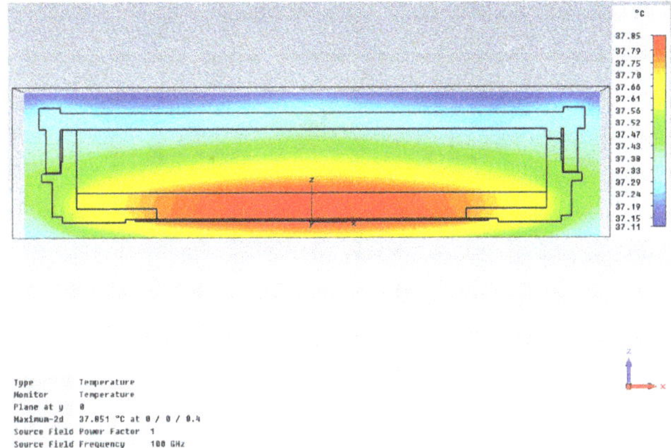

Fig. 7. Results of the heat distribution calculation at 100 GHz in the cell container for a plane wave with 2 mW/cm^2 travelling in the z-direction. The cut plane is a cross section through the center. Open boundary conditions have been chosen above and underneath the sample container.

expect considerable heating of the cells. Field exposition experiments above the power flux density limit were therefore performed at 2 mW/cm^2 to exclude possible thermal effects to influence the outcome of the experiments. Figures 4 and 5 show the SAR calculation for this power flux density. The maximum SAR value is now reduced to 13.34 W/kg. Within the numerical accuracy the maximum SAR value scales linearly with the applied power flux density.

While the SAR calculation is independent of the boundary conditions chosen for bottom and lid, the temperature distribution in the cell container will strongly depend on them. Figure 6 shows the temperature distribution in a cross section for the isothermic case, whereas Fig. 7 shows the results for the open boundary conditions, both for an excitation of 2 mW/cm^2. While the calculation with isothermic boundary conditions predicts a maximum temperature rise of 0.38 K (incubator temperature was considered to be 36 °C), the calculations with open boundary condition predicts up to 1.85 K.

4 Verification measurements

In order to verify the predictions of the model, we performed measurements of the time-dependent temperature in the DMEM culture medium during exposition using a metal-free and fiber-coupled electro-optic thermometer Fotemp4 (OptoCon, Dresden, Germany) and the probe TK5/2 (0.55 mm thickness) already used in a similar context, before (Schrader et al., 2008). Figure 8 shows time traces measured for two different power flux densities of approximately 30 mW/cm^2 and 3 mW/cm^2. The incubator temperature was set to 37 °C in this case. Both experiments show that the culture medium heats up from room temperature to a steady-state value well above the incubator temperature that depends on the power flux density. For the power flux density of 3 mW/cm^2 the THz beam was blocked temporarily during the experiment causing the temperature to decrease to the incubator temperature.

In Fig. 8 the acceptable temperature range between 36 °C and 40 °C in which no detrimental effect on the cell growth is expected is indicated as hatched area. The temperature curve in Fig. 8 indicates a temperature below 30 °C when the experiment is started. This is due to the fact that the sample had to be removed from another incubator before being measured in the exposition chamber. It can be seen, that a power flux density of approximately 30 mW/cm^2 during exposure leads to overheating of the cells, whereas a power flux density of 3 mW/cm^2 causes the temperature to stay in the acceptable range. In the case of irradiation with 3 mW/cm^2 the steady-state value of the temperature lies approximately 2.5 K above the incubator temperature. Compared to the calculation results shown in Figs. 6 and 7, this indicates that the model is valid.

5 Conclusions

In this contribution we have shown the experimental setups, the working plan and first dosimetric calculations for a comprehensive study on possible genotoxic effects of THz radiation as initiated by the German Federal Office for Radiation Protection. The dosimetric calculations show that SAR values up to 13.34 W/kg and a heating of the cell by 1.85 K have to be expected when irradiating the cell containers with 2 mW/cm^2. However, these values represent upper limits since the illumination of the samples with a Gaussian beam will transfer less heat into the sample and the sample

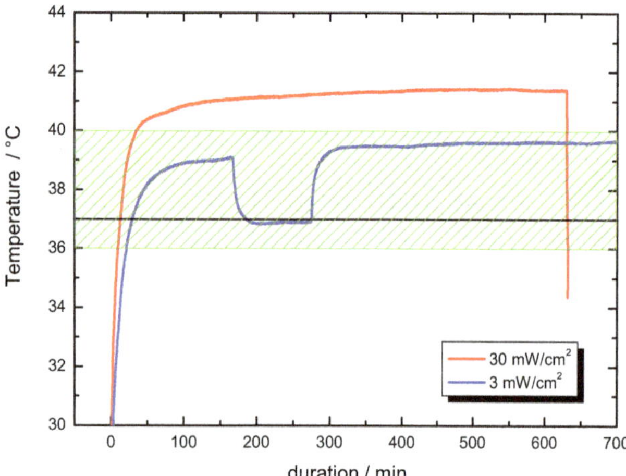

Fig. 8. Measurements of the time dependent temperature in the culture medium for two different irradiation power densities of approximately 30 mW/cm^2 and 3 mW/cm^2 using a fibre-optic semiconductor thermometer. The room-temperature sample has been put into the incubator set at 37 °C. For 3 mW/cm^2 the THz beam was blocked temporarily during the experiment. The hatched area indicates the temperature region in which no detrimental effect on the cell growth is expected.

container is not thermally insulated to the surrounding air as assumed by open boundary conditions. Temperature measurements in the culture medium during exposition show that the dosimetric model approximates realistic results. Field expositions are currently underway. Results of the experiments are expected in 2010.

Acknowledgements. The authors would like to thank the project team that is working closely together to perform the field exposition experiments: B. Heinen, K. Baaske, and M. Koch, all with the Institut für Hochfrequenztechnik at Technical University Braunschweig for performing the field exposition experiments above 600 GHz and H. Hintzsche and H. Stopper from the Department of Toxicology at the University of Würzburg and U. Kärst and J. Wehland from the Helmholtz Centre for Infection Research in Braunschweig for preparation and evaluation of the biological samples.

Furthermore the authors would like to thank E. Schmid from the Institute for Cell Biology at University Munich and A. Enders from the Institute for Electromagnetic Compatibility at Technical University Braunschweig for scientific advice.

References

ANSI Z136.1-2007, American National Standard for Safe Use of Lasers, 2007.

Berry, E., Walker, G. C., Fitzgerald, A. J., Zinov'ev, N. N., Chamberlain, M., Smye, S. W., Miles, R. E., and Smith, M. A.: Do in-vivo terahertz imaging systems comply with safety guidelines?, J. Laser Appl., 15, 192–198, 2002.

CST Microwave Studio, http://www.CST.de/, 2009.

ICNIRP Guidelines for limiting exposure to time varying electric, magnetic, and electromagnetic fields (up to 300 GHz), Health Phys., 74, 494–522, 1998.

Kleine-Ostmann, T., Münter, K., Spitzer, M., and Schrader, T.: The electromagnetic environment above 100 GHz: Electromagnetic compatibility, personal safety and regulation issues, Joint 31st International Conference on Infrared and Millimeter Waves and 14th IEEE International Conference on THz Electronics (IRMMW-THz 2006), Shanghai, China, 2006.

Kleine-Ostmann, T., Schrader, T., Bieler, M., Siegner, U., Monte, C., Gutschwager, B., Hollandt, J., Steiger, A., Werner, L., Müller, R., Ulm, G., Pupeza, I., and Koch, M.: THz Metrology, Frequenz, special issue on Terahertz Technologies and Applications, 62, 137–148, 2008.

Kleine-Ostmann, T., Jastrow, C., Salhi, M., Schrader, T., Hintzsche, H., Stopper, H., Kärst, U., Heinen, B., Baaske, K., and Koch, M.: In Vitro Field Exposition of Skin Cells between 100 GHz and 2.52 THz, Procedings of the 34th International Conference on Infrared, Millimeter and Terahertz Waves (IRMMW-THz 2009), Busan, Korea, 2009.

Korenstein-Ilan, A., Barbul, A., Hasin, P., Eliran, A., Gover, A., and Korenstein, R.: Terahertz Radiation Increases Genomic Instability in Human Lymphocytes, Rad. Research, 170, 224–234, 2008.

Matter, B. and Schmid, W.: Trenimon-induced chromosomal damage in bone-marrow cells of six mammalian species, evaluated by the micronucleus test, Mutat. Res., 12, 417–425, 1971.

NiSG-Gesetz zum Schutz vor nichtionisierender Strahlung bei der Anwendung am Menschen (NiSG), Bundesgesetzblatt Jahrgang 2009 Teil I Nr. 49, 2433–2435, 2009.

Physical Agents Directive 2004/40/EC: Directive of the European Parliament and of the Council on the minimum health and safety requirements regarding the exposure of workers to the risks arising from physical agents (electromagnetic fields), 2004.

Piesiewicz, R., Kleine-Ostmann, T., Krumbholz, N., Mittleman, D., Koch, M., Schöbel, J., and Kürner, T.: Short-Range Ultra Broadband Terahertz Communications: Concept and Perspectives, IEEE Antenn. Propag. M., 49, 24–39, 2007.

Pupeza, I., Wilk, R., and Koch, M.: Highly Accurate Optical Material Parameter Determination with THz Time Domain Spectroscopy, Opt. Express, 15, 4335–4350, 2007.

Schrader, T., Münter, K., Kleine-Ostmann, T., and Schmid, E.: Spindle Disturbances in Human-Hamster Hybrid (AL) Cells Induced by Mobile Communication Frequency Range Signals, Bioelectromagnetics, 29, 626–639, 2008.

Siegel, P. H.: THz Technology, IEEE Trans. Microwave Theory Tech. 50th Anniversary Issue, 50, 910–928, 2002.

Singh, N. P., McCoy, M. T., Tice, R. R., and Schneider, E. L.: A simple technique for quantitation of low levels of DNA damage in individual cells, Exp. Cell Res., 175, 184–191, 1988.

Thomas Keating Instruments, http://www.terahertz.co.uk/, 2009.

THz-Bridge: Tera-Hertz radiation in Biological Research, Investigation on Diagnostics and study of potential Genotoxic Effects, Final Report, http://www.frascati.enea.it/THz-BRIDGE, 2004.

Werner, L., Hübers, H.-W., Meindl, P., Müller, R., Richter, H., and Steiger, A.: Towards traceable radiometry in the terahertz region, Metrologia, 46, 160–164, 2009.

A torsional sensor for MEMS-based RMS voltage measurements

J. Dittmer[1,2], R. Judaschke[2], and S. Büttgenbach[1]

[1]Institute for Microtechnology, Technische Universität Braunschweig, Germany
[2]Physikalisch-Technische Bundesanstalt, Braunschweig, Germany

Abstract. RF voltage measurement based on electrostatic RMS voltage-to-force conversion is an alternative method in comparison to the conventional thermal power dissipation method. It is based on a mechanical force induced by an RF voltage applied to a micro-mechanical system. For a theoretically adequate resolution and high precision measurements, the necessary geometrical dimensions of the sensor require the application of micro machining. In this contribution, the dependence between electrical and geometrical properties of different sensor designs is investigated. Based on these results, problems related to practical micro-machining and solutions with respect to possible sensor realizations are discussed. The evolution of different sensor generations is shown.

1 Introduction

RF voltage measurement based on electrostatic RMS voltage-to-force conversion can be performed by a mechanically tunable capacitor having one elastically suspended electrode plate. By applying a voltage, a mechanical force proportional to the square of its high frequency voltage is generated between the plates and consequently, the plate moves to the equilibrium position between spring counterforce and electrostatic force. The position of the plate is measured using a second capacitor. Taking into account the necessary mechanical dimensions of a micro-machined sensor, several generations of devices have been developed. In this contribution we first discuss the theoretical basis for MEMS-based RF voltage measurement, with an emphasis on the design choices. We present an overview of first (Beissner et al., 2003), second (Dittmer et al., 2007a), and third (Dittmer et al., 2007b) generation devices consecutively and analyze their performance from a fabrication process point of view. Similar devices have been developed by Bartek et al. (2000) in bulk- and Fernandez et al. (2003) in surface-micro machining technology.

2 Theory

Electrostatic voltage measurements are based on the principle of attractive forces induced by charge carriers of opposite polarity. Applying a voltage V to a capacitance C results in charge carriers Q on the electrodes. Taking account the virtual work analogy along with the stored energy in a capacitor $W=\frac{1}{2}CV^2$, the work needed to separate the electrodes to a distance h is known to be $W=F_V h$. Therefore the force F_V acting on the electrodes is given by

$$F_V = \frac{1}{2}\frac{C}{h}V^2. \tag{1}$$

If one of the capacitor plates is suspended elastically with a linear elasticity coefficient k, the necessary force F_k for a displacement x is

$$F_k = kx. \tag{2}$$

Combining Eqs. (1) and (2) yields the equilibrium position for displacements $x \ll h_0$ to be

$$x = \frac{1}{2}\frac{C}{kh_0}V^2. \tag{3}$$

Figure 1 illustrates this basic interaction. Concluding Eq. (3), it is obvious that a high capacitance with a small working distance along with a low stiffness of the suspension are necessary for a good device resolution. The conversion of a RMS voltage to a displacement depends on the inertia and damping of the system. Having an accelerated proof mass m and a velocity depended gas damping with a constant η leads to

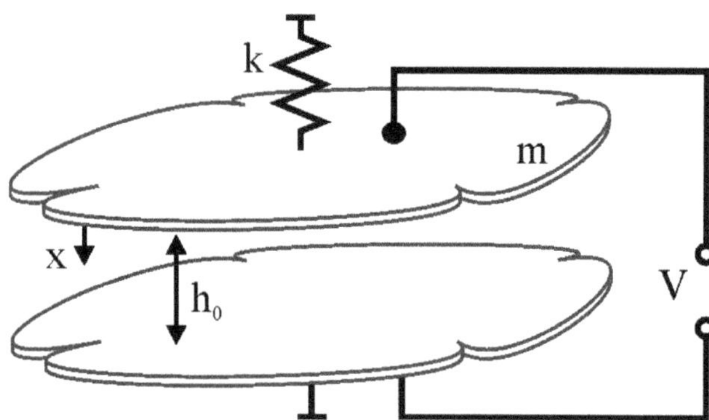

Fig. 1. Coupled spring-mass system with voltage excitation.

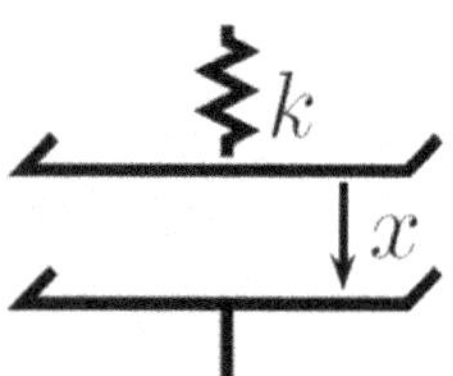

Fig. 2. Parallel plate capactitance.

a 2nd order dynamical system. The basic condition for measuring RF voltages is that the lowest measurement frequency ω is beyond any mechanical resonance frequency ω_r of the system. Having a high proof mass is therefore critical for a good performance. An in-depth discussion for a rotational system is given e.g. by Sattler et al. (2002).

2.1 Parallel plate vs. interdigital sensors

In a first step, a choice has to be made which kind of capacitor to be used. State-of-the art micro machining provides two well-understood kinds of geometries: the first one, popular in silicon-on-insulator (SOI) based systems, uses interdigital or comb-like structures (Fig. 3). It is commonly used in accelerometers or gyroscopes and is well understood regarding manufacturing and calculating. From an electrical point of view, problems of the structure are unpredictable fringing fields not allowing a reliable direct calculation of the displacement resulting from an applied voltage. Therefore, a parallel-plate structure as shown in Fig. 2 has been chosen. Having a large capacitor area extensively suppresses the fringing fields except for the edges of the structure. Having big electrode areas, this effect can be neglected in a first order approximation. The equilibrium position for small displacements of a parallel plate capacitor with electrodes of an area A with an initial distance h_0 can be found to be

$$x = \frac{1}{2}\frac{\varepsilon_0 A}{k h_0^2} U^2. \tag{4}$$

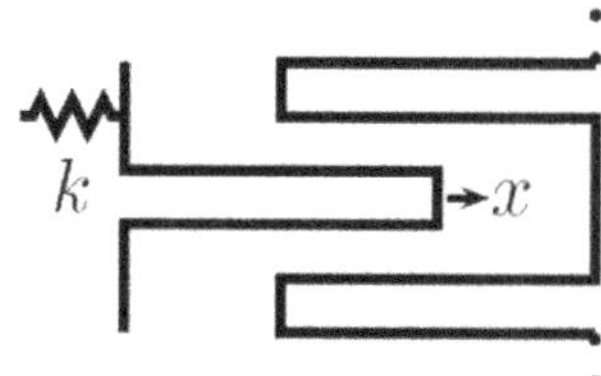

Fig. 3. Interdigital, comb-like capacitance.

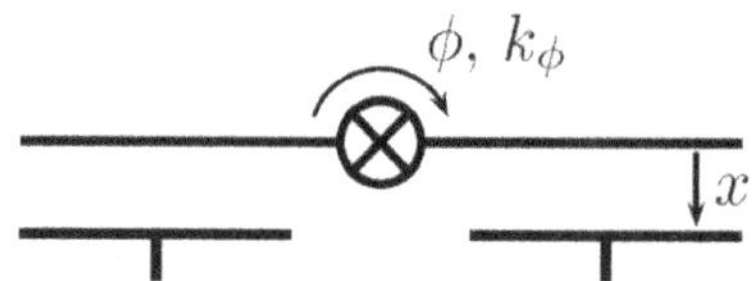

Fig. 4. Rotational design with opposing capacitances.

Equation (4) shows that for a high sensitivity the initial distance h_0 is the dominant quantity describing the the system. Parallel-plate capacitors in micro machining applications are used for example in RF switches or phase shifters fabricated in surface micro-machining technology. Thus, for RMS voltage measurements purposes, the low thickness of the actuated plate and therefore its light mass, and the low air damping coefficient due to the necessary openings and the resulting fringing fields make this technology unusable for the devices presented here. In bulk micro machining technology, a high mass and a closed flat area are achievable by combining multiple substrates.

2.2 Linear vs. rotational sensors

The second design choice for the sensing system, after selecting a parallel plate geometry, has to be made between a strictly linear parallel plate system, as shown in Fig. 3 and a rotational design, as illustrated in Fig. 4. Additionally a rotational design allows compensation measurements in comparison to a linear design. Unknown RF voltages can be directly balanced to known and precise DC voltages. As this is an important aspect for precision metrology applications, a rotational layout is chosen for the sensors discussed here. For large deflection angles, the electrical field is distored. However, as compensation measurements will primary be performed, where only the zero-deflection point is of interest, this disadvantage is not critical.

3 Devices

For the rotational design, different fabrication methods have been tested. Main part of each design is a bulk silicon torsional actuator. As mentioned before, bulk silicon is used due to its elastic behavior and the big achievable mass. It

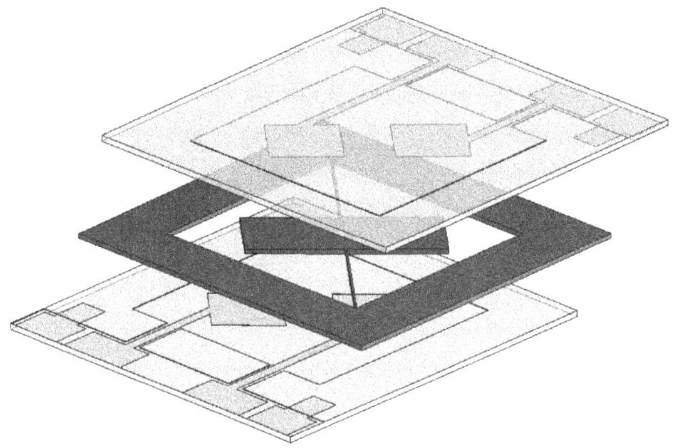

Fig. 5. Schematic of first generation device.

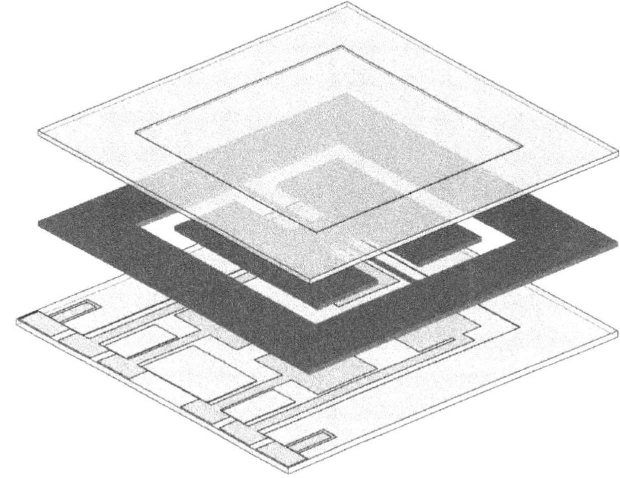

Fig. 6. Schematic of second generation device.

is also a readily available material in micro fabrication with good surface properties. Furthermore, several etchants are applicable, either for dry or wet processing. The actuator is finally embedded between two glass plates, providing the electrodes for readout and excitation connections and for environmental shielding.

3.1 First generation

The first generation devices (Beissner et al., 2003) have electrodes on both glass plates as shown in Fig. 5. The center actuator forms a common electric potential for all four electrodes. It is KOH wet etched, producing vertical side-walls by using the anisotropic crystallographic structure of silicon. The glass plates are anodically bonded to the actuator in a batch process, and afterwards the whole structure is diced. Is is a common problem during the bonding process that the actuator sticks to the glass wafers, producing non-functional devices. A second problem is the unreliable bond process due to the big open areas of the whole layout. Consecutively, this leads to water and particles in the cavities during the dicing process. Anodic bonding works best for closed surfaces. A third major problem are the bottom-up electrodes on the top glass wafer which makes the connections to surrounding electronics impractical. All these observations lead to the development of a second generation of devices.

3.2 Second generation

In the second device generation (Dittmer et al., 2007a), all electrodes are put on one side of the actuator without loss of functionality compared to the first generation. A cross section of the device is shown in Fig. 6. This solves the problem of wire-bonding of electrodes on two opposite sides. The glass cover serves only as protection sealing to prevent unwanted environmental influences. The second generation main actuator is dry-etched in an ICP enhanced process allowing more freedom in the geometrical design and thus better space exploitation. Experiments were made by using electroplated copper instead of gold electrodes with unsatisfactory results due to uneven, bowl-like surfaces. The actuator is anodically bonded to the bottom plate after dicing the parts. This improved the yield as a trade-off to a higher manufacturing overhead. As only four devices fit on one wafer the total work to yield ratio can still be improved.

3.3 Third generation

In the third device generation (Dittmer et al., 2007b), the structure is shrunk for a more efficient usage of the wafer area. The shrinking is mainly done by carefully reducing the bonding area and by replacing two big electrodes by smaller ones. The excitation electrodes still have the same size, as in the second generation. Thirteen 3rd generation devices can be placed on one wafer. To improve the isolation between the actuation and sensing electrodes, the common silicon electrode is replaced by multiple electrodes which are isolated from each other by using gold conductors on silicon oxide. These are connected to the base glass wafer by conductive glue. However, as this connection suffers from uneven glue distribution, unwanted interconnects, and short circuits occur frequently, reducing the yield. Typical geometrical dimensions of third generation devices are given in Table 1. Additional side electrodes are realized to detect skewed motion of the actuator. A cross section of the device is shown in Fig. 7 and an assembled device in Fig. 8. To improve the long-term reliability of the devices, bumpers have been added in the cavities of the sensing electrodes. These small bumpers are pyramidically shaped structures with a peak height below the gap height. They restrict the movement range of the actuator and hence prevent short-circuits between actuator and opposite electrodes as well es stiction of the actuator, which has been a major drawback of the previous device generation.

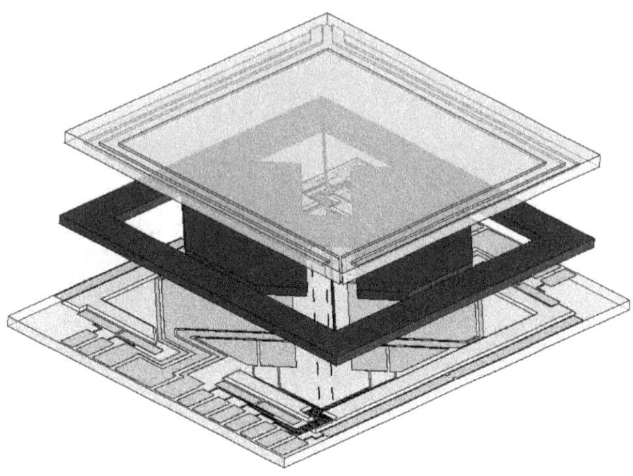

Fig. 7. Schematic of third generation device.

Fig. 8. Assembled third generation device.

Measurement results of third generation devices show the expected behavior for DC voltages. Due to the non-automated assembly process, the absolute and relative measurement error is still too high. The RF behavior of the devices has also been investigated. A low pass behavior of the whole structure can be observed due to the high ohmic resistance and inductance of the long slender torsional beams. Also cross coupling between the excitation and measurement electrodes increase the capacitance measurement error, in the worst case making the results unusable.

4 Conclusions

In this contribution we discuss why an rotational parallel plate design is superior to a linear or interdigital design for RF voltage metrology. Different generations of designs are presented and their shortcomings are discussed. The fourth generation of devices is already under development, improving the electrical connection between base and body wafer by using press-on contacts. Also, the whole process is optimized for batch fabrication allowing for a higher automation degree. The fabrication process is adapted to avoid any wet etching steps after etching the actuators.

Table 1. Major geometrical sensor dimensions of a second generation device.

Main electrode area	A	$5\,\text{mm}\times5\,\text{mm}$
Side electrodes area	A_s	$5\,\text{mm}\times2.5\,\text{mm}$
Gap distance	h_0	$10, 15, 20, 25\,\mu\text{m}$
Beam length	l	$8\,\text{mm}$
Beam width	w	$30, 40, 50, 60\,\mu\text{m}$
Lever	$\alpha_1 L$	$1\,\text{mm}$
Base thickness	t_e	$500\,\mu\text{m}$
Cover thickness	t_c	$500\,\mu\text{m}$
Body thickness	t_b	$370\,\mu\text{m}$

Acknowledgements. This research has been undertaken as a joined research project of the Institute of Microtechnology at the Technische Universität Braunschweig and the High-Frequency Measuring Group of the Physikalisch-Technische Bundesanstalt, Braunschweig, Germany.

References

Bartek, M., Xiao, Z., van Mullem, C., and Wolffenbuttel, R.: Bulk-micromachined electrostatic RMS-to-DC converter: Design and fabrication, in: Tech. digest MME 2000, 1–3 October, Uppsala, Sweden, p. A14, 2000.

Beissner, S., Wogersien, A., Buttgenbach, S., Schrader, T., and Stumper, U.: Micromechanical device for the measurement of the RMS value of high-frequency voltages, Sensors, 2003, Proc. IEEE, 1, 631–635, 2003.

Dittmer, J., Judaschke, R., and Büttgenbach, S.: Aufbau und Charakterisierung eines mikro-elektromechanischen Torsionssensors für die Hochfrequenzspannungsmessung, in: Mikrosystemtechnik Kongress 2007, Dresden, pp. 775–758, VDE Verlag GmbH, 2007a.

Dittmer, J., Judaschke, R., and Büttgenbach, S.: A Miniaturized RMS Voltage Sensor Based on a Torsional Actuator in Bulk Silicon Technology, pp. 769–770, Micro- and Nano Engineering, Kopenhagen, 2007b.

Fernandez, L. J., Visser, E., Sese, J., Wiegerink, R., Jansen, H., Flokstra, J., and Elwenspoek, M.: Radio frequency power sensor based on MEMS technology, Proc. IEEE Sensors, pp. 549–552, 2003.

Sattler, R., Plötz, F., Fattinger, G., and Wachutka, G.: Modeling of an electrostatic torsional actuator: demonstrated with an RF MEMS switch, Sensors and Actuators A, 97–98, 337–346, 2002.

Calibration-measurement unit for the automation of vector network analyzer measurements

I. Rolfes[1], B. Will[2], and B. Schiek[2]

[1]Institut für Hochfrequenztechnik und Funksysteme, Leibniz Universität Hannover, Appelstraße 9A, 30167 Hannover, Germany

[2]Institut für Hochfrequenztechnik, Ruhr-Universität Bochum, Universitätsstraße 150, 44801 Bochum, Germany

Abstract. With the availability of multi-port vector network analyzers, the need for automated, calibrated measurement facilities increases. In this contribution, a calibration-measurement unit is presented which realizes a repeatable automated calibration of the measurement setup as well as a user-friendly measurement of the device under test (DUT). In difference to commercially available calibration units, which are connected to the ports of the vector network analyzer preceding a measurement and which are then removed so that the DUT can be connected, the presented calibration-measurement unit is permanently connected to the ports of the VNA for the calibration as well as for the measurement of the DUT. This helps to simplify the calibrated measurement of complex scattering parameters. Moreover, a full integration of the calibration unit into the analyzer setup becomes possible. The calibration-measurement unit is based on a multiport switch setup of e.g. electromechanical relays. Under the assumption of symmetry of a switch, on the one hand the unit realizes the connection of calibration standards like one-port reflection standards and two-port through connections between different ports and on the other hand it enables the connection of the DUT. The calibration-measurement unit is applicable for two-port VNAs as well as for multiport VNAs. For the calibration of the unit, methods with completely known calibration standards like SOLT (short, open, load, through) as well as self-calibration procedures like TMR or TLR can be applied.

1 Introduction

For the automation of the calibration procedure of vector network analyzers, complex calibration units based on calibration methods with completely known calibration standards, like SOLT (Short, Open, Load, Thru) are already commercially available (Henkel, 2006; Krekels and Schiek, 1995; Krekels, 1996). In Fig. 1 a possible setup of a calibration unit for a 4-port-analyzer is shown. As can be deduced from the number of necessary calibration standards in Fig. 1, an automation of the calibration procedure is desirable and also recommendable in order to avoid calibration errors caused by e.g. connecting errors. The various standards are connected consecutively to the ports of the analyzer by a switch. For this purpose, the calibration standards have to be known exactly. After calibration the device under test (DUT) can be connected to the ports of the analyzer, and the scattering parameters of the DUT can be determined error-corrected in relation to the chosen phase reference plane, e.g. in the phase plane of the DUT's connectors.

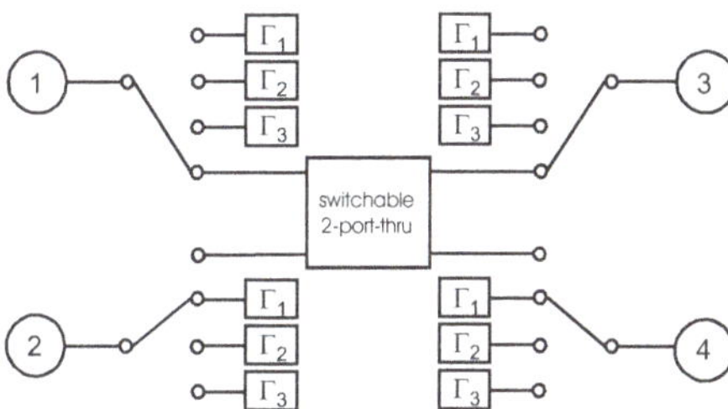

Fig. 1. Example of a calibration unit for a 4-port-analyzer.

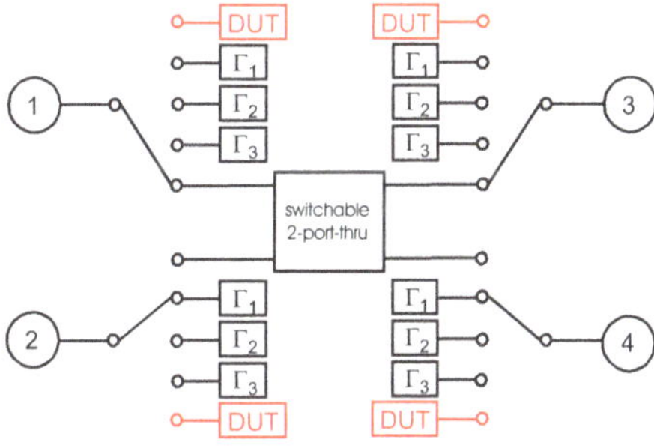

Fig. 2. Setup of the calibration-measurement unit.

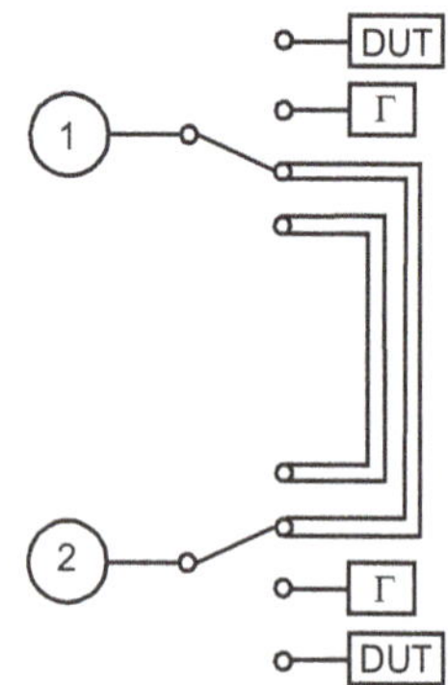

Fig. 3. Combined calibration measurement setup based on a TLR-calibration.

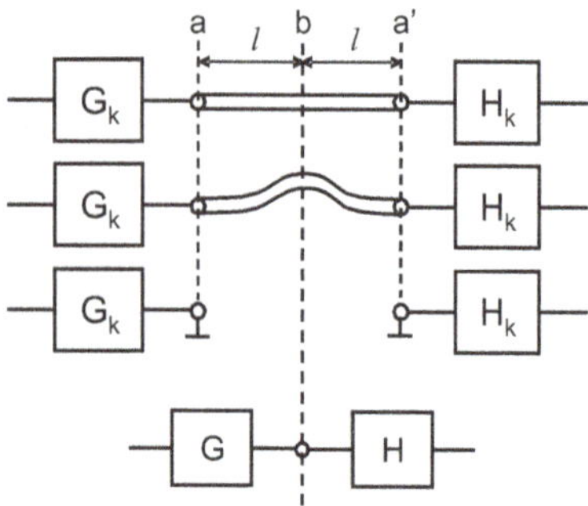

Fig. 4. Two-port error model of the VNA with TLR-calibration standards.

2 Calibration-measurement unit

For the automation of both the calibration of the analyzer and the measurement of the scattering parameters of a DUT, a combined setup, a so-called calibration-measurement unit is introduced. A setup for 4-port-measurements is shown in Fig. 2. The ports 1 to 4 are connected to the vector network analyzer and the DUT is connected to the ports named DUT. Thus, a calibration-measurement unit results, where the calibration standards do not have to be removed by the user in order to connect the DUT to the analyzer. The user can directly connect the DUT to the resulting measurement ports, thus leading to a user-friendly setup, which can easily be recalibrated without having to replace the DUT. However, as a precondition for this combined setup, the different switch positions of a switch have to be symmetrical, so that the phase reference planes of the measured scattering parameters lie behind the switches in the phase plane of the DUT. As a consequence, the non-idealities of the switch are included in the error terms representing the systematic errors of the network analyzer. The calibration can either be performed based on calibration methods with completely known calibration standards as e.g. SOLT as shown in Fig. 2 or with partly unknown calibration standards like TLR (Engen and Hoer, 1979; Eul and Schiek, 1991) as shown in Fig. 3 for a two-port setup. As the phase reference planes are preferably chosen in the plane

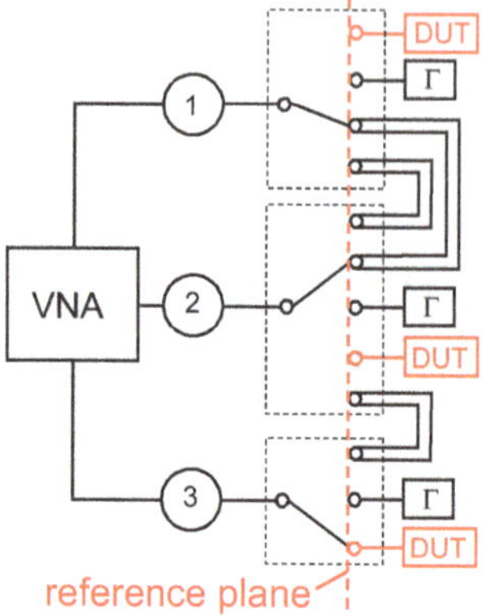

Fig. 5. Extension of setup for multiport analyzers.

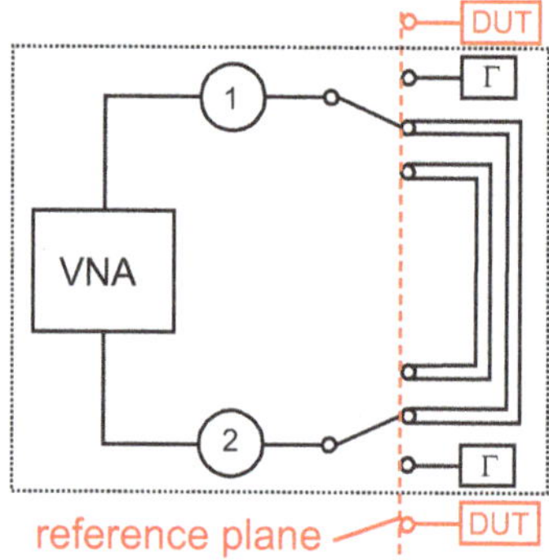

Fig. 6. Network analyzer with integrated calibration unit and measurement port for the contacting of the DUT.

of the DUT, for the TLR calibration a transformation of the phase reference plane is necessary.

This transformation can most easily be realized by choosing the reflection standard Γ for the TLR calibration as a short or open standard in the desired reference planes a, a' as illustrated in Fig. 4.

The error two-ports G and H in Fig. 4 are known after a TLR calibration. The error two-ports G_k and H_k for the transformed reference planes a, a' are calculable by

$$[G_k] = [G] \begin{bmatrix} e^{-\gamma l} & 0 \\ 0 & e^{\gamma l} \end{bmatrix} \tag{1}$$

with

$$e^{2\gamma l} = \frac{\rho}{\Gamma} \tag{2}$$

where ρ is the reflection coefficient of the TLR reflection standard in the reference plane b, which is known from self-calibration after the TLR calibration, and Γ is the reflection coefficient in reference planes a and a', which here is chosen as $\Gamma = -1$ for the termination with a short. Based on Eq. (1) for G_k and a similar equation for H_k the phase reference plane for the measurements can thus be moved into the phase plane of the DUT. The measured scattering parameters refer directly to the DUT without need for a deembedding.

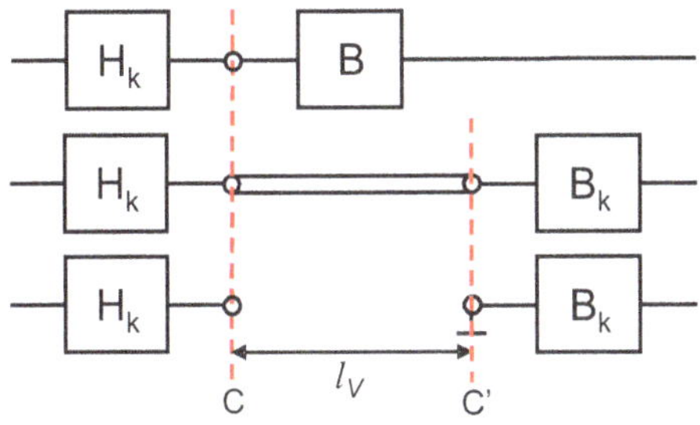

Fig. 7. Two-port setup for elimination of influence of additional cables.

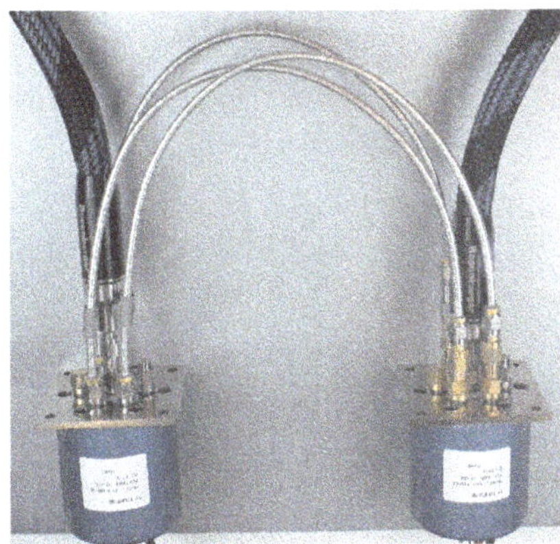

Fig. 8. Photo of the calibration measurement unit.
%vspace-3mm

The TLR-calibration setup of Fig. 4 as well as the previously presented calibration setups based on SOLT can easily be adopted to multiport analyzer systems by adding further line connections to the calibration measurement unit as shown in Fig. 5. In case that these lines are of unknown length but of identical geometry and line impedance as the previously used line standards for TLR calibration, it is possible to determine this unknown line length from an additional reflection measurement.

3 Integration into an analyzer system

The calibration standards can be integrated completely into a network analyzer system, so that only the measurement port for the contacting of the DUT remains as an interface to the user. This is demonstrated in Fig. 6. This setup allows an automated calibration of the measurement system without being in need for a manuel interaction of the user. On the one hand, this helps to reduce the required knowledge about calibration technology of the user and on the other hand, this possibility of automated calibration and combined measurement can furthermore be useful for applications where no user interaction is possible in the measurement plane, as e.g. in industrial drill hole applications where the measurement plane can be deep under earth's surface.

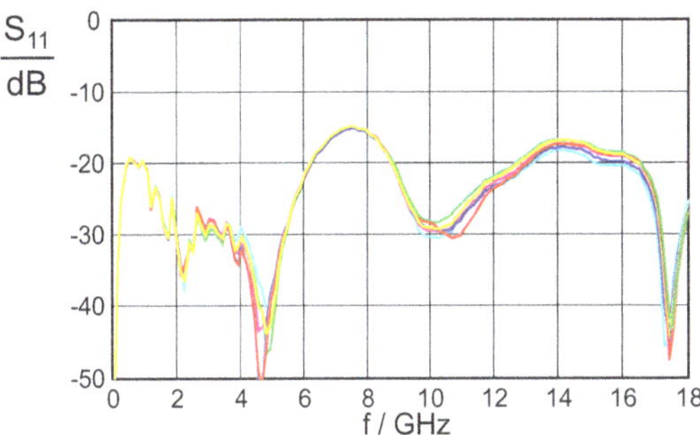

Fig. 9. Verification of the symmetry of the switch. Comparison of S_{11} measurements for all 6 positions of the switch.

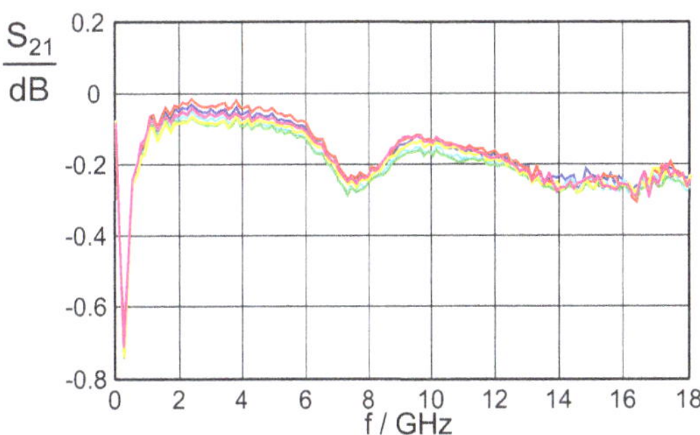

Fig. 10. Verification of the symmetry of the switch. Comparison of S_{21} measurements for all 6 positions of the switch.

For the case that additional cables are necessary for the contacting of the DUT, it is possible to eliminate the influence of the cables by performing two additional measurements: first, the measurement of the transmission of the cables and secondly, the measurement of the cables with a reflection standard placed in the phase reference plane. The setup is shown exemplarily in Fig. 7. Similarly to the previously described multiport extension, the influence of the cables can be corrected.

4 Measurement results

In order to verify the functionality of the proposed calibration-measurement unit, a demonstrator was realized on the basis of electromechanical relays as shown in Fig. 8. For verification of the symmetry of a switch, the different positions of a switch are compared. The measurements of the scattering parameters S_{11} and S_{21} in Figs. 9 and 10 show a good agreement of the high-frequency behavior for the different switching positions.

In Figs. 11 and 12, some further results for the verification of the switches symmetry are given. In this case, the scattering parameters are measured for varying terminations like open, short, match and the repetition of contacting.

In Figs. 13 and 14, the calibrated measurements of the scattering parameters S_{11} and S_{21} of a 3dB-attenuator are

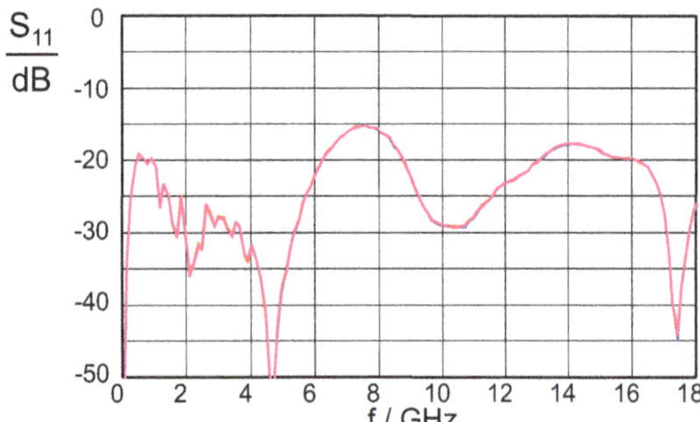

Fig. 11. Verification of the symmetry of the switch. Comparison of measurements of S_{11} while contacting varying terminations like open, match and repetition of contacting.

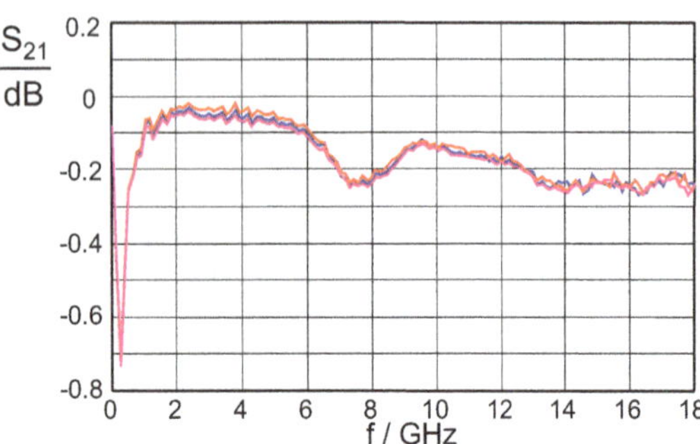

Fig. 12. Verification of the symmetry of the switch. Comparison of measurements of S_{21} while contacting varying terminations like open, match, and repetition of contacting.

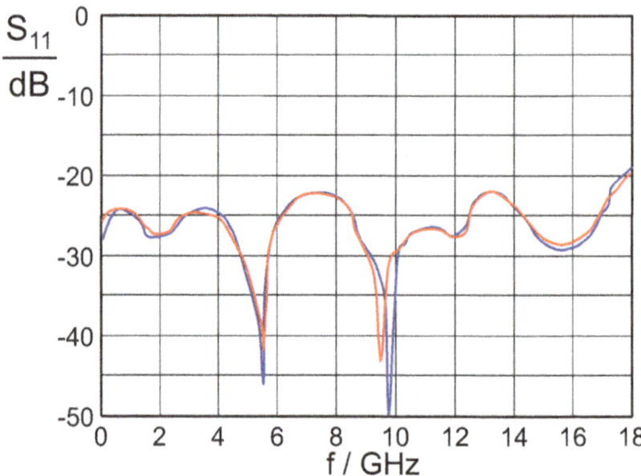

Fig. 13. Comparison of the measurement of S_{11} of a 3dB-attenuator based on a direct TLR-calibration (blue line) and based on the calibration-measurement-setup (red line).

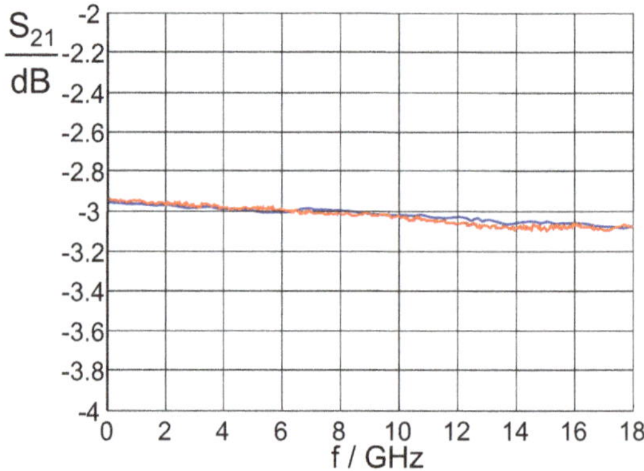

Fig. 14. Comparison of the measurement of S_{21} of a 3dB-attenuator based on a direct TLR-calibration (blue line) and based on the calibration-measurement-setup (red line).

shown exemplarily. A comparison of a direct measurement of the DUT and the measurement with the help of the calibration-measurement unit proves the robust functionality of the developed method.

5 Summary

A calibration-measurement unit is presented which enables on the one hand automated, repeatable calibrations of vector network analyzers and on the other hand the measurement of scattering parameters of DUTs. Well-known calibration methods like SOLT as well as self-calibration methods like TLR are applicable. The possibility of integration of the whole setup is described, leading to a measurement system where the user can directly connect the DUT to the measurement ports of the system, while the calibration standards can be left in the system so that a repetition of calibration is possible at every time. The setup is based on switching multiports, which have to be symmetrical individually for the proposed setup. For the verification of the robust functionality of the calibration-measurement unit a demonstrator and measurement results are presented.

References

Engen, G. F. and Hoer, C. A.: Thru-Reflect-Line: An improved technique for calibrating the dual six port automatic network analyzer, IEEE T. Microw. Theory, 27, 987–993, 1979.

Eul, H. J. and Schiek, B.: A Generalized Theory and New Calibration Procedures for Network Analyzer Self-Calibration, IEEE T. Microw. Theory, 39, 724–731, 1991.

Henkel, A.: Unrivaled – up to eight test ports in a single unit, News from Rohde & Schwarz, 189, 26–29, 2006.

Krekels, H.-G.: AutoKal: Automatic Calibration of Vector Network Analyzer ZVR, Rohde & Schwarz Application Note 1EZ30 IE, 1996.

Krekels, H.-G. and Schiek, B.: A novel Procedure for an automatic Network-Analyzer Calibration, IEEE T. Instrum. Meas., 44, 291–294, 1995.

25

Remote atomic clock synchronization via satellites and optical fibers

D. Piester[1], M. Rost[1], M. Fujieda[2], T. Feldmann[1], and A. Bauch[1]

[1]Physikalisch-Technische Bundesanstalt (PTB), Bundesallee 100, 38116 Braunschweig, Germany
[2]National Institute of Information and Communications Technology (NICT), Tokyo, Japan

Abstract. In the global network of institutions engaged with the realization of International Atomic Time (TAI), atomic clocks and time scales are compared by means of the Global Positioning System (GPS) and by employing telecommunication satellites for two-way satellite time and frequency transfer (TWSTFT). The frequencies of the state-of-the-art primary caesium fountain clocks can be compared at the level of 10^{-15} (relative, 1 day averaging) and time scales can be synchronized with an uncertainty of one nanosecond. Future improvements of worldwide clock comparisons will require also an improvement of the local signal distribution systems. For example, the future ACES (atomic clock ensemble in space) mission shall demonstrate remote time scale comparisons at the uncertainty level of 100 ps.

To ensure that the ACES ground instrument will be synchronized to the local time scale at the Physikalisch-Technische Bundesanstalt (PTB) without a significant uncertainty contribution, we have developed a means for calibrated clock comparisons through optical fibers. An uncertainty below 40 ps over a distance of 2 km has been demonstrated on the campus of PTB. This technology is thus in general a promising candidate for synchronization of enhanced time transfer equipment with the local realizations of Coordinated Universal Time UTC.

Based on these experiments we estimate the uncertainty level for calibrated time transfer through optical fibers over longer distances. These findings are compared with the current status and developments of satellite based time transfer systems, with a focus on the calibration techniques for operational systems.

1 Introduction

Clock comparisons are one of the essential tasks of international time metrology, e.g. for the harmonization of national standards, for enabling the interoperability between satellite navigation systems, and for the dissemination of time to the public. As an internationally agreed reference the Coordinated Universal Time UTC and, more specific, the underlying International Atomic Time TAI are computed by the Bureau International des Poids et Mesures (BIPM) by using data from 391 atomic clocks distributed all over the world in 69 different institutes (as of October 2010). Most of them are National Metrology Institutes (NMIs) (Arias and Panfillo, 2009; Circular T).

The data involved consist of two sets to be delivered by the institutes to the BIPM. The first one is from measurements between all available clocks in each laboratory k with respect to the local "physical" realization of UTC called UTC(k). The second set is from time comparisons between the 69 UTC(k) laboratories (see above). For these comparisons mainly two techniques are employed: in the first instance, signals from the global navigation satellite systems (GNSS) GPS and GLONASS (and in future also the European Galileo) are received and processed in different modes for global comparisons. It is a one-way technique in which time signals propagate from the satellites in known orbits through the atmosphere until they reach dedicated receivers at fixed positions. The used comparison modes depend on the equipment available at the laboratories and are called (in the order of increasing performance): "Single Channel", "Multi Channel", "P3", and "PPP". Details about the techniques can be found in the respective literature, e.g. see the papers from Arias (2007) and Petit and Ziang (2008) and references therein. All modes offer the possibility of true global comparisons, which means that clocks located at nearly every location on Earth can be compared with each other. The second technique in use is two-way satellite time and frequency transfer (TWSTFT) (Kirchner, 1999). It has the advantage that modulated signals are exchanged between two sites, and thus propagate simultaneously through the same medium along the same path, just in opposite direction. This leads to a cancellation of most effects which have an impact on the signal delay, in particular those caused by the propagation through the troposphere and ionosphere. TWSTFT usually employs a geostationary telecommunication satellite as a relay for the signals in space. This limits the choice of locations for ground terminal installations and the operational distance to roughly 10 000 km. An advantage of TWSTFT is that results of clock comparisons are available instantly after a link between two ground sites has been established. No post processing is required as compared to GNSS-based time and frequency transfer.

With these currently operational time and frequency transfer techniques, the frequencies of state-of-the-art primary caesium fountain clocks can be compared relatively at the level of 10^{-15} and time scales can be synchronized with an uncertainty of one nanosecond. At present experiments for satellite based time and frequency transfer are performed or designed with the aim to reduce the so far achieved uncertainty significantly. For example a new system for time and frequency transfer will be available in 2014: the Atomic Clock Ensemble in Space (ACES) will be installed at the International Space Station (ISS) revolving as time reference about every 90 minutes around the Earth. Time transfer with an uncertainty of as low as 100 ps will be possible by using the integrated ACES-Microwave-Link (MWL) (Cacciapuoti and Salomon, 2009). Compared to currently operational TWSTFT the ACES MWL uses a broader bandwidth for time transfer signals. Thus also on ground dedicated technologies are needed to supply these new experiments with ultra stable reference frequency and time. Recently optical fiber based techniques show very promising results in both frequency and time transfer. Distances from campus solutions (1 km) to about 500 km have been investigated and demonstrated. Distances up to 1500 km are discussed (Piester and Schnatz, 2009a). In this paper the possibility of a combination of these techniques is discussed for the application of global time and frequency transfer.

2 Comparison of time scales and atomic clocks

Facilities for satellite based time and frequency comparisons are standard equipment in every time laboratory. They are operated in a continuous mode and allow the determination of frequency differences between atomic clocks and phase (time) differences between time scales maintained at these laboratories. The latter requires a dedicated calibration of propagation delay in the equipment involved. Before measuring time differences via optical fibers the involved equipment has to be calibrated as well. In this section we briefly describe the two standard satellite based techniques for time transfer and finally discuss the achieved results of optical fiber time and frequency transfer based on recently performed experiments.

2.1 GPS time and frequency transfer

GPS signal reception has become the standard tool for time and frequency transfer between time laboratories (see e.g. Piester and Schnatz, 2009a). The GPS maintains a minimum of 24 satellites, in a way that at nearly each location on Earth more than four satellites are simultaneously visible. On each satellite an atomic clock serves as the onboard reference, which is related to the system time reference. From the onboard reference two carriers are generated and transmitted. These signals are phase modulated by two characteristic pseudorandom noise codes, uniquely associated with each space vehicle, the so called coarse acquisition and the precise codes. As part of the GPS operation practice the positions of the satellites are precisely known and reported inter alia in the data stream from the satellites (see e.g. Kaplan and Hegarty, 2006). With known distance from the receivers' antennae to the satellites ground-based clocks can be compared to GPS time with a typical precision of 10 ns or 1 ns using the coarse or precise code, respectively. Exchange of the collected data between two stations via file transfer enables the a posteriori calculation of the time differences between the two local clocks involved. If the carrier itself is used, the precision is in the range of some 10 ps (Gotoh et al., 2003), but typically subject to unknown absolute offsets.

The precision of operational GPS time and frequency transfer has been improved significantly during the last years. It is characterized by the statistical uncertainty u_A for the techniques "single channel" (4.5 ns), "multi channel" (2.5 ns), "P3" (1.0 ns), and "PPP" (0.3 ns) as entitled by BIPM and is carefully examined by the BIPM and documented in the monthly published Circular T, see e.g. issue October 2010. The accuracy of GPS time transfer strongly depends on the calibration procedure. The long-term instability of signal delays in the equipment making up the link determines the necessity of recalibrations, and on the other hand the accuracy of long-term frequency comparisons. For many links a systematic uncertainty u_B of 5 ns has been stated, reflecting the long-lasting conservative practice. This has been put in question when Esteban et al. (2010) estimated a significantly reduced uncertainty of calibrations using a traveling GPS receiver. Recently this has been refined and uncertainties below 1 ns have been reported by Feldmann et al. (2010). There is still a need for confirmation of such small values, but an uncertainty of 1.6 ns should be feasible. It is clear that these small values for u_B can be maintained over long times only when periodic re-calibrations are made.

2.2 Two-way satellite time and frequency transfer

TWSTFT is the second satellite based technique which is regularly used for generating TAI. The main advantage of the two-way technique is that unknown delays along the signal path cancel out to first order because of the path reciprocity of the transmitted signals (Kirchner, 1999). At present TWSTFT makes use of established communication satellite services mainly in the Ku-band (10.7 GHz to 14.5 GHz). Time signals are transmitted by relating the phase of a pseudorandom noise modulation on the carrier signal to the one-pulse-per-second input from a clock. A dedicated code is allocated to each transmitting station. The receive equipment correlates the received signal with a local replica of the signal expected from the transmitting site, and determines the time of arrival of the received signal with respect to the local clock.

With currently operational satellite link budget parameters, frequency instabilities (in terms of Allan deviation, see e.g. Riley, 2008) following a $\sigma_y(\tau) = 10^{-9}(\tau/\mathrm{s})^{-1}$ law can be achieved, which enables frequency comparisons at the level of 10^{-15} at reasonable averaging times τ (Bauch et al., 2006). The precision at an averaging time of 1 s is about 0.5 ns. TWSTFT time comparisons can be performed with an uncertainty of about 1 ns if the internal delays in the ground station equipment are determined by calibration. As an example, for one time link this is usually achieved by circulating a portable reference TWSTFT station to determine the overall relative delay difference between the two local ground stations' transmit and receive equipment. Such calibrations have been done repeatedly between European stations and between PTB and the U.S. Naval Observatory (Piester et al., 2008). Breakiron et al. (2004) report for a series of intra U.S. calibration campaigns overall uncertainties down to 0.38 ns. On the other hand also values of slightly above 1 ns have been reported for recent European campaigns (Bauch et al., 2009) which lead us to conclude that using currently available equipment link uncertainties between 0.4 ns and 1.2 ns are achievable.

2.3 Optical fiber time and frequency transfer

The most recent developments in optical fiber time and frequency transfer have demonstrated that a further and significant reduction of uncertainty is possible. Transfer of frequency signals on fiber lengths up to more than 100 km have been demonstrated (see e.g. Grosche et al., 2009). The introduction of Brillouin amplification for the use in frequency transfer has the potential to bridge more than 250 km without intermediate amplifier stations (Terra et al., 2010). Also very promising results have been achieved for frequency transfer using standard signals for a broader application. We give two examples: a 10 MHz signal as used widely in timing laboratories (Ebenhag et al., 2008) and a 1.5 GHz signal for radio astronomy applications (McCool et al., 2008). Both approaches demonstrate frequency distributions with uncertainties below 10^{-15}.

To cancel long term fiber length variations in time and frequency transfer the two-way method for exchanging optical signals has been investigated (Amemiya et al., 2006). Such an approach was initially proposed in the framework of network synchronization (Kihara and Imaoka, 1995).

Currently, analysis of time transfer through optical fibers (TTTOF) is investigated by different groups. Stability analyzes prove the measurement precision to be at a level of 10 ps (Czubla et al., 2006) or below (Śliwczyński et al., 2010). Smotlacha et al. (2010) demonstrated their two-way system using a 740 km link. In our work we have focused on the capability of calibrated time transfer (Piester et al., 2009b) using equipment similar to the TWSTFT scheme described before. In the two latter systems code division multiple access (CDMA) signals are employed and precisions below 100 ps and 10 ps, respectively, have been reported. The capability for calibrated time transfer on the level of 40 ps has been demonstrated in our study. A further improvement and the extension to longer distances is discussed in the next section.

3 Calibration and uncertainty evaluation of an optical fiber time transfer link

For the operation of a time transfer link in the strict sense a suitable calibration of the propagation delays is obligatory. Here we describe a procedure which can be executed at a single laboratory before the equipment will be installed at the locations where the systems will be operated. We use the nomenclature of existing literature (ITU-R, 2003; Bauch et al., 2011) and deviate only if necessary. The measurement setup is depicted in Fig. 1. We use in our setup modems which are also used for TWSTFT. Generally, we want to compare two time scales TA(1) and TA(2) generated at different distant laboratories. The modems' time-of-arrival measurements TW(i) are given by

$$\mathrm{TW}(1) = \mathrm{TA}(1) - \mathrm{TA}(2) + \mathrm{TX}(2) + \mathrm{SP}(2) + \mathrm{RX}(1), \quad (1)$$

and

$$\mathrm{TW}(2) = \mathrm{TA}(2) - \mathrm{TA}(1) + \mathrm{TX}(1) + \mathrm{SP}(1) + \mathrm{RX}(2). \quad (2)$$

TX(i) and RX(i) represent the complete internal transmission and receive delay of setup i. SP(1) is the transmission path delay from the local setup (1) to the remote site (2). SP(2) is the propagation delay in the opposite direction through the same fiber. For a single optical fiber connection between two sites we assume complete reciprocity of this part of signal path setting SP(1) = SP(2), and thus

$$\mathrm{TA}(1) - \mathrm{TA}(2) = 1/2[\mathrm{TW}(1) - \mathrm{TW}(2)] \\ +1/2[\mathrm{DLD}(1) - \mathrm{DLD}(2)], \quad (3)$$

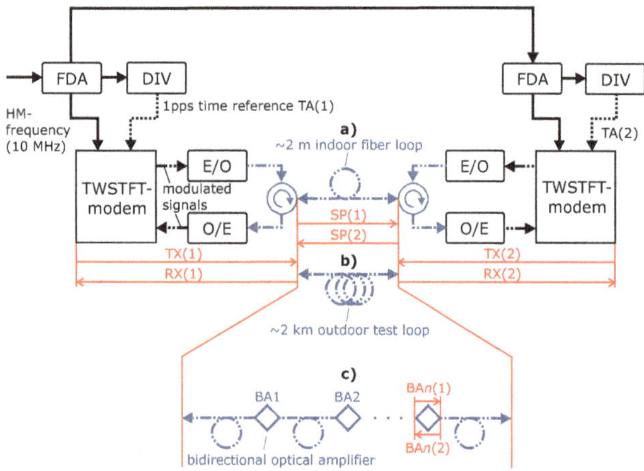

Fig. 1. Calibration setup for testing the independence of time transfer results from the length of the optical fiber. The two fibers used are depicted as **(a)** a short fiber and **(b)** a long fiber. On long distance connections **(c)** a number of bidirectional amplifiers is inserted into the signal path.

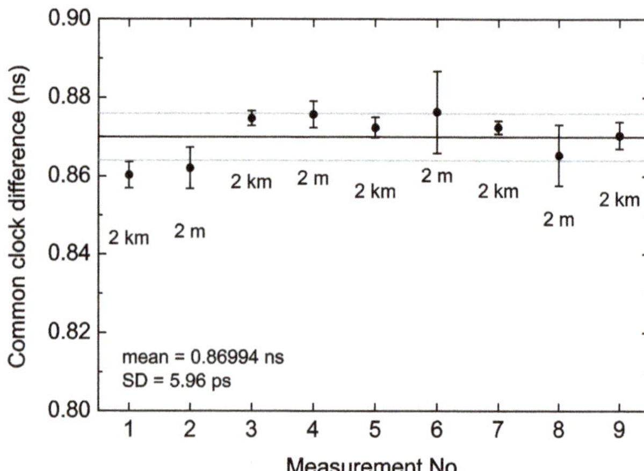

Fig. 2. Display of the sequence of measurements with short fiber loop and long outdoor fiber loop (see Rost et al., 2010). The measurements are labeled with the used fiber length. Mean and standard deviation (SD) are depicted as black and gray lines, respectively.

where $\mathrm{DLD}(i) = \mathrm{TX}(i) - \mathrm{RX}(i)$ is the delay difference between the TX path and the RX path of one setup. For accurate time transfer we want to determine the last term in Eq. (3) as the calibration result. For this calibration we use a common clock configuration and adjust TA(1) = TA(2), which leads to

$$0 = 1/2[\mathrm{TW}(1) - \mathrm{TW}(2)] + 1/2[\mathrm{DLD}(1) - \mathrm{DLD}(2)]. \quad (4)$$

The first term is the so called common clock difference CCD(1,2) = 1/2 [TW(1) − TW(2)]. It is measured to determine the delay difference between both systems corresponding to a calibration value defined as

$$\mathrm{CALR}(1,2) = 1/2[\mathrm{DLD}(1) - \mathrm{DLD}(2)] = -\mathrm{CCD}(1,2). \quad (5)$$

For the calibrated time transfer link between the two setups, finally installed at remote sites, we get

$$\mathrm{TA}(1) - \mathrm{TA}(2) = 1/2[\mathrm{TW}(1) - \mathrm{TW}(2)] + \mathrm{CALR}(1,2). \quad (6)$$

As noted before, a prerequisite for accurate comparisons of remote time scales is the possibility for delay calibration of the whole system. Because the optical fiber length in the final setup is unknown, a calibration test is needed to ensure the independence of the setup from the length of the used fiber which connects both laboratories. For this purpose we connected the modems to reference frequency and 1pps (1-pulse-per-second) from signal distribution equipment (FDAs) and pulse generators (DIVs) as illustrated in Fig. 1. The 1 pps cable connectors represent generally the time scales TA(1) and TA(2). Both TTTOF setups were then connected by two fibers which were subsequently exchanged: a 2 m short indoor fiber (Fig. 1a) and a 2 km long fiber (Fig. 1b) buried on the PTB campus. The attenuation of the two optical fibers was adjusted to be at the same level by inserting a variable optical attenuator into the bidirectional fiber path SP(i), in order to minimize the impact of receive power dependent delay variations in the modems. The optical power was kept constant within ±0.1 dB.

The results of measurements CCD(1,2) = 1/2 [TW(1) − TW(2)] when switching over between long and short fibers is depicted in Fig. 2 (see Rost et al., 2010 for details of the experimental setup). The sequence comprises eight switches between the long and the short fiber. The error bars in Fig. 2 represent the standard deviation of single secondly recorded measurements. The standard deviation of the single CCD(1,2) values around the mean is only 6 ps. However, two effects were observed which have to be investigated in more detail in future: variations at the beginning of the sequence might be a result of temperature variations and the higher standard deviation of the measurements with the short fiber may be due to instabilities caused by interference of the optical signals partly reflected at fiber connectors. A reduction of interference could be achieved if different optical wavelengths are used in the local and remote setup. Nevertheless, the small variations of below 40 ps (including error bars) obtained under different experimental conditions are promising results to comply with the aim of enabling time transfer with an uncertainty well below 100 ps to supply next generation satellite based time and frequency transfer techniques.

The operational distance between the two connected sites is limited to about 100 km if no amplifiers are used (Amemiya et al., 2010). If longer distances should be bridged, the bidirectional signal path requires also bidirectional amplifiers. Because their internal delay is generally different for each direction a calibration of the amplifiers is also necessary. For a link with n amplifiers n additional calibrations have to be made. Equation (3) is extended as follows

$$\mathrm{TA}(1)-\mathrm{TA}(2)=1/2[\mathrm{TW}(1)-\mathrm{TW}(2)] +1/2[\mathrm{DLD}(1)-\mathrm{DLD}(2)]+1/2[\mathrm{dBA}(1,2)+\mathrm{dBA2}(1,2) +...+\mathrm{dBA}n(1,2)] \quad (7)$$

where dBAn(1,2) is the delay difference of the n-th bidirectional amplifier BAn(1) − BAn(2) (see Fig. 1c). The values for dBA$n(i)$ can be determined as follows. The necessary amplifiers for a link are inserted into the calibration setup (common clock TA(1) = TA(2)). From Eq. (7) we get

$$0=1/2[\mathrm{TW}(1)-\mathrm{TW}(2)] +1/2[\mathrm{DLD}(1)-\mathrm{DLD}(2)]+1/2[\mathrm{dBA1}(1,2)+\mathrm{dBA2}(1,2) +...+\mathrm{dBA}n(1,2)] \quad (8)$$

After changing the direction of the first amplifier the measurement is repeated:

$$0=1/2[\mathrm{TW}(1)-\mathrm{TW}(2)]+1/2[\mathrm{DLD}(1)-\mathrm{DLD}(2)] +1/2[-\mathrm{dBA1}(1,2)+\mathrm{dBA2}(1,2)+...+\mathrm{dBA}n(1,2)] \quad (9)$$

Then the direction of each of all amplifiers is subsequently changed:

$$0=1/2[\mathrm{TW}(1)-\mathrm{TW}(2)]+1/2[\mathrm{DLD}(1)-\mathrm{DLD}(2)] +1/2[-\mathrm{dBA1}(1,2)-\mathrm{dBA2}(1,2)-...-\mathrm{dBA}n(1,2)] \quad (10)$$

We get DLD(1) − DLD(2) from adding Eqs. (8) and (10). We get the differential delay e.g. for the first amplifier by subtracting Eqs. (8)–(9). If we only want to calibrate the total link as a whole, only measurements Eqs. (8) and (10) need to be combined. Thus only the uncertainties u of two measurements will contribute to the overall uncertainty

$$U=\sqrt{2}\cdot u, \quad (11)$$

if we assume the same u for both measurements. However, this will not allow the exchange of a single device without loosing the link calibration information. For the calibration of a whole link including n bidirectional amplifiers one will need $n+1$ common clock difference measurements. The overall uncertainty U is in this case

$$U=\sqrt{n+1}\cdot u. \quad (12)$$

If we consider about $u=40$ ps for each delay difference determination, we expect for a 900 km link with 8 bidirectional erbium doped fiber amplifiers an uncertainty of about $U=60$ ps following Eq. (11) and $U=120$ ps after Eq. (12).

4 Summary and outlook

At present three methods exist (or are experimentally evaluated) to synchronize remote atomic clocks and time scales on the nanosecond level and below. The results for these time synchronization methods are summarized in Table 1. If one needs a time synchronization at the level of 5 ns or slightly below, GPS is the choice. GPS receivers are rather inexpensive compared to the other techniques presented, and their operational performance is almost site independent. Using the so called "all-in-view" data computation approach one can compare two sites regardless where they are located on Earth. Most laboratories maintaining a local realization of UTC use GPS links for their connection to the international network of timing laboratories.

TWSTFT offers a lower calibration uncertainty together with a link stability both at the level of one nanosecond. Beside the superior performance some drawbacks have to be taken into account: the operational distance is limited to about 10 000 km and a geostationary satellite has to be available. It must be equipped with transponders providing the required connectivity in the visibility range of both participating stations. So the establishment of a network is more demanding. Also the number of simultaneous clock comparisons in a network is limited by the hardware used. TWSTFT ground stations are elaborate and expensive, and furthermore transponder bandwidth and time need to be purchased from the satellite operating agency, resulting in substantial running costs.

Optical fiber connections offer the best performance characteristics due to the wider bandwidth of the transmission, when a dedicated dark fiber is available. Limiting factors of time transfer via optical fibers are the repeaters needed, when exceeding 100 km distance. In the near future the very high costs for long distance fiber connections will hamper applications on the continental scale. On the other hand, campus solutions will surely be established supporting deep-space network and antenna arrays in astronomy (Calhoun et al., 2007) or in accelerators for particle physics (Kärtner et al., 2010). Connections between cities surely will be established at least for experimental purposes, e.g. to compare optical frequency standards. Broad applications requiring calibrated time synchronization at an uncertainty level well below 100 ps have not been identified so far. Nevertheless, future space experiments will require dedicated ground infrastructure for time synchronization. As mentioned above, in 2014 the ACES mission will offer the possibility to perform two-way time and frequency transfer at enhanced precision. Compared to operational TWSTFT the ACES microwave link (MWL) will allocate a broader bandwidth and therefore improve the precision to 4 ps at an averaging time of 1 s. It will be a demonstrator for enhanced TWSTFT capabilities reducing the accuracy for time transfer to 100 ps on a global scale.

Table 1. Achievable calibration uncertainties of time transfer links. The stated uncertainties do not comprise the long-term stability of the calibrated links.

	Distance (km)	Uncertainty (ns)	References
GPS	$\sim$20000	$<$1.6 to 5	Esteban et al., 2010; Feldmann et al., 2010; Circular T, 2010
TWSTFT	$\sim$10000	$\sim$0.4 to $\sim$1.2	Breakiron et al., 2005; Piester et al., 2008; Bauch et al., 2009
Optical Fibers	$\sim$1500	$<$0.1 to 0.2	Rost et al., 2010; this work

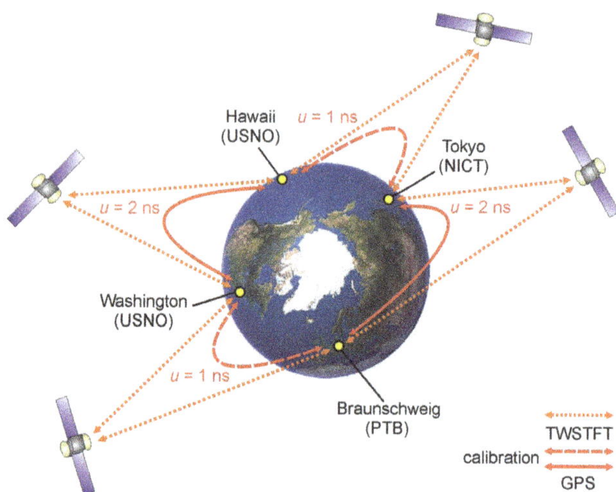

Fig. 3. An example for TWSTFT around the Earth and prospective calibration uncertainties.

A future scenario could be a combination of the presented techniques and might be used for global time comparisons: on shorter distances of several hundred kilometers laboratories might be connected via optical fibers, while these clusters are connected by intercontinental TWSTFT links and enhanced GNSS time links. The capability for TWSTFT calibrations on the global scale will be tested by circulating reference GPS receivers and mobile TWSTFT stations between the United States Naval Observatory USNO (Washington DC), the National Institute of Information and Communications Technology NICT (Tokyo) and the Physikalisch-Technische Bundesanstalt (Braunschweig, Germany). An example is depicted in Fig. 3. First steps have been done during some hours in 2010 by performing a first TWSTFT test for time comparisons around the world with the inclusion of Telecommunication Laboratories of Taiwan (W.-H. Tseng, private communication, 2010). An operational global link will offer the possibility to test the accuracy of a round trip calibration to better than 4 ns, if the two GPS and the two TWSTFT calibrations' results will have uncertainties of 2 ns and 1 ns, respectively.

References

Amemiya, M., Imae, M., Fujii, Y., Suzuyama, T., Ohshima, S., Aoyagi, S., Takigawa, Y., and Kihara, M.: Time and Frequency Transfer and Dissemination Methods Using Optical Fiber Network, IEEJ Trans. FM, 126, 458–463, 2006.

Amemiya, M., Imae, M., Fujii, Y., Suzuyama, T., Hong, F.-L., and Takamoto, M.: Precise Frequency Comparison System Using Bidirectional Optical Amplifiers, IEEE Trans. Instr. Meas., 59, 631–640, 2010.

Arias, E. F. and Panfillo, G.: International time scales at the BIPM: impact and applications, Proc. 14th International Metrology Congress, Paris, 2009.

Arias, E. F.: Time scales and relativity, Proc. International School of Physics "Enrico Fermi" Course CLXVI Metrology and Fundamental Constants, edited by: T. W. Hänsch et al., IOS Press, Amsterdam, 367–392, 2007.

Bauch, A., Achkar, J., Bize, S., Calonico, D., Dach, R., Hlavać, R., Lorini, L., Parker, T., Petit, G., Piester, D., Szymaniec, K., and Uhrich, P.: Comparison between frequency standards in Europe and the USA at the 10^{-15} uncertainty level, Metrologia, 43, 109–120, 2006.

Bauch, A., Piester, D., Blanzano, B., Koudelka, O., Kroon, E., Dierikx, E., Whibberley, P., Achkar, J., Rovera, D., Lorini, L., Cordara, F., and Schlunegger, C.: Results of the 2008 TWSTFT Calibration of Seven European Stations, Proc. European Frequency and Time Forum – IEEE Frequency Control Symposium Joint Conference, Besançon, France, 1209–1215, 2009.

Bauch, A., Piester, D., Fujieda, M., and Lewandowski, W.: Directive for operational use and data handling in two-way satellite time and frequency transfer (TWSTFT), Rapport BIPM-2011/01, 2011.

Breakiron, L. A., Smith, A. L., Fonville, B. C., Powers, E., and Matsakis, D. N.: The Accuracy of Two-Way Satellite Time Transfer Calibrations; Proc. 36th Annual Precise Time and Time Interval (PTTI) Systems and Applications Meeting, Reston, VA, USA, 139–148, 2004.

Cacciapuoti, L. and Salomon, C.: Space clocks and fundamental tests: The ACES experiment, Eur. Phys. J. Special Topics, 172, 57–68, 2009.

Calhoun, M., Huang, S., and Tjoelker, R. L.: Stable Photonic Links for Frequency and Time Transfer in the Deep-Space Network and Antenna Arrays, Proc. IEEE, 95, 1931–1946, 2007.

Circular T.: monthly publication of the BIPM, URL: http://www.bipm.org/jsp/en/TimeFtp.jsp, December 2010.

Czubla, A., Konopka, J., Górnik, M., Adamowicz, W., Struś, J., Pawszak, T., Romsicki, J., Lipiński, M., Krehlik, P., Śliwczyński, L., and Wolczko, A: Comparison of Precise Time Transfer with

Usage of Multi-Channel GPS CV Receivers and Optical Fibers over Distances of About 3 Kilometers, Proc. 38th Annual Precise Time and Time Interval (PTTI) Systems and Applications Meeting, Reston, Virginia, USA, 337–345, 2006.

Ebenhag, S.-C., Hedekvist, P. O., Rieck, C., Skoogh, H., Jarlemark, P., and Jaldehag, K., Evaluation of Output Phase Stability in an Fiber-Optic Two-Way Frequency Distribution System, Proc. 40th Annual Precise Time and Time Interval (PTTI) Systems and Applications Meeting, Reston, VA, USA, 117–124, 2008.

Esteban, H., Palacio, J., Galindo, F. J., Feldmann, T., Bauch, A., and Piester, D.: Improved GPS-Based Time Link Calibration Involving ROA and PTB, IEEE Trans. UFFC, 57, 714–720, 2010.

Feldmann, T., Bauch, A., Piester, D., Rost, M., Goldberg, E., Mitchell, S., and Fonville, B.: Advanced GPS Based Time Link Calibration with PTB's New Calibration Set-Up; to be published in Proc. 42th Annual Precise Time and Time Interval (PTTI) Systems and Applications Meeting, Reston, Virginia, USA, 2010.

Gotoh, T., Kaneko, A., Shibuya, Y., and Imae, M.: GPS Common View; Journal of the NICT, 50, 113–123, 2003.

Grosche, G., Terra, O., Predehl, K., Holzwarth, R., Lipphardt, B., Vogt, F., Sterr, U., and Schnatz, H.: Optical frequency transfer via 146 km fiber link with 10^{-19} relative accuracy, Opt. Lett., 34, 2270–2272, 2009.

ITU Radiocommunication Sector: The operational use of two-way satellite time and frequency transfer employing PN codes; Recommendation ITU-R TF.1153-2, Geneva, Switzerland, 2003.

Kärtner, F. X., Kim, J., Cox, J., Chen, J., and Nejadmalayeri, A. H.: Femtosecond Precision Timing Distribution for Accelerators and Light Sources, Proc. IEEE Int. Frequency Control Symposium, Newport Beach, CA, USA, 564–568, 2010.

Kaplan, E. D. and Hegarty, C. J. (eds.): Understanding GPS: Principles and Applications, Artech House, Boston, 2006.

Kihara, M. and Imaoka, A.: SDH-Based Time and Frequency Transfer System, Proc. 9th European Frequency and Time Forum, Besançon, France, 317–322, 1995.

Kirchner, D.: Two-Way Satellite Time and Frequency Transfer (TWSTFT): Principle, Implementation, and Current Performance, Review of Radio Sciences 1996–1999, Oxford University Press, 27–44, 1999.

McCool, R., Bentley, M., Garrington, S., Spencer, R., Davis, R., and Anderson, B.: Phase Transfer for Radio Astronomy Interferometers, over Installed Fiber Networks, Using a Round-Trip Correction System, Proc. 40th Annual Precise Time and Time Interval (PTTI) Systems and Applications Meeting, Reston, VA, USA, 107–116, 2008.

Petit, G. and Jiang, Z.: Precise Point Positioning for TAI Computation, Int. J. Nav. Obs., 562878, doi:10.1155/2008/562878, 2008.

Piester, D., Bauch, A., Breakiron, L., Matsakis, D., Blanzano, B., and Koudelka, O.: Time transfer with nanosecond accuracy for the realization of International Atomic Time; Metrologia, 45, 185–198, 2008.

Piester, D. and Schnatz, H.: Novel Techniques for Remote Time and Frequency Comparisons, PTB-Mitteilungen Special Issue, 119, 33–44, 2009a.

Piester, D., Fujieda, M., Rost, M., and Bauch, A.: Time Transfer Through Optical Fibers (TTTOF): First Results of Calibrated Clock Comparisons, Proc. 41th Annual Precise Time and Time Interval (PTTI) Systems and Applications Meeting, Santa Ana Pueblo, NM, USA, 2009b.

Riley, W. J.: Handbook of Frequency Stability Analysis, Natl. Inst. Stand. Technol. Spec. Publ. 1065, Washington, 2008.

Rost, M., Fujieda, M., and Piester, D.: Time Transfer Through Optical Fibers (TTTOF): Progress on Calibrated Clock Comparisons, Proc. 24th European Frequency and Time Forum, Noordwijk, The Netherlands, 2010.

Śliwczyński, L., Krehlik, P., and Lipiński, M.: Optical fibers in time and frequency transfer Meas. Sci. Techn., 21, 075302, 2010.

Smotlacha, V., Kuna, A., and Mache, W.: Time Transfer Using Fiber Links, Proc. 24th European Frequency and Time Forum, Noordwijk, The Netherlands, 2010.

Terra, O., Grosche, G., and Schnatz, H.: Brillouin amplification in phase coherent transfer of optical frequencies over 480 km fiber, Opt. Exp., 18, 16102–16111, 2010.

26

Application of postured human model for SAR measurements

M. Vuchkovikj[1,2], **I. Munteanu**[1,2], **and T. Weiland**[1,2]

[1]Graduate School of Computational Engineering, TU Darmstadt, Darmstadt, Germany
[2]Computational Electromagnetics Laboratory, TU Darmstadt, Darmstadt, Germany

Correspondence to: M. Vuchkovikj (vuchkovikj@gsc.tu-darmstadt.de)

Abstract. In the last two decades, the increasing number of electronic devices used in day-to-day life led to a growing interest in the study of the electromagnetic field interaction with biological tissues. The design of medical devices and wireless communication devices such as mobile phones benefits a lot from the bio-electromagnetic simulations in which digital human models are used.

The digital human models currently available have an upright position which limits the research activities in realistic scenarios, where postured human bodies must be considered. For this reason, a software application called "BodyFlex for CST STUDIO SUITE" was developed. In its current version, this application can deform the voxel-based human model named HUGO (Dipp GmbH, 2010) to allow the generation of common postures that people use in normal life, ensuring the continuity of tissues and conserving the mass to an acceptable level. This paper describes the enhancement of the "BodyFlex" application, which is related to the movements of the forearm and the wrist of a digital human model.

One of the electromagnetic applications in which the forearm and the wrist movement of a voxel based human model has a significant meaning is the measurement of the specific absorption rate (SAR) when a model is exposed to a radio frequency electromagnetic field produced by a mobile phone. Current SAR measurements of the exposure from mobile phones are performed with the SAM (Specific Anthropomorphic Mannequin) phantom which is filled with a dispersive but homogeneous material. We are interested what happens with the SAR values if a realistic inhomogeneous human model is used. To this aim, two human models, a homogeneous and an inhomogeneous one, in two simulation scenarios are used, in order to examine and observe the differences in the results for the SAR values.

1 Introduction

Bioelectromagnetics as an interdisciplinary science that investigates the interaction between the biological systems and the electromagnetic fields offers new and important opportunities for development of medical devices for diagnosis and therapeutic purposes. This science is very attractive and interesting nowadays, as electromagnetic devices, in particular mobile phones, play an increasing role in everyday life. During the development of the electromagnetic devices it is very important to understand the field distribution inside the human body, since direct measurement of the electromagnetic field inside the tissues and organs of the living organisms is almost impossible. As a result of the rapid development of computer science, measurements can be replaced by simulations on the human body models to predict the electromagnetic radiation effects and the macroscopic effects such as heating and specific absorption rate (SAR) distribution.

The upright position of all of the digital human models is a limiting factor in the electromagnetic research activities for realistic situations, and imposes the need for the generation of postured human models. Some deformed human models already exist (Allen et al. 2003, 2005; Findlay and Dimbylow et al., 2005, 2006; Dawson et al., 1999, 2002; Nagaoka and Watanabe, 2008), but the posing techniques are often very time consuming, and the positions of the models are limited. One approach for generation of a postured human model where continuity of the tissues and mass conservation are considered was proposed by Gao (2011), and was implemented within the software tool "BodyFlex". The software is based on an improved version of the well known free form deformation technique (FFD) introduced by Sederberg and Perry (1986), combined with the marching cubes algorithm (MC) introduced by Lorensen and Cline (1987). The "BodyFlex" application deforms and postures the voxel-based human model named "HUGO" (Dipp GmbH, 2010) which is built from the Visual Human Project data (Ackerman, 1998). However, some parts of HUGO's body such as wrist and the fingers cannot be moved with the current version of the software application.

In this paper, we describe the enhancement of the "BodyFlex" application in order to allow a proper movement of the forearm and the wrist, which is necessary for evaluation of the electromagnetic effects from mobile phones. To this aim, non-axis aligned control lattices and a special treatment for the forearm and hand deformation were introduced, for coping with the non-standard position of these body parts, which are bent over the lower part of the abdomen in the original model. Additionally, we analyze and compare the SAR values obtained by electromagnetic simulations, in which the inhomogeneous HUGO model in original and deformed position and the homogeneous model SAM are exposed to the electromagnetic field produced by a mobile phone.

2 "BodyFlex" application overview and enhancement

2.1 "BodyFlex" application overview

The powerful and quick application "BodyFlex" developed by Gao (2011) can generate common postures that people use in normal life, by deforming the voxel-based human model named HUGO. The combination of the free form deformation technique (FFD) for 3-D solid geometric models deformation introduced by Sederberg and Perry (1986) and the marching cubes (MC) algorithm introduced by Lorensen and Cline (1987) allow the generation of different positions for the HUGO model, while ensuring continuity of the tissues and conservation of the mass to an acceptable level.

In order to generate a new position of the HUGO model, the "BodyFlex" application imports a voxel dataset file with a specific resolution which can be chosen by the user. After this step, the user can choose to see the model and its tissues by a rendering technique based on the MC algorithm. The movement of particular body part is defined on a simplified human model in which only the joints of the model and their positions are considered. In this step, control FFD lattices around all parts of the body are created with the improved version of the FFD technique. Next, after the definition of the movement for one or few body parts has been completed, the posturing of the model is performed. The last step is the export of the deformed model in a new voxel dataset file with the same model resolution as the original imported model. Finally, the mass conservation for both the individual tissues and the entire deformed model can be checked.

Within the first version of the "BodyFlex" application, deformations of almost all parts of the body are possible. Exceptions are the wrist and fingers movements which require a change in the current algorithm because of the position of the forearm and the hand of the HUGO model which are bent on the lower part of the abdomen. This enhancement of the "BodyFlex" application is described in the next section.

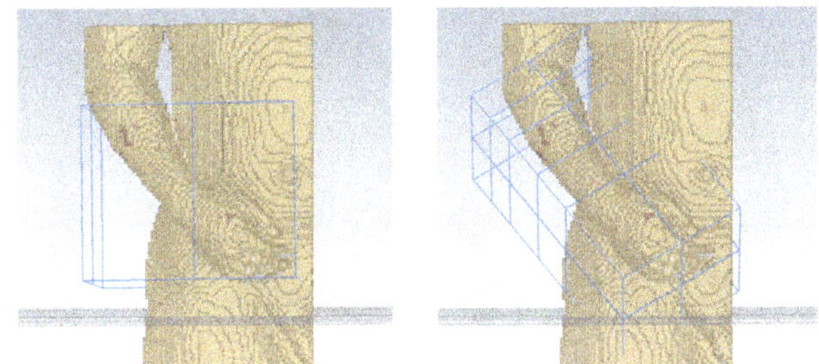

Fig. 1. Old version (on the left side) and new version (on the right side) of the FFD lattices for the forearm and the wrist.

2.2 Enhancement of the "BodyFlex" application

In order to allow an evaluation of the electromagnetic radiation from mobile phones, a movement of the forearm and wrist joint needs to be considered. The first version of the "BodyFlex" application does not deal with the wrist movements. As shown in Fig. 1 left, the FFD control lattices which embed the forearm and the wrist part do not treat these parts separately. Furthermore, the FFD control lattices are aligned to the global coordinate system axes, while the forearm has a tilted position in the original HUGO model.

In the new version of the "BodyFlex" application, as shown in Fig. 1 right, the FFD control lattices in which the forearm and the wrist are embedded for deformation have a rotated starting position. To determine a position of a control point or a voxel data point in the rotated lattice set in local coordinates, the cosine matrix was used. The direction cosine matrix is widely used in computer graphics for a transformation from one to another coordinate system and is defined by the following formula:

$$\mathbf{R} = \begin{bmatrix} \cos\theta_{x'x} & \cos\theta_{x'y} & \cos\theta_{x'z} \\ \cos\theta_{y'x} & \cos\theta_{y'y} & \cos\theta_{y'z} \\ \cos\theta_{z'x} & \cos\theta_{z'y} & \cos\theta_{z'z} \end{bmatrix}. \tag{1}$$

In Eq. (1), θ is the angle between the global coordinate system axis x, y or z and the axis x', y' or z' of the rotated coordinate system, respectively. To explain the determination of a point in the translated and rotated local coordinate system, we take the following example. We consider two coordinate systems, such that the first coordinate system is the Cartesian coordinate system and the second one is attached to the FFD control lattice set on the wrist. The second coordinate system has axes aligned to the FFD control lattices and center in point O (a,b,c). We want to determine the local coordinates x_2, y_2, z_2 of a point M in the second coordinate system, with respect to its coordinates x_1, y_1, z_1 in the first coordinate system. The dimensions of the FFD lattice set is *len_x*,*len_y* and *len_z*. To this aim, we use the direction cosine matrix and the position of the point M in the second coordinate system can

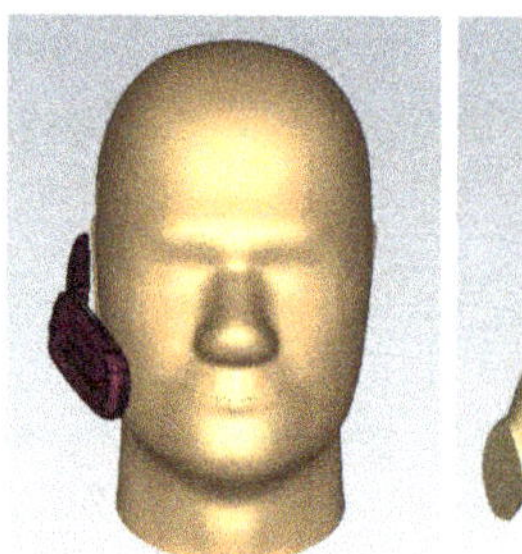

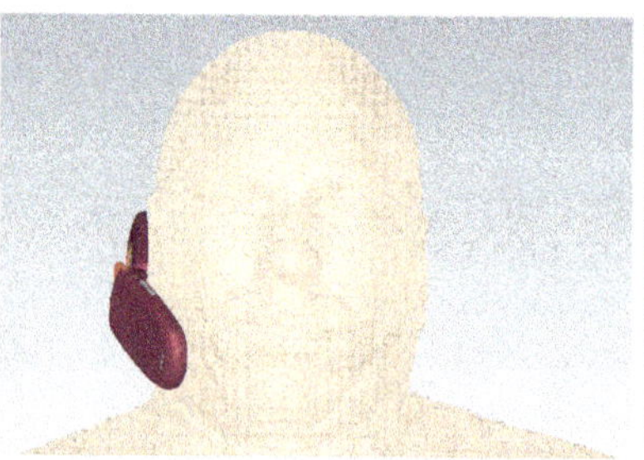

Fig. 2. SAM (on the left side) and HUGO (on the right side) models used for simulations.

be obtained by solving the Eq (2):

$$\begin{pmatrix} x_2 \\ y_2 \\ z_2 \end{pmatrix} = \begin{pmatrix} x_1 - a \\ y_1 - b \\ z_1 - c \end{pmatrix} * \mathbf{R} * \begin{pmatrix} len\_x^{-1} \\ len\_y^{-1} \\ len\_z^{-1} \end{pmatrix}. \tag{2}$$

After the voxels of the HUGO model are placed into the desired position by applying the trivariate tensor product Bernstein polynomial, the local coordinates of the voxels are transformed back to global coordinates by the following formula:

$$\begin{pmatrix} x_1 \\ y_1 \\ z_1 \end{pmatrix} = \begin{pmatrix} a \\ b \\ c \end{pmatrix} + \mathbf{R}^{-1} * \begin{pmatrix} x_2 * len\_x \\ y_2 * len\_y \\ z_2 * len\_z \end{pmatrix}. \tag{3}$$

Another problem which arises is the proper rotation of the wrist. Because the hand of the HUGO model is bent on the lower part of the abdomen, the rotation of the wrist cannot be performed around the axes aligned with the global coordinate system. Therefore, an algorithm for a rotation around arbitrary axes, which in this case are the axes of the local coordinate system, should be implemented. The calculation of a new position of a point described by Brouke (1992) makes series of transformations and rotations on the original position of the point, which are defined by the following formula:

$$\begin{pmatrix} x' \\ y' \\ z' \\ 1 \end{pmatrix} = \mathbf{T}^{-1}\mathbf{R}_x^{-1}\mathbf{R}_y^{-1}\mathbf{R}_z\mathbf{R}_y\mathbf{R}_x\mathbf{T} \begin{pmatrix} x \\ y \\ z \\ 1 \end{pmatrix}. \tag{4}$$

In Eq. 4), x, y, z and x', y', z' are the coordinates of the original and the new position of the point respectively. The translation of the point is performed by using the translation matrix $\mathbf{T}$ and its inverse $\mathbf{T}^{-1}$, while the rotations and the inverse rotations around the x, y and z axes are denoted by $\mathbf{R}_x, \mathbf{R}_y, \mathbf{R}_z$ and $\mathbf{R}_x^{-1}, \mathbf{R}_y^{-1}$ respectively.

3 Application: Analysis of the SAR distribution in simulations with mobile phones

3.1 Specific Absorption Rate (SAR)

Two international bodies ICNIRP (2009) and IEEE (2005) have developed guidelines in which the radiofrequency exposure limits for mobile phone users are expressed in terms of specific absorption rate (SAR). The SAR is defined as a measure of the rate at which energy is absorbed by the body which is exposed to a radio frequency electromagnetic field. Usually it is averaged either over the whole body or over a small sample volume (typically 1 g or 10 g of tissue) and can be expressed by the following formula:

$$\mathrm{SAR} = \frac{\iiint \frac{\sigma(r)|E(r)|^2 dV}{\rho(r)}}{V} \left[\frac{W}{kg}\right]. \tag{5}$$

In Eq. (5), σ represents the electrical conductivity and ρ is the mass density of the sample tissue. The electric field strength E magnitude is given in terms of the root mean square value.

For radiofrequency exposure to mobile phones, governments define the maximal allowed exposure in terms of energy absorbed in the head and the limbs. Namely, FCC in USA requires the phones on the market to have an SAR level of maximum $1.6\,\mathrm{W\,kg^{-1}}$ taken over a volume which contains 1 g of tissue, while CENELEC in Europe requires maximal SAR value of $2\,\mathrm{W\,kg^{-1}}$ averaged over 10 g of tissue. The SAR values mentioned previously refer to the maximal value which may appear in the head, while for the limbs, the maximal SAR value is $4\,\mathrm{W\,kg^{-1}}$.

3.2 Simulation scenarios and setup

In order to analyse the SAR distribution in the human body resulting from an exposure to radiofrequency electromagnetic fields, simulations were performed with the commercial software CST MICROWAVE STUDIO (CST AG, 2012) on two models: a homogenous one (SAM) and the realistic HUGO model. Two scenarios with mobile phone placed in a talk position were considered. In the first scenario, the mobile

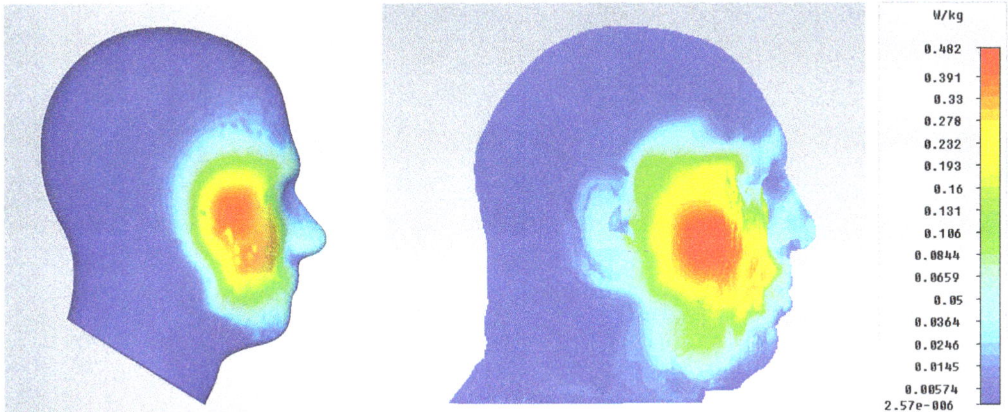

Fig. 3. SAR distribution in SAM (on the left side) and HUGO (on the right side) models on 1.8 GHz when no hand is present.

phone is not held by the hand (there is no hand in the simulation) and in the second scenario, the hand is placed behind the mobile phone as shown in Fig. 2. In the case of the SAM model, the mobile phone is held in the hand, while in the case of the HUGO model the hand is placed behind the mobile phone, since the finger movement is currently still under development.

The simulations were performed on a hexahedral mesh by using the transient solver. For the excitation, a discrete face port with an impedance of 50 Ohm was defined. The frequency range was between 0 and 2.5 GHz and field monitors for farfields, electric fields and power losses at 0.9 and 1.8 GHz were defined. At the frequency of 0.9 GHz the reference power was set to 0.125 W and at the frequency of 1.8 GHz the reference power was set to 0.25 W. As a post processing result, SAR values were obtained according to the IEEE C95.3 standard averaging method: at the frequency of 0.9 GHz the averaging mass was 10 g, while at the frequency of 1.8 GHz, the averaging mass was 1 g. The results for the SAR values for the SAM and the HUGO model, for the two scenarios, are presented in the next section.

3.3 Simulation results

As expected, a difference between the SAR values computed on the SAM and the HUGO model can be noticed.

Figure 3 shows the SAR distribution in the SAM and the HUGO models at 1.8 GHz, when no hand is present. It can be noticed that the maximal SAR value is higher in the realistic HUGO model than in the SAM model. However, this SAR value does not exceed the maximum allowed SAR value in the head prescribed by FCC and CENELEC. The realistic inhomogeneous human model allows an accurate analysis of the SAR distribution "hot spots" inside the model. As shown in Fig. 4, the parts of the head which absorb the most energy can be identified. The maximum SAR values occur within the skin and fat tissue near the chick, but also some increased SAR values occur in the muscles in the chick and in the mucous membrane in the nose.

In reality, a hand is always present near the head when talking on the phone. The plots of the SAR for this scenario are shown for both SAM and HUGO models in Fig. 5. The results show that in this case the SAR values in the HUGO's head are less than the ones in the SAM's head. Moreover, it is important to emphasize that when using a voxel human model, such as HUGO, the SAR distribution in the hand can be computed and compared to the maximal SAR values allowed in the limbs.

Figure 6 presents a comparison of the SAR values in the HUGO model at 1.8 GHz with and without hand. The maximal SAR value in HUGO's head decreases in case when the hand is present: in this case, a large part of the energy is absorbed by the hand.

4 Conclusions

In this paper we describe the recent advances of the posing program "BodyFlex", which allow the movement of the forearm and wrist of the inhomogeneous voxel human model HUGO. The existing algorithm based on the powerful free form deformation technique was enhanced to allow the definition and use of non-axis aligned control lattices, better fitted to the position of the forearm in the original voxel model. Besides that, an algorithm for the rotation of the wrist around arbitrary axes was implemented in order to enable the anatomically correct movement of this body part.

The presented application was the analysis of the SAR values obtained by electromagnetic simulations, where the inhomogeneous model HUGO in original and deformed positions and the homogeneous model SAM were exposed to the electromagnetic field produced by a mobile phone. As expected, a difference between the SAR values measured on the SAM and the HUGO model was noticed. The SAR values in the head of the realistic model, although larger than in

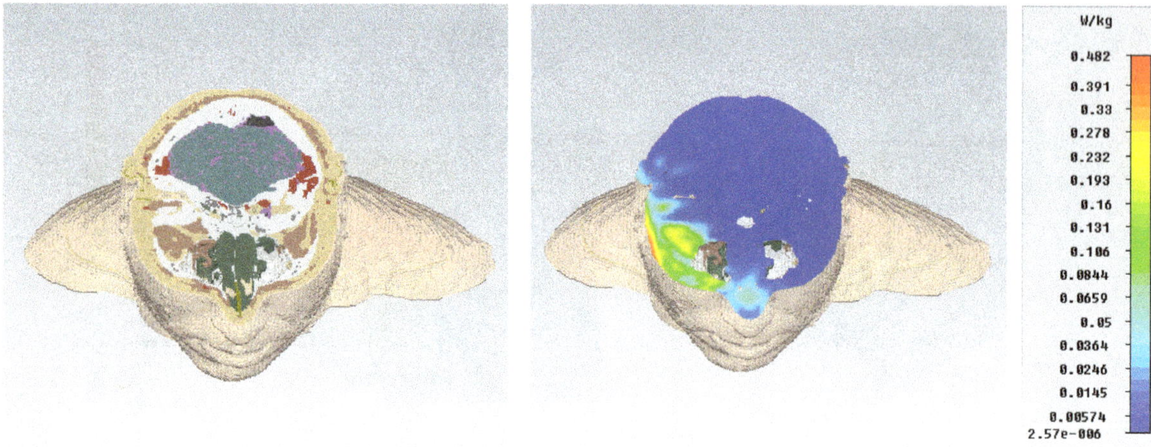

Fig. 4. Cut view of the HUGO model with SAR distribution observed on 1.8 GHz when no hand is present.

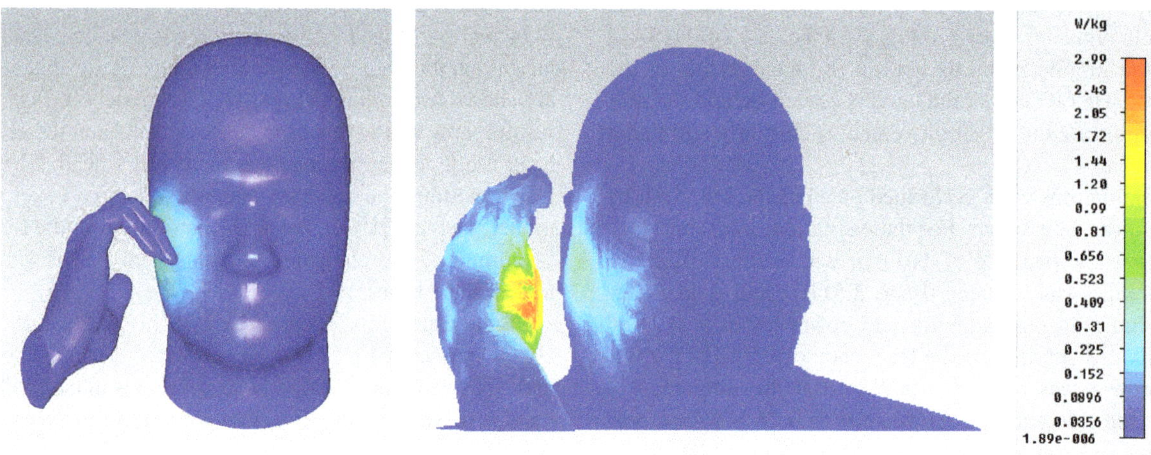

Fig. 5. SAR distribution in SAM (on the left side) and HUGO (on the right side) models on 1.8 GHz when hand is present.

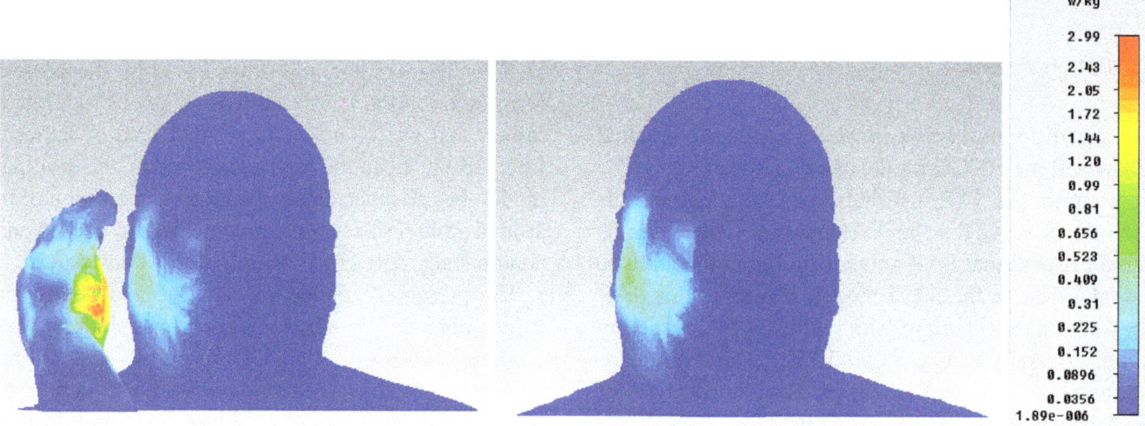

Fig. 6. SAR distribution in HUGO models on 1.8 GHz.

the homogeneous model, did not exceed the limits prescribed in the standards. Additionally, a comparison of the results between two scenarios (with and without hand) on the HUGO model showed that the presence of the hand in the model leads to a decrease of the maximal SAR value in the head. It can be concluded that the hand has a "protective role", reducing the radio-frequency absorption in the head. Moreover, it is important to emphasize that when using a voxel human model, such as HUGO, the SAR values measured in the hand can be evaluated and compared to the maximal SAR values allowed in the limbs. The analysis of the maximal SAR values obtained by these simulations leads to the conclusion that the prescribed limits for the radiofrequency exposure to mobile phones do not underestimate the SAR values.

Acknowledgements. This work is supported by the 'Excellence Initiative' of the German Federal and State Governments and the Graduate School of Computational Engineering at Technische Universität Darmstadt.

References

Ackerman, M. J.: The visible human project, Proceedings of the IEEE, 86, 504–511, 1998.

Allen, S. J., Adair, E. R., Mylacraine, K. S., Hurt, W., and Ziriax, J.: Empirical and theoretical dosimetry in support of whole body resonant RF exposure (100 MHz) in human volunteers, Bioelectromagnetics, 24, 502–509, 2003.

Allen, S. J., Adair, E. R., Mylacraine, K. S., Hurt, W., and Ziriax, J.: Empirical and theoretical dosimetry in support of whole body radio frequency (RF) exposure in seated human volunteers at 220 MHz, Bioelectromagnetics, 26, 440–447, 2005.

Bourke, P.: Rotation about an arbitrary axis (3 dimensions), online available at: http://paulbourke.net/geometry/rotate/, last access: 19 September 2012, 2012.

CST AG: CST – Computer Simulation Technology, online available at: http://www.cst.com/, last access: 18 September 2012, 2012.

Dawson, T. W., Caputa, K., and Stuchly, M. A.: Numerical evaluation of 60 Hz magnetic induction in the human body in complex occupational environments, Phys. Med. Biol., 44, 1025–1040, 1999.

Dawson, T. W., Caputa, K., and Stuchly, M. A.: Magnetic field exposures for UK live-line workers, Phys. Med. Biol., 47, 995–1012, 2002.

Dipp GmbH: Anatomical volume data sets, online available at: http://www.vr-laboratory.com/, last access: 25 July 2012, 2012.

Findlay, R. P. and Dimbylow, P. J.: Effects of posture on FDTD calculations of specific absorption rate in a voxel model of the human body, Phys. Med. Biol., 50, 3825–3835, 2005.

Findlay, R. P. and Dimbylow, P. J.: FDTD calculations of specific energy absorption rate in a seated voxel model of the human body from 10 MHz to 3 GHz, Phys. Med. Biol., 51, 2339–2352, 2006.

Gao, J., Munteanu, I., Müller, W. F. O., and Weiland, T.: Generation of postured voxel-based human models for the study of step voltage excited by lightning current, Adv. Radio Sci., 9, 99–105, doi:10.5194/ars-9-99-2011, 2011.

Institute of Electrical and Electronics Engineers (IEEE), IEEE standard for safety levels with respect to human exposure to radio frequency electromagnetic fields, 3 kHz to 300 GHz, IEEE Std C95.1, 2005.

International Commission on Non-Ionizing Radiation Protection (ICNIRP), Statement on the Guidelines for limiting exposure to time-varying electric, magnetic and electromagnetic fields (up to 300 GHz), 2009.

Lorensen, W. E. and Cline, H. E.: Marching cubes: A high resolution 3-D surface construction algorithm, in: Proceedings of the 14th annual conference on Computer graphics and interactive techniques (ACM Siggraph 87), 24, 163–169, 1987.

Nagaoka, T. and Watanabe, S.: Postured voxel-based human models for electromagnetic dosimetry, Phys. Med. Biol., 53, 7047–7061, 2008.

Sederberg, T. W. and Parry, S. R.: Free-form deformation of solid geometric models, Proceedings of the 13th annual conference on computer graphics and interactive techniques (ACM Siggraph 86), 20, 151–160, 1986.

Compact mode-matched excitation structures for radar distance measurements in overmoded circular waveguides

G. Armbrecht[1], E. Denicke[1], I. Rolfes[1], N. Pohl[2], T. Musch[3], and B. Schiek[3]

[1]Inst. für Hochfrequenztechnik und Funksysteme, Leibniz Universität Hannover, Appelstraße 9A, 30167 Hannover, Germany
[2]Lehrstuhl für Integrierte Systeme, Ruhr-Universität Bochum, Universitätsstraße 150, 44801 Bochum, Germany
[3]Institut für Hochfrequenztechnik, Ruhr-Universität Bochum, Universitätsstraße 150, 44801 Bochum, Germany

Abstract. This contribution deals with guided radar level measurements of liquid materials in large metal tubes, so-called stilling wells, bypass or still pipes. In the RF domain these tubes function as overmoded circular waveguides and mode-matched excitation structures like waveguide tapers are needed to avoid higher order waveguide modes. Especially for high-precision radar measurements the multimode propagation effects need to be minimized to achieve submillimeter accuracy. Therefore, a still pipe simulator is introduced with the purpose to fundamentally analyze the modal effects. Furthermore, a generalized design criterion is derived for the spurious mode suppression of compact circular waveguide transitions under the constraint of specified accuracy levels. According to the obtained results, a promising waveguide taper concept will finally be presented.

1 Introduction

Since the 1980s radar is one of the fastest-growing technologies in the process instrumentation industry (Parker, 2002), e.g. for tank level control. Due to its robustness, flexibility and the simultaneously falling market prices of microwave components, radar systems are advantageous in comparison to other level technologies such as differential pressure level, ultrasonic, capacitance, and displacer measurement systems (Kielb and Pulkrabek, 1999). Basically, there are two different techniques of radar setups which can be subdivided by the way the RF signal propagates: free-space radiating and guided wave setups. In free-space applications, state-of-the-art radar systems can be utilized not only for general process gauging, but also for calibratable high-precision level detection in an industrial environment (Weiss, 2001). Especially many commercial radar systems are based on frequency modulated continuous wave (FMCW) technologies (Brumbi, 1995). These systems provide an accuracy of the distance error within the submillimeter range (Musch, 2003). In the other case, the radar signal is guided along a transmission line probe. Coaxial probes are commonly used for exploiting the dispersion-free TEM waveguide mode for accurate level detection. Another upcoming important application for industrial radar systems is the application of permanently built-in still pipes, as depicted in Fig. 1, consisting of metal tubes that are large with respect to the radar wavelength. The usage of such configurations can be superior compared to the free-space case, due to the well-defined conditions of the waveguide suppressing parasitic reflections of tank internals. Even level turbulences that are caused by stirrers, resulting in measurement inaccuracies by waves and foam, can be neglected.

If the level detection in still pipes is conducted by using the same antennas as in the free-space radiating system, the accuracy of the measurements is significantly deteriorated compared to the free-space application (Pohl and Gerding, 2007). In contrast to Sai and Kastelein (2006), for our purpose solely conventional detection and signal processing algorithms are applied due to the limited processing power and power consumption in standardized industrial environment, e.g. the popular HART protocol (Brumbi, 2000). The still pipe functions as an overmoded circular hollow metallic waveguide. The deterioration is caused by intermodal dispersion due to the appearance of higher order multimode propagation and additionally due to chromatic waveguide dispersion of every single mode itself. The observed effects are strongly related to the behavior of multimode fiber optical transmission lines. Hence, this contribution investigates novel taper approaches and design criterions for a broadband mode-matched transition between waveguides with different diameters of the cross section, that can be utilized as feeding sections for still pipes (see Fig. 1). The structures are

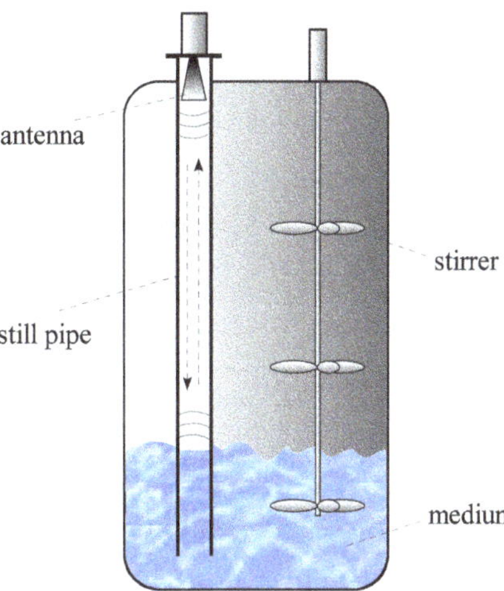

Fig. 1. Tank configuration incorporating a still pipe for accurate level detection.

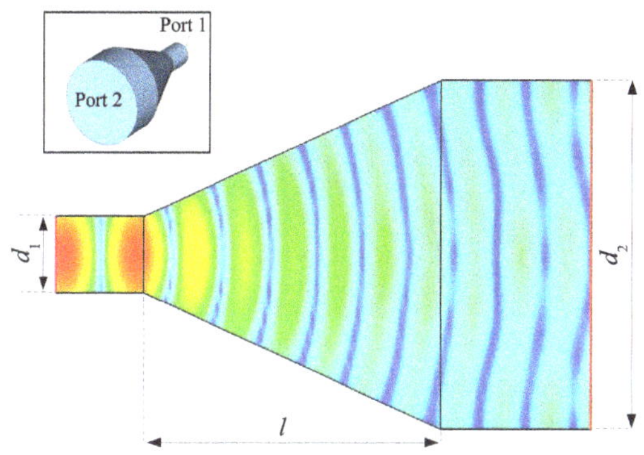

Fig. 2. Waveguide taper model of a two-port circular linear horn including the electric field at f=9.5 GHz (red color $\widehat{=}$max. amplitude) and typical geometrical settings.

optimized with respect to the conservation of the fundamental H_{11} field distribution between a single-moded small circular waveguide and an overmoded large circular waveguide by providing solely the excitation of the fundamental mode, i.e. the H_{11} mode, in an overmoded waveguide over a wide frequency range from 8.5 up to 10.5 GHz.

Characteristic optimization values are either the modal transmission coefficients, given in terms of scattering parameters of a two-port model derived from a commercial 3-D FIT field simulator, or the resulting FMCW distance error that is provided by a still pipe simulator incorporating all the modal and dispersive effects inside a still pipe. Realistic applications limit the space requirements of such transition geometries which results in a high demand for compact mode-matched structures. Thus, the maximal geometrical dimensions are limited. The lengthwise extension as well as the diameter at the second port are set to a value of l=d_2=80 mm to realize a transition starting from a feeding waveguide incorporating a fixed diameter of d_1=22 mm.

In the following, the structure of this article will be given: in Sect. 2 the general properties of a prominent waveguide taper will be introduced. Section 3 deals with the complete radar system and its simulation by utilizing a MATLAB implemented still pipe simulator, incorporating the mentioned dispersion and distortion effects. Subsequently, an analytical model of a waveguide transition is used for an in-depth analysis of the intermodal dispersion effects, providing the fundamentals for the derivation of an adequate design criterion in Sect. 3.3 that is directly related to the obtained measurement deterioration. Finally, according to this criterion, in Sect. 4 a novel subreflector-based taper concept will be shown.

2 Compact waveguide tapers

The simplest way to realize a compact waveguide transition is to use a conical horn with circular cross-section, as depicted in Fig. 2. Such a linear horn consists of a cone structure and is excited at the apex of the cone at a small diameter d_1. The impinging fundamental H_{11} field distribution, which exhibits planar phase fronts at the small cross section, is transformed into the corresponding spherical eigenmode inside the cone (Narasimhan and Balasubramanya, 1974). The H_{11} field distribution is mapped to spherical phases fronts with marginal higher order mode excitation within the cone itself. In general, the longer the horn structure the more the curvature of the phase fronts decreases. For the limit of an infinitely long cone, the fundamental H_{11} mode distribution will be completely conserved, when passing into a cylindrical eigenmode system at port 2 (see also Table 1).

In accordance with Fig. 2, a circular horn of *finite* length acting as a transition between cylindrical circular waveguides with different diameters is considered. A multiplicity of modes will be excited, due to the spherical phase fronts in the large waveguide. This kind of multimode excitation is primarily caused by the geometrical discontinuity at the transition to the large circular waveguide, due to the change in the corresponding eigenmode system. Even if discontinuities are similar to those at the feeding section (port 1), the influence on the mode conversion behavior of the taper is much smaller. All of the higher order modes are below their cut-off frequencies f_c. Figure 3 shows the scattering parameters S_{21} for the proposed circular linear horn in the frequency range from 8.5 up to 10.5 GHz, where $|S_{2(\text{Mode x}),1(H_{11})}|$ denotes the magnitude of the mode conversion between the fundamental mode at port 1 and multimode excitation at port 2.

Theoretically, a total number of 41 cylindrical modes are able to propagate in such a still pipe at the large diameter d_2. However, not all of these modes have to be taken into account for the taper investigation. According to Tang (1966), in case of distinct polarization properties of the incident H_{11} mode

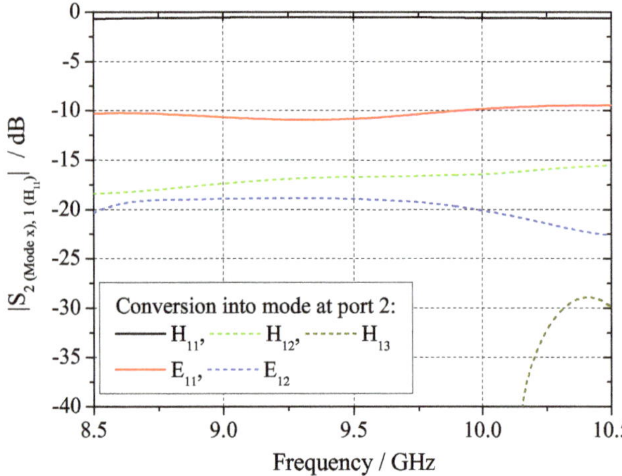

Fig. 3. Mode conversion from the fundamental mode H_{11} at port 1 into a mixture of different modes at port 2.

and under the assumption of axial straight and radial symmetrical waveguide tapers, the total number of modes with noticeable excitation reduces significantly to five. Solely H_{1p} and E_{1p} modes are excitable, if the respective cut-off frequencies f_c are passed. As shown in Fig. 3, the first higher order mode E_{11} is constantly suppressed by approximately 10 dB over the whole frequency range of interest. The other eigenmodes are more attenuated, so E_{11} is identified as the dominant higher order mode. The results given in Fig. 3 are representative for any kind of waveguide taper that is to be considered in this paper, regardless of the exact shape or filling. Hence, in the following sections the E_{11} mode is anticipated to have the major effects on potential intermodal dispersion inside a still pipe. If the total horn length would be a degree of freedom, Table 1 denotes the distinct improvements in the derived mode conversion behavior depending solely on length extension, showing the steady reduction of the parasitic eigenmodes at the center frequency of f=9.5 GHz. This trivial solution is well-known and will therefore not be considered in our evaluation.

3 Still pipe simulation

In this section the features and the underlying analytical formulations of a MATLAB implemented still pipe simulator are introduced, yielding the possibility to verify the waveguide tapers directly by means of the overall gauging performance. The accuracy prediction is an important figure of merit, especially for the design and optimization process of novel taper structures. Thus, an existing measurement setup is emulated in software by utilizing analytical waveguide equations to account for the loss-free mode-dependent wave propagation behavior inside the metal pipe (Barrow, 1984). The setup is schematically depicted in Fig. 4. The setup approximates the expected reflections caused by arbitrary material properties of liquids in terms of a movable plane sliding short within the still pipe. Thus, modal coupling and conversion at this reflector can be neglected (Katsenelenbaum and Mercader, 1998) and the reflection coefficients for all $N-1$ eigenmodes are equally set to Γ_n=-1 (cp. Eq. 3). Without loss of generality it is advantageous to choose magnitudes of $|\Gamma_n|<1$ to attenuate the radar signal. Therefore, multiple signal reflections and resulting ringing can be reduced that could cause aliasing effects due to undersampling by exceeding the range of unambiguity. Generally, the range of unambiguity is determined by the number of frequency samples N_s, whereas N_s=1001 samples over a total bandwidth of Δf=2 GHz were evaluated. Signal interpolation in time domain can be considered via zeropadding in the frequency domain.

Table 1. Magnitudes of the conversion of the H_{11} mode into the spurious cylindrical waveguide modes E_{11}, H_{12}, E_{12} and H_{13} in dependence of the total linear horn length at the center frequency of f=9.5 GHz.

l/mm	80	160	240	320
$S_{2(E_{11}),1(H_{11})}$/dB	−10.8	−15.5	−19.5	−21.1
$S_{2(H_{12}),1(H_{11})}$/dB	−16.7	−22.6	−25.8	−28.2
$S_{2(E_{12}),1(H_{11})}$/dB	−18.9	−27.9	−33.8	−31.2
$S_{2(H_{13}),1(H_{11})}$/dB		below f_c		

According to Fig. 4 the whole system configuration consists of the excitation structure and the still pipe, whereas the reflection coefficient Γ_{res} represents the transfer function of the complete radar setup. FMCW measurements can be simulated by solely evaluating the real part of the reflection coefficient which results in an image error, as shown by Stolle and Heuermann (1995). Additionally, various window functions can be applied. For our investigations the Hannning window W_{Hann} is used. Subsequently, an inverse Fourier transform provides the impulse response, whereas solely the envelope is evaluated. Common signal processing algorithms, e.g. barycentric pulse detection (Le Huerou and Gindre, 2003), can be utilized, even though the still pipe is a dispersive transmission line. Minimal deteriorations are obtained, if the still pipe is used in single-mode operation considerably beyond the cut-off frequency $f_{c,H_{11}}$ of the desired H_{11} mode. The determined pulse round trip time is then converted into the particular measurement distance by multiplying with the broadband average signal velocity $\overline{v}_{gr,H_{11}}$, as given by Eq. (1). This equation accounts in an appropriate manner for the frequency-dependent and mode-specific propagation properties.

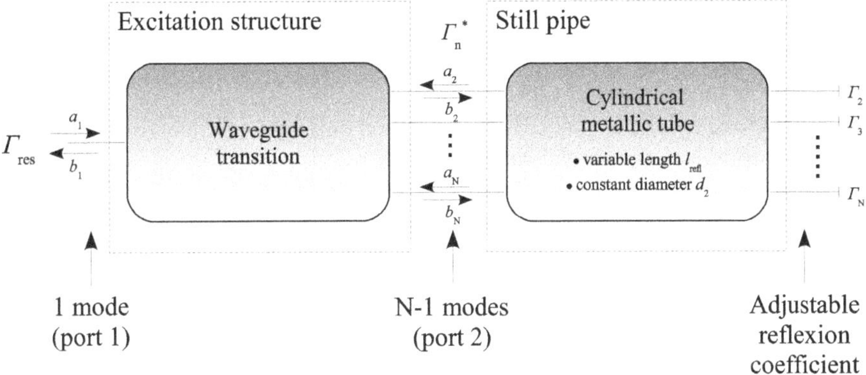

Fig. 4. Scheme of the MATLAB-based still pipe simulation.

$$\overline{v}_{\mathrm{gr,H}_{11}} = \frac{\sum\limits_{k=1}^{N_s} \left(|S_{21}(f_k)| \cdot W_{\mathrm{Hann}}(f_k)\right)^2 \cdot v_{\mathrm{gr,H}_{11}}(f_k)}{\sum\limits_{k=1}^{N_s} \left(|S_{21}(f_k)| \cdot W_{\mathrm{Hann}}(f_k)\right)^2} \tag{1}$$

$$\text{with} \quad v_{\mathrm{gr,H}_{11}}(f_k) = \frac{1}{\sqrt{\varepsilon\mu}} \sqrt{1 - \left(\frac{f_{c,\mathrm{H}_{11}}}{f_k}\right)^2}$$

Equation (1) derives from summing up all monofrequent group velocities $v_{\mathrm{gr,H}_{11}}(f_k)$, each weighted with its individual spectral power, normalizing the entire sum to the total signal power. N_s specifies the frequency samples, whereas $|S_{21}(f_k)|$ denotes the magnitude of conversion into the desired H_{11} mode. ε and μ stand for the material permittivity and permeability, respectively. The formulation is derived in accordance to a similar pulse delay equation given by Hartog (1979). The excitation structure is described as a physical two-port device exhibiting N virtual port modes, one at the first and $N-1$ eigenmodes at the second port. Therefore, the whole waveguide transition can fully be described by a $N \times N$ scattering matrix, as follows:

$$\begin{pmatrix} b_1 \\ \vdots \\ b_N \end{pmatrix} = \begin{bmatrix} S_{11} & \cdots & S_{1N} \\ \vdots & \ddots & \vdots \\ S_{N1} & \cdots & S_{NN} \end{bmatrix} \cdot \begin{pmatrix} a_1 \\ \vdots \\ a_N \end{pmatrix}. \tag{2}$$

In Eq. (2) a_n and b_n denote the nth incident and reflected wave quantities, respectively. Whereas the reflected waves b_n at the second port can be expressed by the transformed reflection coefficient Γ_n^* of the still pipe, as given by Eq. (3).

$$b_n = \Gamma_n^* \cdot a_n = \frac{1}{\Gamma_n} \cdot e^{2\gamma_n l_{\mathrm{refl}}} \cdot a_n \quad \forall \quad n \geq 2 \tag{3}$$

The parameter γ_n stands for the wave propagation constant of the nth cylindrical waveguide mode. Thus, the reflected wave quantities in Eq. (2) are consequentially substituted according to Eq. (3), leading to a reduction to the wave quantity b_1, as follows:

$$\begin{pmatrix} b_1 \\ 0 \\ \vdots \\ 0 \end{pmatrix} = \overbrace{\begin{bmatrix} S_{11} & S_{12} & \cdots & S_{1N} \\ S_{21} & S_{22} - \Gamma_2^* & \cdots & \vdots \\ \vdots & \vdots & \ddots & \vdots \\ S_{N1} & \cdots & \cdots & S_{NN} - \Gamma_N^* \end{bmatrix}}^{[S_{\mathrm{mod}}]} \cdot \begin{pmatrix} a_1 \\ a_2 \\ \vdots \\ a_N \end{pmatrix}. \tag{4}$$

S_{mod} denotes the modified scattering matrix exhibiting solely one single output coefficient b_1 and combining the unknown incident waves in terms of the vector quantity $\boldsymbol{a}$. The set of linear equations is solved by setting the value of $b_1=1$ and inverting the derived matrix $[S_{\mathrm{mod}}]$.

$$\boldsymbol{a} = [S_{\mathrm{mod}}]^{-1} \cdot (1, 0 \cdots 0)^{\mathrm{T}}, \tag{5}$$

Subsequently, Γ_{res} is given by:

$$\Gamma_{\mathrm{res}} = \frac{b_1}{a_1} = a_1^{-1}, \tag{6}$$

concerning all system components including the full-wave simulation results of the mode conversion behavior of arbitrary waveguide transitions. Hence, on the basis of the still pipe simulator, fundamental evaluation of the intermodal and chromatic dispersion can efficiently be performed. In the following sections, insights to the design constraints will be given with respect to the mode conversion properties of a waveguide transition. As already shown in Sect. 2, due to geometry, polarization and symmetry properties of the regarded cylindrical waveguide transitions, the maximal degree of the mode index is set to $N=6$ for further investigations.

3.1 Analytical model of a waveguide transition

An analytical model of a waveguide transition will be presented leading to a direct relationship between the level of spurious mode excitation at port 2 and the measurement uncertainties and thus providing the fundament of modal design rules for still pipe excitation structures. The choice of modal level can be adjusted by determinating the modal transmission parameters S_{n1} into each mode (with $2 \leq n \leq N$). These scattering parameters are assumed to have purely real values, i.e. the phase is neglected and therefore the spatial extension is set to zero.

Considering loss free, isotropic and passive waveguide structures, the corresponding scattering matrix will become reciprocal (cp. Eq. 7) and unitary (cp. Eq. 9). The complete scattering matrix accounts for the matching properties of such a N-port structure, thus being comparable with a three-port device that cannot be matched at all ports. In this case, the multimode waveguide transition cannot be matched for every single mode, which arises mode-dependent multiple reflection cycles inside the tube. A serious consequence are pulse replica within the system's impulse response, which may exceed the peak amplitudes of the first pulse. This originates from a phenomenon similar to the mode beating effect well-known from the theory of optical transmission line (Fernandez Casares and Balle, 1994). As a result, the applied barycentric processing algorithm unlatches due to the inaccurate pulse maximum detection on pulse replica.

Exemplary, $N=3$ is considered for the following evaluation, resulting in solely one parasitic mode, i.e. the incident mode power is purely split up in the modes H_{11} and E_{11}, respectively. By incorporating the principal of reciprocity, given by:

$$[S] = [S]^{\mathrm{T}}, \tag{7}$$

the corresponding scattering matrix is given by Eq. (8), whereas the assumption $S_{11}=0$ is chosen to approximate a perfect matched excitation port, as follows:

$$[S_{\mathrm{ana}}] = \begin{bmatrix} 0 & S_{21} & S_{31} \\ S_{21} & \boldsymbol{S_{22}} & \boldsymbol{S_{23}} \\ S_{31} & \boldsymbol{S_{23}} & \boldsymbol{S_{33}} \end{bmatrix}. \tag{8}$$

Thus, the three unknown quantities are the marked scattering parameters S_{22}, S_{33} and S_{23}. Hence, utilizing the assumed unitarity, a set of nonlinear equations is derived from the equations

$$[S]^{\mathrm{T}} [S]^{*} = [S]^{*\mathrm{T}} [S] = [I] \quad \Leftrightarrow \quad [S]^{-1} = [S]^{*\mathrm{T}} \tag{9}$$

and is subsequently solved numerically. In Eq. (9), $[I]$ denotes the identity matrix, $(^{*})$ accounts for the complex conjugate operation and $(^{\mathrm{T}})$ stands for the transpose of matrix. The phase of the unknown three parameters reaches merely two discrete values of 0° and 180°, due to absent spatial extension. In general, solving the nonlinear system numerically for $N>3$ becomes more complex, because it is overdetermined, however, also includes redundancies among the equations itself. This causes ambiguities in the solution

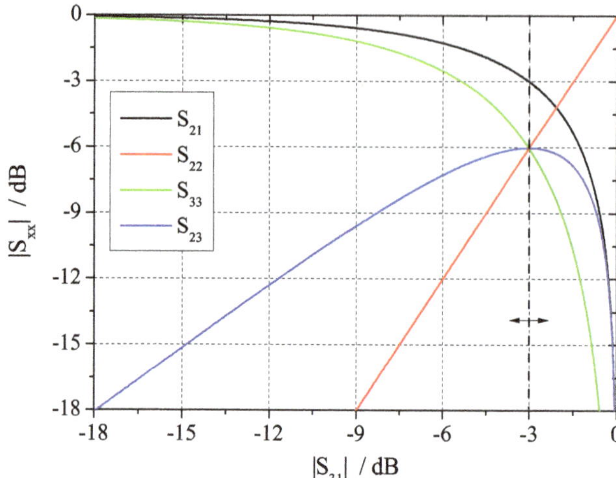

Fig. 5. Scattering parameters for a virtual 3×3 transition in dependence of the parasitic transmission level of $|S_{31}|$.

strongly depending on the choice of the initial values for the unknowns. The total number of unknowns is given by $(N^2-N)/2$.

Finally, the derived analytical waveguide transition can be utilized for the generation of scattering matrices having user-defined spurious modal transmission levels, as depicted in Fig. 5 for the special case of $N=3$. Here, the parameter S_{31} accounts for the excitation of the parasitic mode E_{11} that entirely defines the remaining parameters, e.g. for a mode level of $S_{31}=-3$ dB, S_{21} takes the same value whereas $|S_{22}|=|S_{33}|=|S_{23}|=-6$ dB is derived. Henceforth, the analytical waveguide transition is utilized to characterize intermodal dispersion effects in still pipes that refers to constant modal conversion levels over arbitrary frequency ranges.

3.2 Intermodal dispersion

This section focuses on the analysis of intermodal dispersion, which significantly determines the influences on the measurement uncertainties. Accurate detection properties are achieved when, due to temporal walk-off, pulse breakup has occurred (Shum, 2004), i.e. the individual modes' pulses have sufficiently separated. The chromatic dispersion is negligible causing rather small pulse distortion due to the low cut-off frequency of the H_{11} operational mode. In Fig. 6 the obtained impulse response for a distinct reflector position at $l_{\mathrm{refl}}=3$ m is depicted, accounting for two different reflection coefficients $|\Gamma_{\mathrm{n}}|=0.5 \vee 1$. The pulse package consists of the two superimposed modes, i.e. H_{11} and E_{11}, equally excited and thus splitted to -3 dB. The subdivision directly refers to Sect. 3.1 and the marked scattering parameters shown in Fig. 5. In both cases the signal energy incorporated by the pulse packages is more and more spread with increasing numbers of reflection, whereas the major difference is observed to be the signal-to-noise ratio (SNR). At $|\Gamma_{\mathrm{n}}|=1$ the

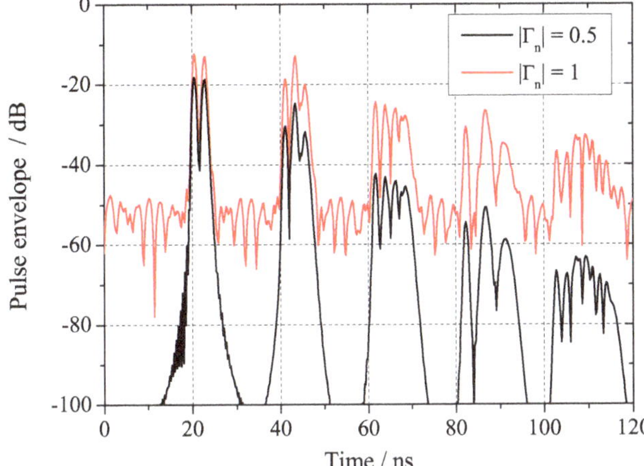

Fig. 6. Multiple reflections of the fundamental H_{11} and the first spurious E_{11} mode in dependence of the reflection coefficient $|\Gamma_n|$

SNR value is raised about 50 dB compared to the reflection coefficient of $|\Gamma_n|=0.5$. As already mentioned in Sect. 3, this is a result of signal ringing. To avoid the arising undersampling phenomenon, an appropriate value of $|\Gamma_n|$ is chosen for further investigations. For the investigation the total number of ports $N=4$ is considered, which specifies two parasitic modes (E_{11}, H_{12}). The adjustable reflector distance of the simulated still pipe is given in an interval of $l_{\text{refl}}=[0\ldots5\,\text{m}]$, while the step width is chosen to $\Delta l_{\text{refl}}=l_{\text{refl,max}}/1000$. Subsequently, the graphs are scaled and normalized in terms of an equivalent non-dispersive pulse propagation without any attenuation, being fully reflected. When measuring with a FMCW radar technology, solely the real part of the derived total reflection coefficient Γ_{res} is evaluated; the peak pulse amplitude becomes bisected to -6 dB. Multimode propagation and thus arising mismatches at the transition site in combination with reflector losses will result in a decrease of the peak pulse amplitudes, if a fixed total incident power level is provided. For an in-depth investigation of the intermodal dispersion of the first pulse package, the same reflector distance $l_{\text{refl}}=3$ m is chosen in combination with a fixed total parasitic mode power of 25%, according to the incident mono-mode power at port 1. At this power level, pulse replica influences are not of any concern. Therefore, the explicit pulse shape deterioration by intermodal dispersion solely of the first incoming package is depicted in Fig. 7a for different mode configurations. The mono-mode H_{11} pulse is well-shaped without any observable indication of chromatic dispersion. In contrast to the curve including the spurious E_{11} mode, the obtained package consists of two pulses, whereas the E_{11} pulse is delayed due to its increased cut-off frequency. If E_{11} is exchanged by H_{12}, that exhibits an even higher cut-off frequency compared to E_{11}, the pulse delay as well as the H_{12} pulse spread is raised and thus leads to an improvement in intermodal pulse separation. Finally, the excitation of both spurious modes is equally combined resulting in attenuated pulses without influencing the absolute pulse position in time domain.

This leads to the assumption, that the excitation of merely the E_{11} mode represents the worst case scenario for the barycentric signal detection algorithm. This pulse package is affected by the longest displacement of the barycenter, that results in an increase in measurement uncertainties. In detail, the described behavior is depicted in Fig. 7b, showing the simulated distance error e over the entire reflector range from $l_{\text{refl}}=0$ up to 5 m. At the distinct position of $l_{\text{refl}}=3$ m, having been already evaluated, the obtained error for all sets of modes is negligible because a sufficient level of pulse separation is obtained. Generally, in the close-up range of $l_{\text{refl}}\leq0.2$ m the distance inaccuracies of all sets are similarly affected by the FMCW image error. Further errors caused by a finite input matching of the analytical waveguide transition are excluded, as given by Eq. (8). Therefore, this range is not considered for further investigations. According to Fig. 7b, all curves have an oscillating characteristic that is caused by the already mentioned pulse beating deterioration. As expected, considering the entire reflector interval l_{refl} the excitation of solely the E_{11} mode exhibits the longest influence on the error curve, followed by H_{12} causing the same peak error of approximately $e_{\max}=20$ mm, however, decaying much faster with increasing reflector distances.

Finally, the combination of these two spurious modes results in a smaller peak error, due to the constant parasitic power level now is split into two modes, which is therefore advantageous for real taper design. If the influences of H_{12} have completely decayed starting from a distinct reflector position of $l_{\text{refl}}\approx1.2$ m, the E_{11} error characteristic remains, obtaining the same decaying length.

3.3 Design criterion for waveguide tapers

The aim of this section is to quantify the level of mode suppression that is necessary to achieve submillimeter accuracy. In accordance with the investigated intermodal dispersion behavior in the previous Sect. 3.2, the exclusive excitation of the E_{11} mode marks the worst case scenario to cope with. Therefore, in the present section a design criterion is derived, especially accounting for the first spurious mode. Based on experience with real waveguide taper, improvement in parasitic mode suppression is limited to a certain frequency range. Hence, the criterion is extracted by assuming a spurious E_{11} mode magnitude of $\left|S_{2(E_{11}),1(H_{11})}\right|=-20$ dB over a bandwidth ranging in an interval of $\Delta f_{\text{notch}}=[0\ldots2\,\text{GHz}]$. Obviously, a certain amount of bandwidth of the proposed suppression level is required. Δf_{notch} is located around the center frequency of $f=9.5$ GHz, where the radar system is supposed to have highest sensitivity compared to the corner frequencies due to the applied signal processing window.

Therefore, an analytical waveguide taper ($N=3$) having a stepwise constant scattering parameter S_{31} accounting for

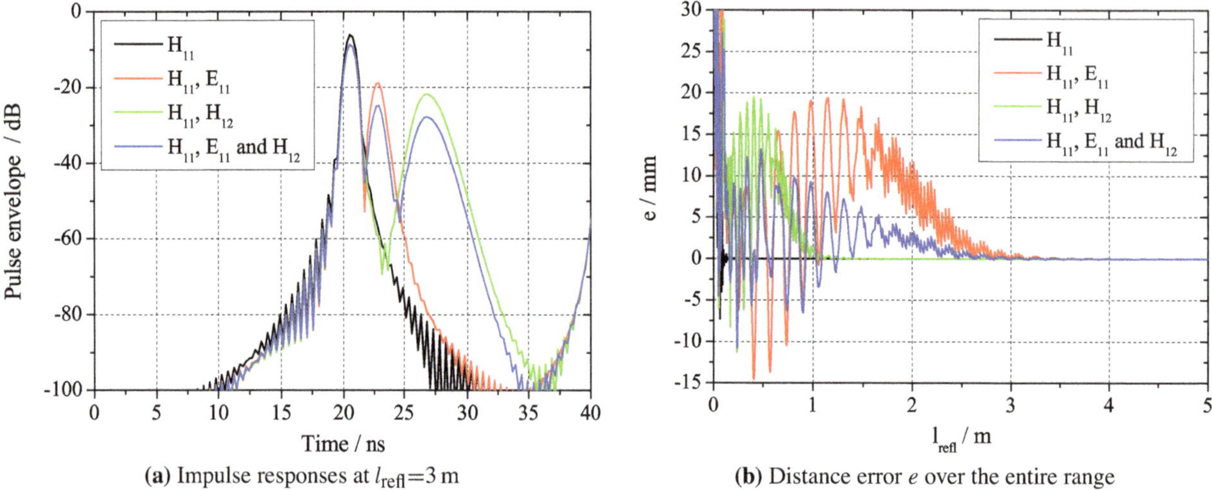

(a) Impulse responses at $l_{\text{refl}}=3\,\text{m}$

(b) Distance error e over the entire range

Fig. 7. Time signal and corresponding distance error for four sets of eigenmodes including $\text{H}_{11}, \text{E}_{11}$ and H_{12}.

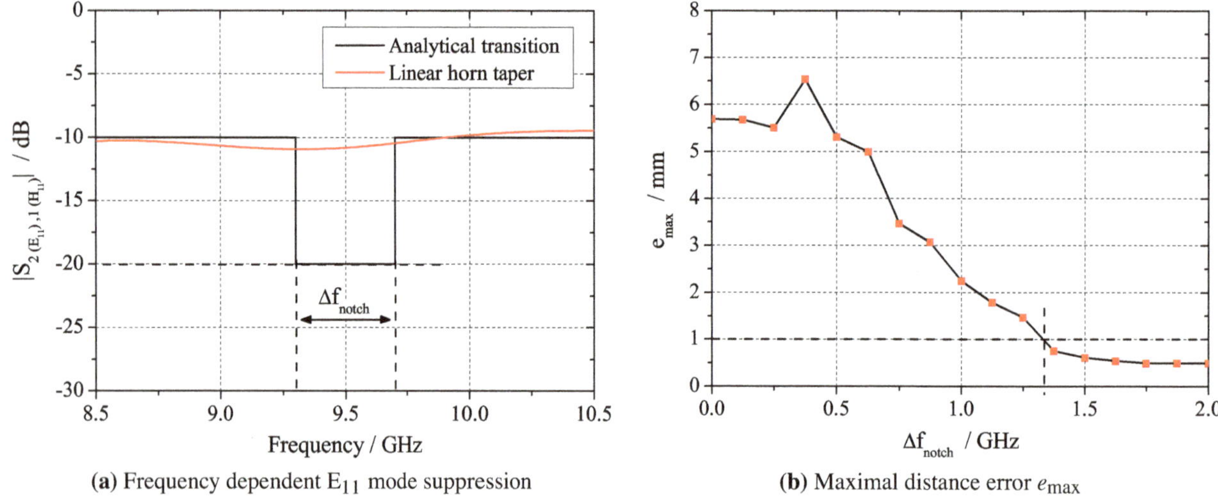

(a) Frequency dependent E_{11} mode suppression

(b) Maximal distance error e_{max}

Fig. 8. Worst case estimation of the level of E_{11} mode suppression in dependence of the required bandwidth.

the transmission properties of $|S_{2(\text{E}_{11}),1(\text{H}_{11})}|$ is utilized, as introduced in Sect. 3.1. The corresponding characteristic is depicted in Fig. 8a starting from the approximation of the circular linear horn at a constant E_{11} magnitude of $|S_{2(\text{E}_{11}),1(\text{H}_{11})}|=-10\,\text{dB}$, as shown in Sect. 2. In Fig. 3.2 the obtained maximal distance error e_{max} is shown depending on the value of Δf_{notch} for $0.2\,\text{m}<l_{\text{refl}}<5\,\text{m}$. As depicted, the error continuously decreases with rising values of the notch bandwidth Δf_{notch}. At a distinct bandwidth of $\Delta f_{\text{notch}}\approx 1.3\,\text{GHz}$ the corresponding maximal distance error e_{max} is falling below the desired value of $e_{\text{max}}<1\,\text{mm}$. Beyond this certain bandwidth the remaining accuracy level keeps within the submillimeter range. In summary, a direct relationship between the spurious mode suppression level and the obtained FMCW radar distance error is accomplished. Henceforth, real waveguide transitions can solely be evaluated by its spurious mode transmission behavior. Thus, the verification of every single step in the design and optimization process of waveguide tapers in terms of distance errors can be omitted, leading primarily to a raise in the design speed of novel taper structures.

4 Mode-matched excitation structures

Based on the derived design criterion according the previous Sect. 3.3, improved real waveguide taper concepts are considered by utilizing a commercial 3-D FIT solver (CST MICROWAVE STUDIO, Vers. 2006B) for numerical evaluation. To briefly review the major consequences drawn by the analysis of Sect. 2, the fundamental mode excitation purity of a circular linear horn taper obviously suffers from spherical phase fronts inside the cone. That results in an increase of transmission into the major spurious mode E_{11},

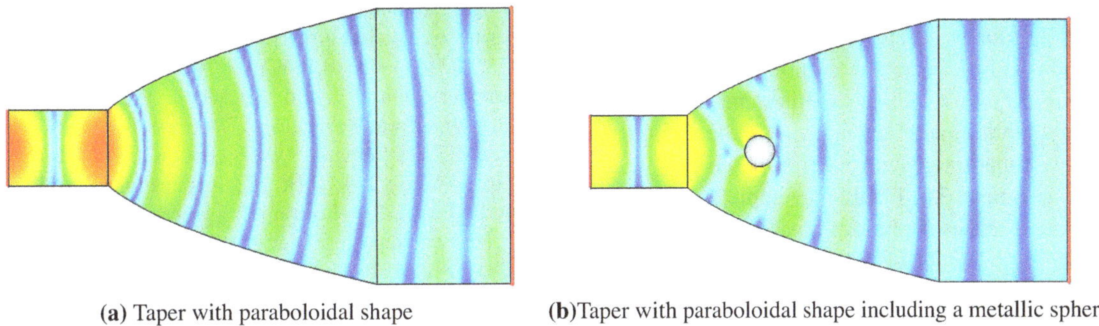

(**a**) Taper with paraboloidal shape (**b**)Taper with paraboloidal shape including a metallic sphere

Fig. 9. Electric field distribution at f=9.5 GHz in two different paraboloidal waveguide transitions.

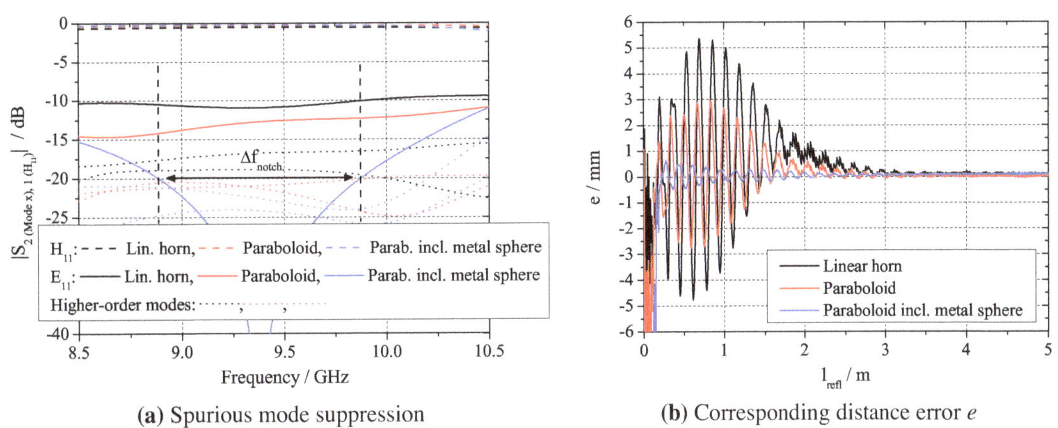

(**a**) Spurious mode suppression (**b**) Corresponding distance error e

Fig. 10. Comparison of three different waveguide tapers in terms of excited mode levels and obtained FMCW measurement uncertainties.

when changing back the eigenmode system from spherical to cylindrical at the second port. At first, well-known analogies from the ray-optics are considered, that may lead to a phase curvature reduction by solely changing the contour of the waveguide taper, so that the phase distribution becomes flattened. Whereas the distinct feature of straight phase fronts denotes for the possible evidence of a single-moded field distribution, it may not be considered as a general proof. Nevertheless, the phase characteristic can be utilized to support the design and optimization process. In the fictional case of a punctual isotropic radiating source in the focal point of a paraboloid, the reflected rays form a straight phase distribution in the plane of the second port. However, this analogy is purely valid when the rays, which are directly propagating in the direction of the second port, are neglected. Hence, our approach consists of a combination of both principles by exploiting the flattening effects of a taper exhibiting a parabolic shape and simultaneously suppressing the direct wave propagation through the whole taper structure by the insertion of an appropriate obstacle. The influence of the applied obstacle on the field distribution inside the taper structure is comparable to a subreflector of a Cassegrainian antenna feed system (Rusch, 1963). For our purpose to design mode-matched waveguide tapers, a metallic sphere was found to provide remarkable improvements.

Figure 9 depicts the achieved improvements regarding the phase distribution for two cases of parabolic transitions. According to the curvature of the phase fronts at the second port, the parabolic shape (see Fig. 9a) is advantageous compared to the distribution obtained by the circular linear horn, depicted in Fig. 2. Furthermore, when the metallic sphere is applied, an additional improvement can be observed. This behavior is verified by the simulated scattering parameter, as given in Fig. 10a. Subsequently, the mode magnitude level $|S_{2(\mathrm{E}_{11}),1(\mathrm{H}_{11})}|$ decreases, offering best results in case of the parabolic structure combined with the metallic obstacle. In this case, a notch bandwidth $\Delta f_{\mathrm{notch}} \approx 1.0$ GHz considering a E_{11} mode suppression level of more than 20 dB can be realized in the range of 8.9 GHz to 9.9 GHz. Other higher order modes are likewise suppressed by approximately more than 20 dB.

Figure 10b shows the derived distance error, in the same manner as introduced in Fig. 7b. The measurement uncertainties are successively decaying between the three structures, starting at a maximal error of about $e \approx 5$ mm in case of a linear horn, for the paraboloid taper a decreased maximal error of $e \approx 3$ mm is obtained. As expected, the best results are achieved by the paraboloid including the proposed metal sphere as a subreflecting obstacle, maintaining a continuous accuracy level of less than one millimeter. Although, the notch bandwidth of this structure is less than the requested $\Delta f_{\mathrm{notch}} \approx 1.3$ GHz, due to a transmission notch of more than 20 dB that extensively exceeds the criterion's requirements, a bandwidth of $\Delta f_{\mathrm{notch}} \approx 1.0$ GHz satisfies the demand for a taper offering submillimeter accuracy. This fact confirms the worst case estimation given by the proposed design criterion according to Sect. 3.3. Finally, the application of such a metallic obstacle causes no major impairments concerning the input matching behavior of the first port. Although not having been an integral part of our investigation, a value of $|S_{1(\mathrm{H}_{11}),1(\mathrm{H}_{11})}| < -10$ dB is reached over the frequency range of operation. Further effort on this quantity could be spent in the design process of a corresponding prototype taper based on this promising concept.

5 Conclusions

In this paper, compact mode-matched excitation structures for the application in FMCW radar distance measurements in still pipes have been fundamentally investigated. These structures are tapers that function as waveguide transitions between circular waveguides of different diameters. By establishing a still pipe simulator, incorporating all effects of multimode propagation, it was shown that intermodal dispersion effects, caused by spurious mode excitation, dominates the measurement uncertainties, if common signal processing algorithms are deployed. This leads to a great demand of a design criterion, that appropriately accounts for the spurious mode attenuation as well as for the suppression bandwidth to accomplish certain measurement specifications, e.g. for augmenting the accuracy level to the submillimeter domain. Finally, a novel subreflector-based taper concept was introduced, exhibiting promising mode suppression levels. By verifying the corresponding distance error, it was clarified that this concept meets the requirements for FMCW high-precision level detection conducted in large overmoded circular waveguides.

References

Barrow, W.: Transmission of electromagnetic waves in hollow tubes of metal, Proceedings of the IEEE, 72, 1064–1076, 1984.

Brumbi, D.: Measuring process and storage tank level with radar technology, Radar Conference, 1995., Record of the IEEE 1995 International, pp. 256–260, doi:10.1109/RADAR.1995.522555, 1995.

Brumbi, D.: Low power FMCW radar system for level gaging, Microwave Symposium Digest., 2000 IEEE MTT-S International, 3, 1559–1562 vol.3, doi:10.1109/MWSYM.2000.862273, 2000.

Fernandez Casares, S. and Balle, S. M.-V. P.: Mode beating and spontaneous emission noise effects in a variable-waveguide model for the dynamics of gain-guided semiconductor laser arrays, IEEE J. Quantum Electronics, 30, 2449–2457, doi:10.1109/3.333695, 1994.

Hartog, A.: Influence of waveguide effects on pulse-delay measurements of material dispersion in optical fibres, Electronics Lett., 15, 632–634, doi:10.1049/el:19790450, 1979.

Katsenelenbaum, B. Z. and Mercader, L. P. M. S. M. T. M.: Theory of Nonuniform Waveguides: The Cross-Section Method, vol. 44, ser. IEE Electromagn. Waves. London, U.K., IEE Press, 1998.

Kielb, J. A. and Pulkrabek, M.: Application of a 25 GHz FMCW radar for industrial control and process level measurement, Microwave Symposium Digest, 1999 IEEE MTT-S International, 1, 281–284, doi:10.1109/MWSYM.1999.779475, 1999.

Le Huerou, J.-Y. and Gindre, M. A. A. U. W. W. M.: Compressibility of nano inclusions in complex fluids by ultrasound velocity measurements, IEEE Transactions on Ultrasonics, Ferroelectrics and Frequency Control, 50, 1595–1600, doi:10.1109/TUFFC.2003.1251143, 2003.

Musch, T.: A high precision 24-GHz FMCW radar based on a fractional-N ramp-PLL, Instrumentation and Measurement, IEEE Trans., 52(2), 324–327, doi:10.1109/TIM.2003.810046, April 2003.

Narasimhan, M. S. and Balasubramanya, K.: Transmission Characteristics of Spherical TE and TM Modes in Conical Waveguides (Short Papers), IEEE Trans. Microwave Theory and Techniques, 22, 965–970, 1974.

Parker, S.: Diverse uses for level radar, InTech, ISA International Society for Measurement and Control, May 2002.

Pohl, N. and Gerding, M. W. B. M. T. H. J. S. B.: High Precision Radar Distance Measurements in Overmoded Circular Waveguides, IEEE Trans. Microwave Theory and Techniques, 55, 1374–1381, doi:10.1109/TMTT.2007.896784, 2007.

Rusch, W.: Scattering from a hyperboloidal reflector in a cassegrainian feed system, IEEE Trans. Antennas and Propagation, 11, 414–421, 1963.

Sai, Bin and Kastelein, B.: Advanced High Precision Radar Gauge for Industrial Applications, International Conference on Radar, 2006. CIE '06., pp. 1–4, doi:10.1109/ICR.2006.343173, 2006.

Shum, M. L. P.: Effects of intermodal dispersion on short pulse propagation in an active nonlinear two-core fiber coupler, Photonics Technology Lett., IEEE, 16, 1080–1082, doi:10.1109/LPT.2004.824994, 2004.

Stolle, R. and Heuermann, H. S. B.: Novel algorithms for FMCW range finding with microwaves, Microwave Systems Conference, 1995. Conference Proceedings., IEEE NTC '95, pp. 129–132, doi:10.1109/NTCMWS.1995.522875, 1995.

Tang, C.: Mode Conversion in Tapered Waveguides At and Near Cutoff, IEEE Trans. Microwave Theory and Techniques, 14, 233–239, 1966.

Weiss, M.: Low-cost, low-power nanosecond pulse radar for industrial applications with mm accuracy, 2001 International Symposium on Electron Devices for Microwave and Optoelectronic Applications, pp. 199–204, doi:10.1109/EDMO.2001.974307, 2001.

Weiss, M. and Knochel, R.: A Highly Accurate Multi-Target Microwave Ranging System for Measuring Liquid Levels in Tanks, 27th European Microwave Conference, 1997, 2, 1103–1112, doi:10.1109/EUMA.1997.337945, 1997.

28

Decreased noise figure measurement uncertainty in Y factor method

I. Gaspard

Hochschule Darmstadt, FB Elektrotechnik und Informationstechnik, Darmstadt, Germany

Correspondence to: I. Gaspard (ingo.gaspard@h-da.de)

Abstract. Almost always noise figure is measured today by using a matched noise source delivering two different but known noise temperatures (Y factor method). In commercially available noise sources these temperatures are characterized by the excess noise ratio (ENR) value, describing the ratio of equivalent noise power when switched on related to the noise power of a resistor at a temperature of 290 K (switched off). For a typical ENR value of 5 dB that means a hot temperature $T_h = 1207$ K and a cold temperature $T_c = 290$ K.

1 Introduction

In this paper the impact of T_c onto the noise figure uncertainty is analysed. There are several advantages when holding T_c as low as possible and not – as usual done – at 290 K. To realize T_c below room temperatures there are several possibilities, like cooling a resistor with liquid nitrogen or using sky noise by a suited antenna, see e.g. Zhutyaew (2010). Drawbacks of these solutions are either that they are expensive/need huge effort or are narrowband.

Therefore in this paper an alternative solution of a broadband noise generator on the basis of a Schottky diode operated in the conduction region will be described. In conduction region the Schottky diode shows a mixture of shot noise as well as thermal noise caused by the metal-semiconductor junction and the bulk resistance. By simple DC (direct current) measurements the overall noise temperature can be precisely determined and is in the range of 150 K...200 K, depending on the diode type used. At the same time the bulk resistance in series with the differential resistance of the proper biased diode shows very good match to 50 Ohms over a wide frequency range. Thus a noise normal can be implemented by simple DC and reflection factor measurement. Application and measurement results will be presented.

2 Noise figure measurement uncertainty

In the standard Y factor measurement method to measure noise figure two main error sources are specified by equipment manufacturers: the uncertainty of the noise source in terms of ΔENR and the uncertainty of the power ratio or instrumentation error ΔY. Typical values range e.g. between $\Delta\text{ENR} = \pm 0.3$ dB for an old Ailtech 7615 or $\Delta\text{ENR} = \pm 0.2$ dB for an HP346A noise source. For the instrumentation error of an HP8970A/B, Swain and Cox (1983), an instrumentation error of $\Delta Y = \pm 0.1$ dB is specified. When analyzing the resulting error in noise figure (NF uncertainty, see annex) it turns out that the resulting NF uncertainty is strongly dependent on the cold temperature T_c of the noise source. In Fig. 1 the NF uncertainty is shown as a function of cold temperature of the source for two different ENR values of the source and for two LNAs which are assumed for simulation purpose to differ in noise figure (NF = 0.1 dB and NF = 0.5 dB). From Fig. 1 it can be concluded that the cold temperature should be as low as possible to keep the resulting NF uncertainty as low as possible. It can also be seen that even for very low ENR (ENR = −4.5 dB) the NF uncertainty is reasonable low if cold temperature is as low as e.g. 150 K. Normally cold temperature of the noise source is at ambient room temperature, e.g. 290K. So it has to be found a (50Ω matched) noise source delivering a cold noise temperature as low as possible.

3 Schottky diode

Schottky diodes are formed by a metal-semiconductor junction and are widely used e.g. in mixers and detectors up to very high frequencies (> 100 GHz). This section will treat the necessary circuit models of a Schottky diode in order to provide a basic understanding of the noise generator implementation described in Sect. 4.

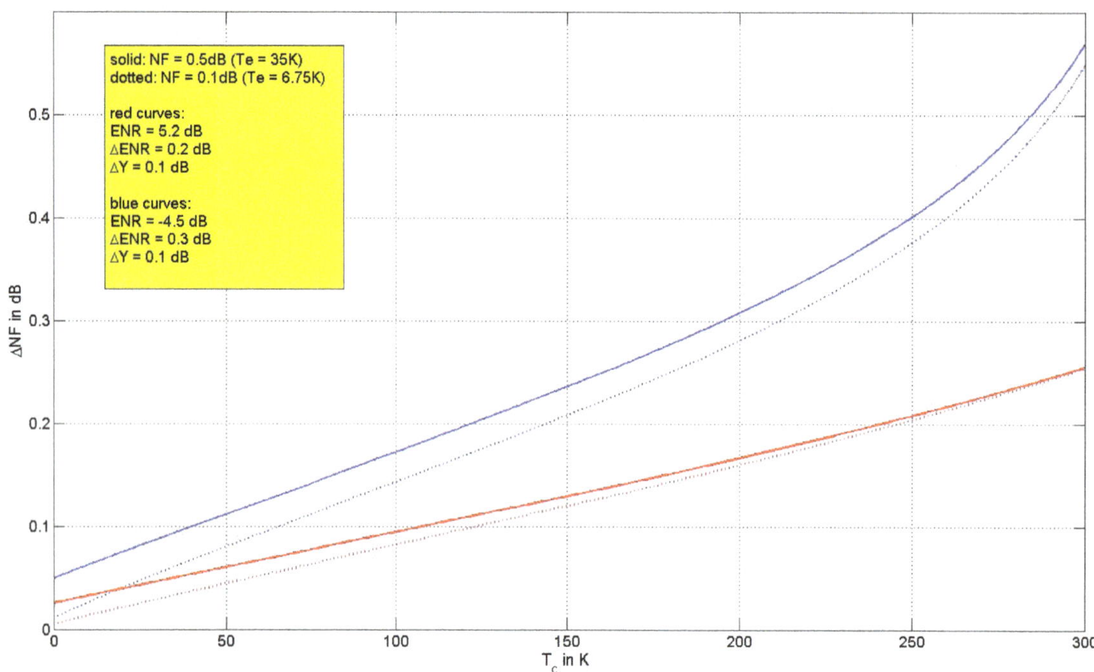

Fig. 1. Noise figure uncertainty in dB as function of T_c, NF of DUT and ENR for ΔENR = 0.2 dB (ENR calibration uncertainty of noise source) and $\Delta Y = 0.1$ dB (instrumentation uncertainty, e.g. HP8970).

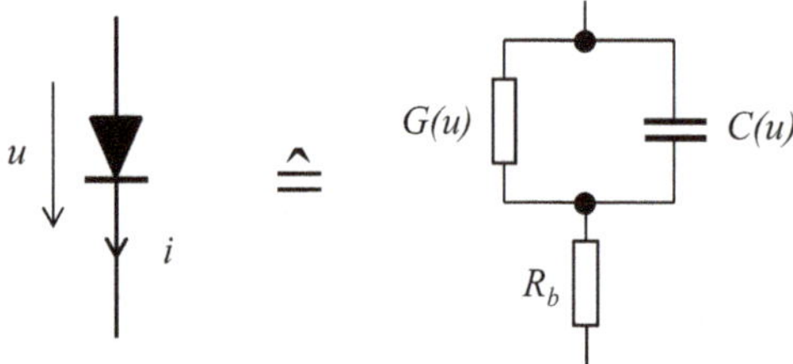

$G(u)$: nonlinear conductance of junction
$C(u)$: nonlinear capacitance of junction
(very small, e.g. < 0.5pF → high f)
R_b: bulk resistance (e.g. < 10Ω)

Fig. 2. Circuit model of Schottky diode.

3.1 DC and small signal properties

A widely used circuit model of the Schottky diode that is valid for both large-signal and small-signal analysis is given in Fig. 2, Maas (1993). It consists of a nonlinear resistance whose value is determined by the operational point and a capacitance representing the junction. Furthermore it contains a bulk resistance of a few Ohms. This model corresponds to the exponential I/U characteristic expressed in the following equation

$$i(u) = I_S \cdot \left(\exp\left(\frac{q \cdot u}{\tilde{n} \cdot kT}\right) - 1\right) \tag{1}$$

where kT/q describes the temperature voltage (26 mV at room temperature), I_S is the saturation current ($< 0.1\,\mu$A) and $\tilde{n}$ is the so-called ideality factor, an empirical factor describing the deviation from the exponential (values range from 1 to 1.2).

By inspecting the DC I/U-characteristic of the Schottky diode (either by taking measurements or by looking to the data sheet) all necessary values to parameterize a suitable model of the Schottky diode can be derived – see Fig 3. That is namely the ideality factor $\tilde{n}$, the bulk resistance R_b and the saturation current I_S.

3.2 Noise model

When operated in forward direction a Schottky diode shows a mixture of thermal noise due to the bulk resistance and shot noise due to the quantized charge carriers which have to climb the metal-semiconductor junction. That is summarized into a noise equivalent circuit with two independent noise sources given in Fig. 4 from Schiek et al. (2006). The conductance G is the differential conductance of the diode at its DC operating point.

It turns out that the ratio of effective noise temperature at the diode's terminals to the ambient (physical) room

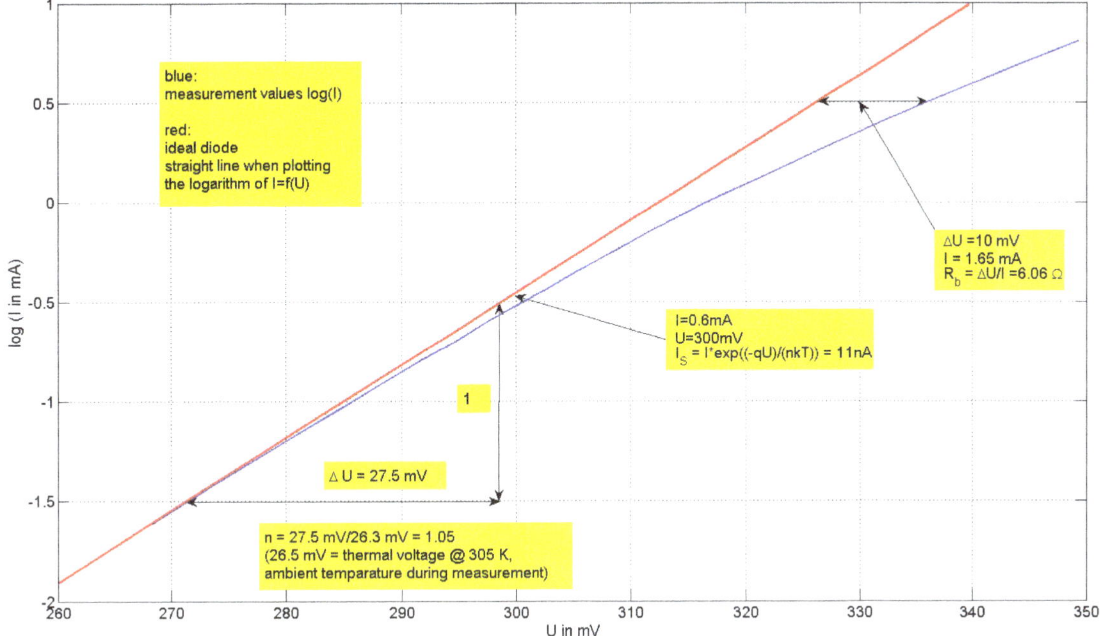

Fig. 3. I/U characteristic of an Agilent HSMS-2823 Schottky diode.

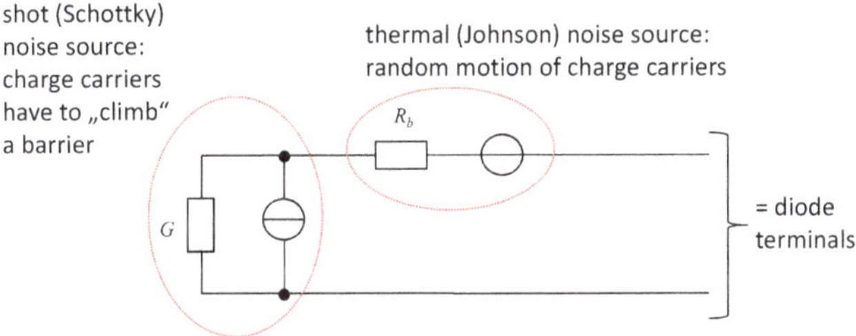

Fig. 4. Noise equivalent circuit model of a Schottky diode with bulk resistance.

temperature as a function of ideality factor $\tilde{n}$, DC current I_0 and saturation current I_S when neglecting the bulk resistance is given to be Schiek et al. (2006)

$$\frac{T^*_{ef}}{T} = \frac{1}{2} \cdot \tilde{n} \cdot \left(1 + \frac{I_S}{I_0 + I_S}\right) \tag{2}$$

Inspecting Eq. (2) it can be seen that because the saturation current is negligible in comparison to the forward current I_0 the noise temperature could be at minimum half of the ambient (physical) temperature.

When including thermal noise from the bulk resistance and taking into account Eq. (2) the overall noise temperature T_{ef} related to the ambient room temperature T is then, see Schiek et al. (2006),

$$\frac{T_{ef}}{T} = \frac{\frac{T^*_{ef}}{T} + R_b \cdot G}{1 + R_b \cdot G} \tag{3}$$

4 Implementation

Based on the diode model introduced in Sect. 3 a noise source by means of a Schottky diode operated in forward direction was implemented, see Figs. 5 and 6.

DC current through the diode can be aligned by the variable 10 kΩ resistor. A precision shunt of 1 kΩ allows DC current measurement. DC voltage across the diode's terminals can be measured directly at the N-male output connector. An additional PT1000 temperature sensor in the same housing placed nearby the diode allows measurement of the ambient temperature of the diode. RF (radio frequency) noise of the diode is decoupled from the DC part by a 10 kΩ chip resistor and 1 nF disk capacitor.

By taking DC measurements in this setup bulk resistance, ideality factor and saturation current were derived according to Fig. 3 and are in good agreement to the data sheet.

5 Application and measurements

In order to provide a matched load the DC current through the diode was aligned to make the bulk resistance in series with the differential junction resistance a pure 50Ω load by means of VNWA (vector network analyzer) measurement. Thus $1/G + R_b$ is equal to 50Ω. Best result was reached for DC current $I_0 = 0.67$ mA at ambient temperature of 305 K (32 °C). In Fig. 7 the measured return loss up to 1300 MHz is shown. The degradation of return loss for increasing frequency is mainly caused by the junction capacitance (~0.7 pF) of the diode. More expensive diodes

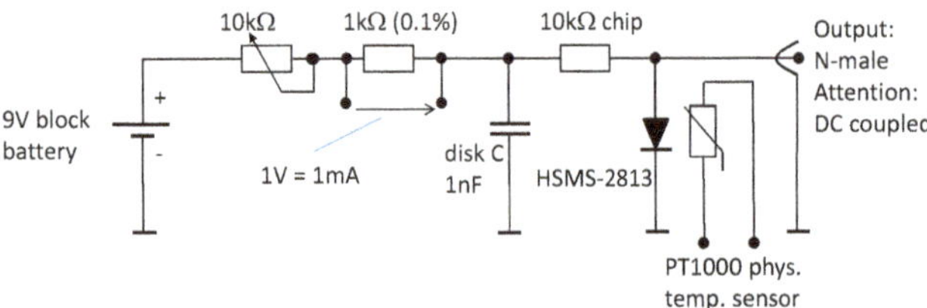

Fig. 5. Cicuit of implemented Schottky diode noise generator.

Fig. 6. Realization of noise generator.

with smaller junction capacitance should work at even much higher frequencies with sufficient return loss (RL). Of course the VNWA measurement has to be done carefully with low stimulus power (e.g. < -20 dBm) in order to avoid nonlinear behavior of the diode.

Applying the noise model and calculus in Sect. 3 delivered an equivalent noise temperature of 180 K at $I_0 = 0.67$ mA and ambient (physical) temperature of 305 K around the diode.

In Fig. 8 the combination of the Schottky noise generator with a standard noise generator is suggested in order to lower the overall cold temperature of the new noise source. This combination can be easily integrated into existing noise figure analyzer setups, e.g. the often used HP8970A/B. By combining the noise powers of the two noise generators by a directional coupler which has to have low loss in main line to avoid too much increasing of the overall cold temperature another advantage results: because of the low coupling (-19.5 dB) there is virtually no change in reflection coefficient for "on" and "off" state which would cause additional errors for mismatched devices under test ("gain error", see Bertelsmeier, 1988).

With the setup shown in Fig. 8 noise figure measurement at an MCL ZEL-1217LN low noise L-band amplifier was taken with an HP8970B noise figure analyzer. The noise figure was measured to be approximately 1 dB at 1296 MHz which is in good agreement with data sheet (typically 1.1 dB) and mea-

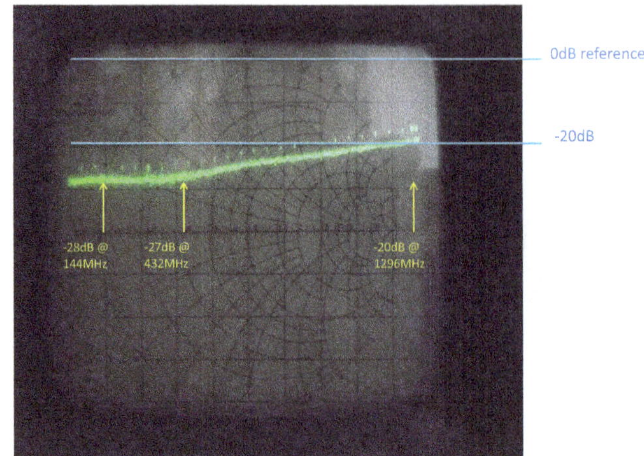

Fig. 7. Return loss (RL) of Schottky noise generator when biased for minimum RL.

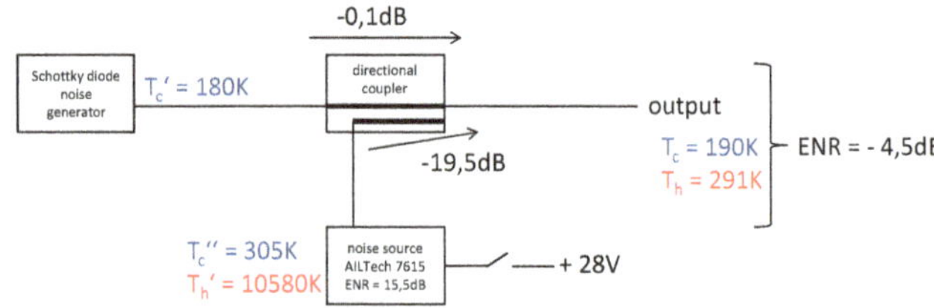

Fig. 8. Combining Schottky noise generator and standard noise source for "new" noise source with decreased overall cold temperature.

surements taken with an HP346A noise source (NF measured 0.97 dB).

6 Conclusions

In this paper a broadband "cold" matched noise source providing noise temperatures below room temperature based on a Schottky diode was described. Its application in a noise figure measurement setup with enhanced accuracy was shown. Because of the decreased cold temperature in comparison to the usually used noise sources a lower hot temperature provided by more attenuation and thus less gain error – see Bertelsmeier (1988) - can be applied in Y factor measurement method. Another advantage of the proposed noise generator is the decreased tolerance in effective ENR value due to decreased cold temperature.

Future work will evaluate in more detail the noise temperature uncertainty of the Schottky noise generator and its impact on NF measurement uncertainty.

Appendix A

Starting with the basic equation which is evaluated in a noise figure measurement setup we have (see Pozar, 2000; Bertelsmeier, 1988)

$$F + \frac{\mathrm{ENR} - Y\frac{T_c}{T_o} - 1}{Y - 1} \tag{A1}$$

where

- F: noise figure of device under test – linear scale,
- ENR: excess noise ratio of noise source – linear scale,,
- Y: power ratio of power at output of DUT when noise source switched on to power at output of DUT when noise source switched off – linear scale,
- T_c: cold noise temperature of noise source (switched off) in K,
- T_o: reference noise temperature, $T_o = 290\,\mathrm{K}$.

Furthermore the relation between equivalent noise temperature T_e given in Kelvin of the DUT and its (linear) noise figure F is given by (see e.g. Pozar, 2000)

$$F = 1 + \frac{T_e}{T_o} \tag{A2}$$

For the noise figure of the DUT in dB it holds

$$NF_{\mathrm{dB}} = 10 \cdot \log(F) \tag{A3}$$

Furthermore the definition of the (linear) ENR is given by

$$\mathrm{ENR} = \frac{T_h - T_c}{T_o} \tag{A4}$$

where T_h is the hot temperature when the noise source is switched on. Mostly ENR is given in dB's, thus

$$\mathrm{ENR}_{\mathrm{dB}} = 10 \cdot \log \frac{T_h - T_c}{T_o} = 10 \cdot \log(\mathrm{ENR}) \tag{A5}$$

The measured Y factor could be expressed in terms of T_e, T_h and T_c according to

$$Y = \frac{T_h + T_e}{T_c + T_e} \tag{A6}$$

or in dB's

$$Y_{\mathrm{dB}} = 10 \cdot \log(Y) \tag{A7}$$

As can be seen from equation Eq. (A1) two major sources of error in a noise figure measurement setup are the ENR calibration uncertainty and Y factor measurement uncertainty. For state of the art noise sources, e.g. a HP346A an ENR = 5...6 dB and an uncertainty of $\Delta\mathrm{ENR} = \pm 0.2$ dB are specified. The uncertainty of Y factor measurement is specified in terms of instrument uncertainty, e.g. $\Delta Y = \pm 0.1$ dB for an HP8970.

Obviously these two variables are independent and thus differential calculus in the form of Taylor series and combining the uncertainties in a root-sum-of-squares fashion can be applied to find the uncertainty of F:

$$\Delta F = \sqrt{\left(\frac{\delta F}{\delta Y} \cdot \Delta Y\right)^2 + \left(\frac{\delta F}{\delta \mathrm{ENR}} \cdot \Delta\mathrm{ENR}\right)^2} \tag{A8}$$

The partial derivatives of the function given in Eq. (A1) to the respective variables needed in equation Eq. (A8) are given as follows:

$$\frac{\delta F}{\delta Y} = -\frac{\mathrm{ENR} + 1}{(Y-1)^2} + \frac{T_c}{T_o} \cdot \frac{1}{(Y-1)^2} \tag{A9}$$

and

$$\frac{\delta F}{\delta \mathrm{ENR}} + \frac{1}{Y - 1} \tag{A10}$$

with

$$\Delta\mathrm{ENR} = 10^{\frac{\mathrm{ENR_{dB}}}{10}} \cdot \left(10^{\frac{\mathrm{ENR_{dB}}}{10}} - 1\right) \tag{A11}$$

and

$$\Delta Y = Y \cdot \left(10^{\frac{\Delta Y_{\mathrm{dB}}}{10}} - 1\right) \tag{A12}$$

The linear ΔF according to Eq. (A8) can thus be calculated as a function of ΔENR, ΔY and T_c – or in terms of dB it holds

$$\Delta NF_{\mathrm{dB}} = \frac{10}{\log(10)} \cdot \frac{\Delta F}{F} \tag{A13}$$

which is the resulting NF uncertainty.

References

Zhutyaew, S.: 1296 MHz Small EME Station with Good Capability (part 1–4), available at: www.vhfdx.ru, 2010.

Maas, S. A.: Microwave Mixers, Artech House, 23–27, 1993.

Schiek, B., Rolfes, I., and Siwers, H.-J.: Noise in High-Frequency Circuits and Oscillators, Wiley, 133–136, 2006.

Pozar, D.: Microwave and RF Design of Wireless Systems, Wiley, 87–97, 2000.

Bertelsmeier, R.: Low Noise GaAs-FET Preamps for EME: Construction and Measurement Problems, DUBUS 4, 1988.

Swain, H. L. and Cox, R. M.: Noise Figure Meter Sets Records for Accuracy, Repeatability, and Convenience, Hewlett-Packard J., 4, 23–24, 1983.

29

Traceable calibration of a horizontally polarised reference antenna with omnidirectional pattern at VHF frequencies for ILS field strength validation

T. Schrader[1], **T. Kleine-Ostmann**[1], **and J. Bredemeyer**[2]

[1]Physikalisch-Technische Bundesanstalt (PTB), Bundesallee 100, 38116 Braunschweig, Germany
[2]Flight Calibration Services FCS GmbH, Hermann-Blenk-Straße 32 A, 38108 Braunschweig, Germany

Correspondence to: T. Schrader (thorsten.schrader@ptb.de)

Abstract. We present a traceable calibration of a specially designed horizontally polarised reference antenna with an omnidirectional pattern in the E-plane for the frequency range between 105 MHz and 120 MHz. This antenna is used as a validation tool for absolute field strength measurements at the localizer transmitter of an instrument landing system (ILS) at airports and is carried by a helicopter. We investigate whether we can treat it as a dipole-like antenna in the calibration setup despite its disk-shape body. We also investigate the suitability of an anechoic chamber for antenna calibration though it was not designed for that purpose. The measurements are based on scattering parameters (S-parameters) which we apply in the *3-antenna-method* (TAM or 3-AM) to obtain the antenna gain and the antenna factor, respectively. An uncertainty budget for the antenna gain calibration is derived. We also report on the first practical application of the calibrated reference antenna.

1 Introduction

Terrestrial instrument landing systems (ILS) supporting air traffic management and navigation close to airports are subject to regular flight inspection (FI), where, in addition to other values of interest, absolute electric field strength values have to be determined according to the International Civil Aviation Organisation (ICAO 2000 and 2006). For this purpose the flight inspection service providers utilize medium size FI aircrafts equipped with navigation receivers and omnidirectional antennas mounted on the upper fuselage of the aircraft.

In order to validate the measured field strength values obtained during flight inspection, we developed a new method being totally independent of the aircrafts typically used. In our method, we use a helicopter carrying an autonomous payload on its external load hook, which consists of the new reference antenna and the receiving/recording system (Bredemeyer et al., 2012). The latter contains a Rohde & Schwarz EVS 300 navigation receiver, a global positioning system (GPS) receiver, a data storage system, and a battery unit. Nylon ropes with 8 m length provide the required clearance between the lower fuselage of the helicopter and the reference antenna and between the antenna and the receiving/recording system, respectively. As the whole equipment is likely to swing underneath the helicopter during the flight and, moreover, may also rotate around the load hook, a special antenna was designed which provides an omnidirectional pattern in the E-plane. The antenna also shows a null in its sensitivity diagram along the vertical axis. Thus, it reduces the fringing effects on the electromagnetic field caused by the helicopter, the coaxial cable, and by the instrumentation box, respectively. The desired frequency range is 105 MHz to 120 MHz, the co-polar receive mode of the antenna matches the horizontally polarized ILS LOC signal (localizer, indicates the lateral displacement of the aircraft to the landing runway).

The outer shape of the antenna is a flat disk with a diameter of approximately 70 cm and a height of approximately 3 cm. The signal picked up by the antenna is fed into a coaxial cable which is routed along the vertical axis down to the receiver box. Thus, the antenna and its diagram are symmetrically with respect to the vertical axis and the coaxial cable does not interfere much with the antenna diagram. The R&S EVS 300 navigation receiver features an additional intermediate frequency (IF) output which provides a full channel band-pass signal. This is sampled at a high data rate and is directly recorded without any preprocessing. Thereby, the raw band-pass signal-in-space covers the complete channel bandwidth and allows for a maximum opportunity for any signal post-processing in order to extract the essential parameters of interest. The sampling electronics was developed by the authors and is based on field programmable gate arrays (FPGAs).

To calculate the electrical field strength from the antenna factor and the input signal of the receiver, the whole instrumentation has to be calibrated. The measurements to be taken later are performed in the far-field of the LOC, so we need to provide the far-field gain of the antenna and the antenna pattern diagrams in the E- and the H-plane. Hence, we discuss how to obtain the antenna gain from which we deduce the antenna factor traceable to the SI units. We also expand on the measurement uncertainty budget as ICAO (2000 and 2006) limits the total measurement uncertainty. The requirements by ICAO (2000 and 2006) state an overall uncertainty of the field strength measurements to be smaller than 3 dB for an absolute value of the electric field strength of $40\,\mu\mathrm{V\,m^{-1}}$. Taking these requirements into account we assume that the uncertainty contribution of the antenna factor should be in the order of 1 dB or less.

2 Measurements

Absolute antenna gain calibrations typically employ the *2-antenna* (2-AM) or *3-antenna method* (3-AM) in free-space and under far-field conditions. The 3-AM may be often found as the TAM in literature. But, in order to introduce a non-ambiguous abbreviation for the 2-AM and 3-AM, we do not use TAM here. Two single or three pairs of similar sized antennas are placed in an echo-free environment and the transmission parameters are measured as a frequency response for each pair. Using Friis' formula (Balanis, 1982) for antennas separated by a distance R with

$$R > 2D^2 f/c, \tag{1}$$

where D is the largest dimension of either antenna. Here $D = 1.4\,\mathrm{m}$, $f = 110\,\mathrm{MHz}$, and R should be larger than 1.45, which is fulfilled for $R = 3\,\mathrm{m}$), c is the speed of light, and f the frequency. Assuming a polarisation match and maximum reception alignment, the gain $G_i(f)$ in dB of each of the three antennas can be calculated according to Eq. (2a–c). Here, we only have dipole-like or biconical antennas, where their phase center is located in the center axis of the antenna and does not vary with frequency. Using the free-space pathloss $\mathrm{PL}(f) = c^2/(4\pi Rf)^2$ and let $a_j a_k(f)$ be the measured linear S-parameter (forward transmission) between antenna j and antenna k, the gain $G_i(f)$ is calculated using Eq. (2a–c).

$$G_1(f) = 10\,\mathrm{dB}\cdot\log_{10}\left(\sqrt{\frac{a_1a_2(f)\cdot a_1a_3(f)}{a_2a_3(f)\cdot \mathrm{PL}(f)}}\right)\ \text{in dB} \tag{2a}$$

$$G_2(f) = 10\,\mathrm{dB}\cdot\log_{10}\left(\sqrt{\frac{a_2a_3(f)\cdot a_1a_2(f)}{a_1a_3(f)\cdot \mathrm{PL}(f)}}\right)\ \text{in dB} \tag{2b}$$

$$G_3(f) = 10\,\mathrm{dB}\cdot\log_{10}\left(\sqrt{\frac{a_1a_3(f)\cdot a_2a_3(f)}{a_1a_2(f)\cdot \mathrm{PL}(f)}}\right)\ \text{in dB} \tag{2c}$$

When using only two antennas, they have to be identical. In this case, the gain $G_{\mathrm{TX,RX}}$ can be calculated using Eq. (3). The indices TX and RX indicate the transmit and receive mode of the antennas, respectively.

$$G_{\mathrm{TX,RX}}(f) = \frac{1}{2}\left[20\cdot\log_{10}(a_1a_2(f)) - 10\cdot\log_{10}\mathrm{PL}(f)\right] \tag{3}$$

From the results of Eq. (2a–c) we calculate the antenna factor $\mathrm{AF}_i(f)$ in $\mathrm{dB\,m^{-1}}$ according to Eq. (4), where $G_i(f)$ is the gain in dB, $Z_0 = 377\ \Omega$ is the free-space wave impedance, and $Z = 50\Omega$ is the characteristic line impedance. The antenna factor AF is defined as the ratio of the incident electromagnetic field strength to the voltage V on the line connection of an antenna with a specified impedance. For an electric field antenna the antenna factor AF has the unit 1/m.

$$\mathrm{AF}_i = 20\,\mathrm{dB\,m^{-1}}\cdot\log_{10}\left(\frac{2f}{c}\cdot\sqrt{\frac{\pi Z_0}{Z\cdot 10^{G_i(f)|_{\text{in dB}}/10}}}\right) \tag{4}$$

$\mathrm{dB\,m^{-1}}$

Employing S-parameters for the transmission measurements ensures that the antenna's input impedance is taken into account. Proximity effects and multipath interference, mutual coupling and multiple reflections must also be taken into account, when the measurement uncertainty budget is set up.

In order to take into account these parameters influencing the measurement uncertainty, we used the setup in our large anechoic chamber described in the following. Instead of using the center line of the chamber we slightly tilted the boresight axis between the antennas. In addition we positioned the setup in such a way that the distance to all possibly reflective installations including the absorber-lined walls, ground and ceiling of the anechoic chamber is maximized. The distance between the antennas and the supporting masts was at least 1 m. We guided the coaxial cables loaded with ferrites at least 2 m behind the antenna and then routed them down to the ground. Alignment and polarisation of the antennas were carefully checked using vertical and horizontal laser lines. The distance was also measured applying laser lines which indicate the antenna positions on the ground, where we then used a calibrated measuring tape. The minimum distance between the antennas was chosen to 3 m.

After performing a system-error correction on the vector network analyser (VNA), we measured attenuators for which we hold calibration certificates covering the whole dynamic range of interest (e.g. –10 to –50 dB). Comparing the actual measurement results of the attenuation (here scattering parameter $S_{21,\text{act}}$) with those of the certificates ($S_{21,\text{cal}}$) employing the E_n-criteria (Wöger, 1999) in Eq. (5) we found a good agreement within the specified expanded uncertainties U (expansion factor $k = 2$). U_{cal} is taken from the calibration certificate, U_{act} is the measurement uncertainty specified for the actual experiment. Comparing attenuation and mismatch measurements after repeating the setup a week later we found 0.02 dB and 0.03 dB deviations, respectively.

$$E_n \leq \left| \frac{|S_{21,\text{act}}| - |S_{21,\text{cal}}|}{\sqrt{U^2_{S21,\text{act}} + U^2_{S21,\text{cal}}}} \right| \tag{5}$$

Using several mismatches instead of attenuators, also the measured input reflection coefficients were compared to the results from calibration certificates. From these validations we regard the S-parameters measured with the VNA as traceable to the SI units. The uncertainty of S-parameter measurements is within the specified and validated range. After all measurements we finally repeated the validation measurements to ensure the proper functionality of our VNA throughout all experiments described here.

3 Validation

Antenna calibrations in the VHF frequency range are typically carried out for dipole-like structures, e.g. biconical or logarithmic-periodic antennas. The antenna under test (AUT) here is more like a magnetic type of antenna, so we had to investigate (A) whether we could treat this AUT like a dipole or not. The second issue (B) to be solved during this project was the applicability of our anechoic chamber in antenna calibration in the VHF frequency range, though the anechoic chamber was originally not designed for that purpose. In near-field testing a combination of free-space loss and absorber reflectivity should add up to –60 dB (Newell, 1988). According to Hemming (2002) absorbers should provide –30 dB to –40 dB reflectivity level or even better to ensure that the chamber has a negligible effect on the antenna measurements. When we performed the measurements described here, the CISPR 16-1-6 was not published and the guidance about antenna calibration test sites (CALTS) in CISPR 16-1-5 was not applicable. Therefore, we had to come up with our own procedure to validate the results taking into account case (A) and (B). We do not apply the very stringent requirements here which are needed for near-field measurements. We further know from other experiments, that the absorbers do not fulfill fairly high requirements for antenna calibration facilities. But, in some cases antenna measurements are still consistent with free-space measurements, which we verified experimentally. As a general requirement we keep the size of the antennas in the same order of medium mechanical dimensions (e.g. length <1.4 m).

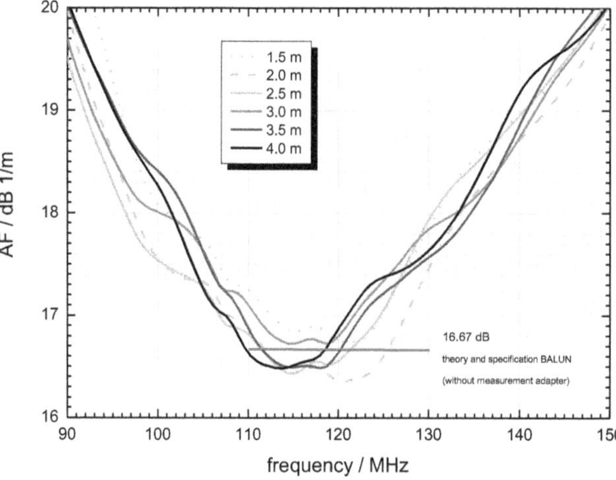

Fig. 1. AF for a set of reference dipoles as a function of distance.

3.1 Validation of the anechoic chamber

In order to deal with case (B) we applied the 2-AM and the 3-AM. For validation of our procedures and the measurement results we chose two identical reference dipoles and two sets of biconical antennas for which we hold several calibration certificates from accredited antenna calibration laboratories. Those employed different methods to obtain the free-space antenna gain and antenna factor, e.g. the 3-AM method (3 m distance applying reference dipoles and biconical antennas) and the standard site method (SSM) using both an open area test site (OATS) and a semi-anechoic chamber (SAC) with 10 m distance. The specifications from the manufacturer obtained for 3 m distance are given as well. Thus, we compared antenna gains determined by measurements (for several measuring distances between 1.5 m and 4 m) in our anechoic chamber with theoretical values of the antenna gain of the reference dipoles and, furthermore, we compared the antenna gain obtained from measurements in our chamber with a set of antenna gains obtained by accredited antenna calibration laboratories for the same identical antenna. Even more, we applied the 2-antenna-method and the 3-antenna-method (including one antenna with calibration certificate) to the reference dipoles and made a consistency check. The deviations found are within the specified uncertainty.

Figure 1 shows the AF as a function of frequency for a set of reference dipoles. We applied the 2-AM with the distance between the antennas as the parameter. Varying the distance from 1.5 m to 4 m, the AF changes a few tenth of a dB. Figure 2 shows a comparison of the AF for a reference dipole obtained from the 2-AM and from the 3-AM. The

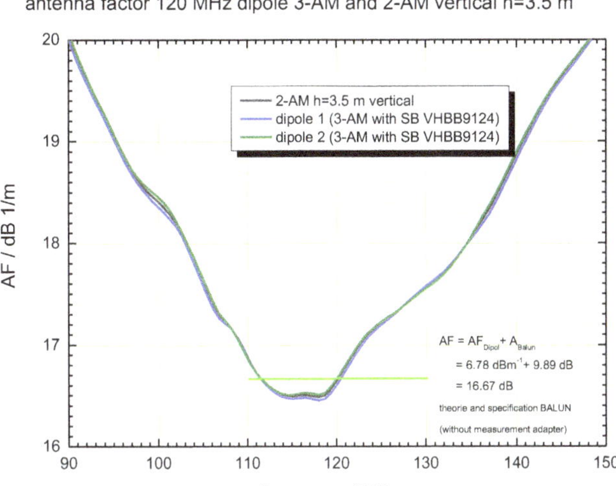

Fig. 2. Deviation of antenna gain of reference dipole obtained from 2-antenna-method and 3-antenna-method. The theoretical value of the AF is applicable at 120 MHz.

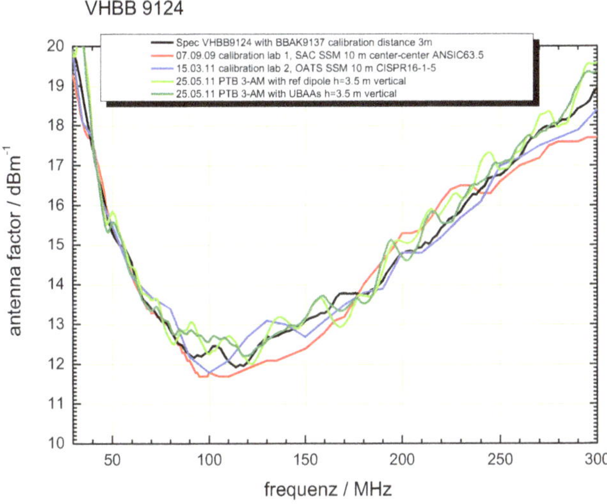

Fig. 3. Comparison of antenna factors from different accredited calibration laboratories and of PTB.

deviation found is about 0.1 dB. Figure 3 shows a comparison of the free-space antenna factor for an antenna VHBB 9124 balun with BBAK 9137 biconical elements (manufacturer Schwarzbeck, Germany) provided from several calibration laboratories and by PTB (measurements obtained in anechoic chamber) employing different methods of antenna calibration. The deviation found for 120 MHz is approximately 1 dB.

As a result of the procedure applied we validated our antenna measurement setup in the frequency range of 105 MHz to 120 MHz. The antenna gain values obtained are consistent to those from accredited laboratories within the specified uncertainty.

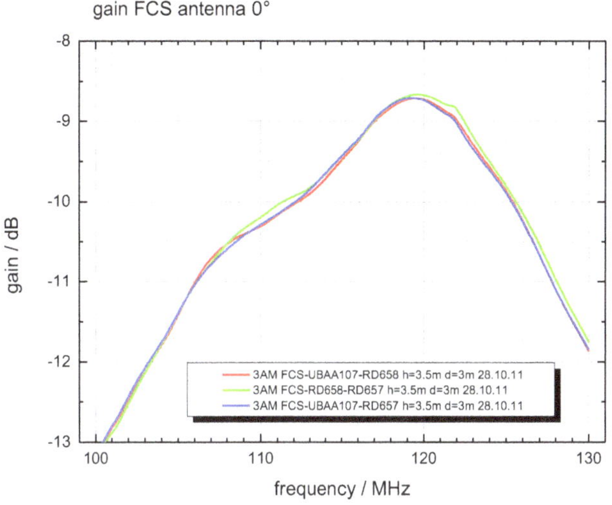

Fig. 4. Antenna gain for the AUT (FCS) obtained using reference dipoles (RD) and Schwarzbeck biconical antennas type UBAA in 3 m distance.

3.2 Validation of the antenna calibration

After this validation we used the 3-AM to determine the gain and subsequentially the antenna factor of the unknown reference antenna under test (AUT; cp. to FCS in Fig. 4) from two sets of three pairs of measurements. The first set included the two reference dipoles and the AUT, the second set included one reference dipole, a biconical antenna, and the AUT. Taking the antenna gain of one of the reference dipoles from (a) the initial validation (cp. Sect. 3.1), (b) from the first set of measurements, (c) from the second set of measurements, and (d) from theoretical calculations, we estimated the influence of our AUT on the measurements. The deviations found are within the specified uncertainty range (cp. Fig. 4). Now, we compared the results from several measurements of antenna gain for a biconical antenna, which is slightly larger than a reference dipole. We used the gain a) from the second set here, (b) from the initial validation (cp. Sect. 3.1), and (c) from the external calibration certificate. Again, the deviations found are within the specified uncertainty range (cp. Fig. 4).

Thus, we have experimentally proven, that the AUT does not change the measurement conditions in an unacceptable manner, and, furthermore, that the antenna gain obtained in this particular setup gives the best estimate for the gain of the AUT.

4 Measurement uncertainty

In order to estimate the overall uncertainty we had to derive the measurement equation first. In a second step we identified the contributions to the uncertainty and determined where they come into effect. We take also into account the steps to ensure traceability to the SI units. The uncertainty budget was derived according to JCGM (2008) using the GUM Workbench Pro tool (Metrodata, 2010). The contributions to uncertainty affect mainly the measurements of the forward transmission. In order to investigate the error propagation we revised the terms $a_j a_k$ in Eq. (2a–c) to

$$\left(a_j a_k\right)' = a_j a_k + d_{\mathrm{SI}} + d_{\mathrm{Cable}} + d_{\mathrm{Misalign}} + d_{\mathrm{Reflex}} + d_{\mathrm{Multi}} + d_{\mathrm{Mismatch}} + d_{\mathrm{Repeat}} \quad (6)$$

In Table 1 we specified all contributions to the overall uncertainty, in which d_{SI} is the uncertainty due to traceability to SI, d_{Cable} is the influence of the cable movement on the measurement of the S-parameter, d_{Misalign} is the misalignment of the antennas, d_{Reflex} is the remaining reflectivity of the environment, d_{Multi} are multiple reflections between the antennas, d_{Mismatch} is the influence of varying impedance levels by cable connections, d_{Repeat} takes into account the repeatability of the used N connector. For the final application, the anisotropy of the sensitivity diagram (d_{Aniso}) has to be taken into account. This will affect the actual gain of the antenna in Eq. (4). Correlations are not taken into account.

The best estimate for the correction terms in Eq. (6) is 0 dB, but we can associate the uncertainty to each contribution. Please note, that all three $a_j a_k$ need to be revised as the influences occur again in each pairing of the antennas. The drift of the VNA was negligible compared to the deviation we introduced for traceability to the SI units (d_{SI}). The reason is that we performed validation measurements before and after the measurement campaigns, which would reveal any drift by the VNA, provided that the artefacts like attenuators and mismatches remain stable. Cable movements during the measurement campaign, e.g. when the antennas are moved up to 3.5 m height, could result in long-term stress and phase shifts, which would also be revealed by measuring the artefacts at the ends of both cables. To obtain as much insight as possible into the measurement setup, these measurements should be taken on the ground and when the cable end including the artefact (mismatch) is moved up along the mast to the actual measurement position. Using some examples for measured data the expanded uncertainty (using $k = 2$) for the antenna factor was calculated to 1 dB m^{-1}.

5 Application of the calibration factor

The reference antenna was designed as a validation tool for absolute electrical field strength measurements on the localizer transmitter of instrument landing systems. In particular, measurements obtained during regular flight inspection are to be verified applying a method which is totally independent from FI aircrafts and their instrumentation.

Therefore, the final step was to compare the reference power density measurements employing the new antenna with those obtained with a computed (MLFMA) 3-D antenna pattern (Bredemeyer, 2007) as of the flight inspection (FI) aircraft.

Two measurement campaigns were carried out at Braunschweig (EDVE) and Bückeburg (ETHB) airports to cover the lower (108.5 MHz) and the upper (111.55 MHz) frequency ranges. The helicopter was deployed at various positions which are subsequently passed by in periodic flight inspection missions. Those power densities gained with the traceable reference antenna were then compared with the most recent flight inspection results, using the aircraft's VHF top dipole for comparison. Measurements at Braunschweig are shown in the above two diagrams. An ILS approach on centerline with the corresponding power density is depicted by Fig. 5. At ILS Point "A" 4 NM (nautic miles) before threshold a value –77.5 dBW m^{-2} can be read. On the orbital flight with the aircraft at 7 NM distance and 1800 ft (feet) altitude a value of about –100 dBW m^{-2} is given in Fig. 6 at –10° offset from the LOC antenna.

Since it is not a fixed assembly, the reference antenna may either rotate horizontally or swing laterally. The former is without influence due to the antenna's omnidirectional pattern. The latter is compensated by monitoring the maximum swings of the received level. This is included as an additional input to the overall measurement uncertainty budget (see section below).

The helicopter measurement values are depicted in the red curves (left Y-axis) in diagrams of Fig. 7. Within the marked areas (blue circles) the helicopter was kept relatively stable along a period of time (X-axis) and clear maximum power densities can be traced. On the right Y-axis the absolute 3-D velocity according to the GPS receiver (green curve) is mapped. Depending on the air speed and the pilot's flight control the absolute speed (vertical and ground) may vary.

From the reference measurements a value of –79.5 dBW m^{-2} is obtained at ILS point A (see Fig. 7, left diagram), which is 2 dB below the aircraft result. From the helicopter measurements on approach 26 and orbit 7 NM we get a maximum of roughly –99 dBW m^{2} in the highlighted area (cp. Fig. 7, right diagram), which is 1 dB above the aircraft measurements.

The results obtained at Bückeburg airport showed deviations in the same order of magnitude.

6 Conclusions

We presented a method for the traceable calibration of a newly designed horizontally polarized and omnidirectional antenna, which is used for the validation of electrical field strength measurements on instrument landing systems (ILS), in particular for the localizer transmitter. ILS' are subject to regular flight inspections (FI), which are performed using midsized aircrafts. In order to obtain absolute values of the field strength levels, aircrafts and their instrumentation need a thorough calibration, e.g. a 3-D-antenna pattern diagram has to be determined as a function of frequency. Of course, this is not an easy task as the calibration would be performed

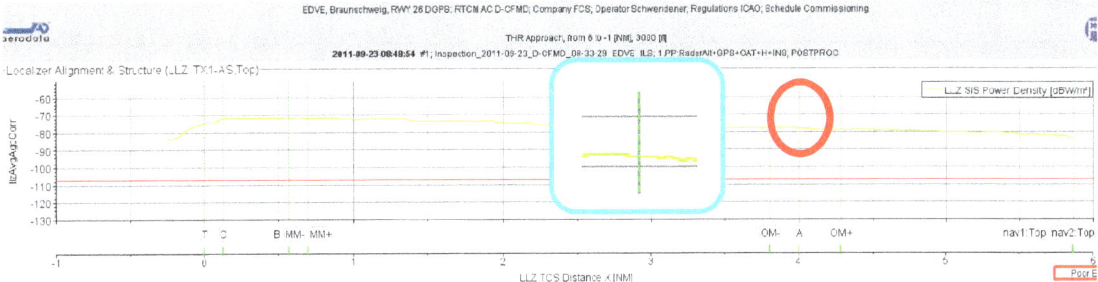

Fig. 5. Flight inspection aircraft: LOC Power Densities on Approach 26 at Braunschweig.

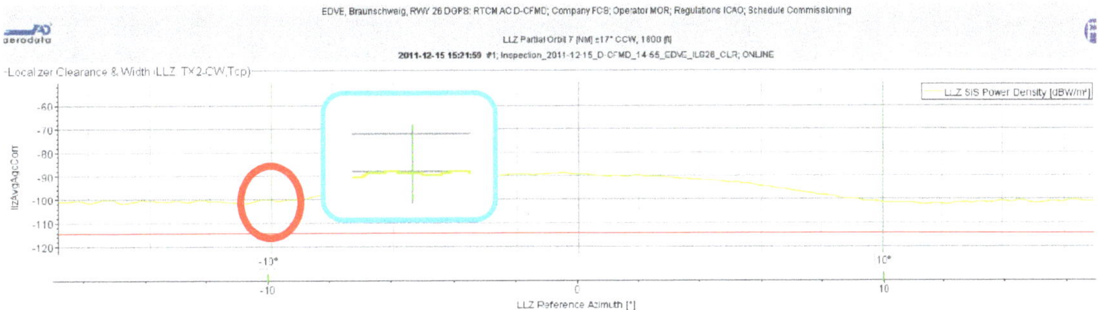

Fig. 6. LOC Power Densities on Orbital Flight 7NM at Braunschweig.

on the ground level, but is actually needed for the case that the aircraft is performing measurements in free-space and at many different bank angles (Bredemeyer, 2007).

To validate FI measurement results obtained with such flight inspection aircrafts, a new method was presented recently (Bredemeyer et al., 2012). This employs a newly designed reference antenna and a receiver system which are carried by a helicopter on its external load hook. Again, for absolute field strength measurements using this reference antenna, a suitable calibration for it as AUT (antenna under test) is required as well. Due to its more magnetic type of functionality, we had to verify that we could perform a typical *3-antenna-method* without sacrificing the results. Another issue to be solved was the applicability of our anechoic chamber. Its design does not meet very stringent and specific requirements for antenna calibrations. In order to investigate these two issues we applied the *2-antenna-method* and the *3-antenna-method* for several pairs of antennas. These comprise a set of reference dipoles and broadband biconical antennas, for which we hold calibration certificates from several ISO/DIN/EN 17025 accredited calibration laboratories. We then compared theoretical and measured data sets of the dipoles and actual measurement results of the biconical antennas with results from calibration certificates applying the E_n-criteria. All measurement results are based on scattering parameter measurements employing a vector network analyser.

Results A

Firstly, we were able to reproduce both the theoretical antenna factors of the reference dipoles and the antenna factors calculated from the antenna gain stated in the calibration certificates of the biconical antennas. Hence, we have experimentally verified that the setup in our anechoic chamber allows for this particular antenna calibration within the specified uncertainty range.

Results B

Secondly, we performed a *3-antenna-method* comprising the new reference antenna, a reference dipole and a biconical antenna, whose calibration factor is known from external – but traceable – calibration. Comparing the actual antenna gains with the theoretical and the external results, we found good agreement within the specified range of uncertainty. Moreover, the antenna gains of the reference dipole and of the biconical antenna were in good agreement comparing the results from measurement campaign A and B (cp. results A and B in this section). Thereby, we experimentally verified the applicability of the *3-antenna-method* on this type and size of the AUT within the specified range of uncertainty. The overall uncertainty was calculated to 1 dB m^{-1}.

Table 1. Uncertainty budget

Quantity	Value	Standard Uncertainty	Distribution	Sensitivity Coefficient	Uncertainty Contribution	Index
lambda	2.7253859818 m	$14.3 \cdot 10^{-9}$ m				
c	$299.792458 \cdot 10^{6}$ m s^{-1}					
f	$110.000000000 \cdot 10^{6}$ 1 s^{-1}	0.577 1 s^{-1}	rectangular	$39 \cdot 10^{-9}$	$23 \cdot 10^{-9}$ dB m^{-1}	0.0 %
PL	$5.2263 \cdot 10^{-3}$ m^{2} s^{2}	$20.1 \cdot 10^{-6}$ m^{2} s^{2}				
PI	3.1415926535898					
R	3.00000 m	$5.77 \cdot 10^{-3}$ m	rectangular	−1.4	$-8.4 \cdot 10^{-3}$ dB m^{-1}	0.0 %
a_1a_2	$199.5 \cdot 10^{-6}$	$37.3 \cdot 10^{-6}$				
$a_1a_2$log	−37.0 dB					
d_{SI}	0.0 dB	0.300 dB	normal	−0.50	−0.15 dB m^{-1}	9.1 %
d_{Cable}	0.0 dB	0.404 dB	rectangular	−0.50	−0.20 dB m^{-1}	16.5 %
$d_{Misalign}$	0.0 dB	0.289 dB	rectangular	−0.50	−0.14 dB m^{-1}	8.4 %
d_{Reflex}	0.0 dB	0.404 dB	rectangular	−0.50	−0.20 dB m^{-1}	16.5 %
d_{Multi}	0.0 dB	0.289 dB	rectangular	−0.50	−0.14 dB m^{-1}	8.4 %
$d_{Mismatch}$	0.0 dB	0.212 dB	U-distr.	−0.50	−0.11 dB m^{-1}	4.5 %
d_{Repeat}	0.0 dB	0.173 dB	rectangular	−0.50	−0.087 dB m^{-1}	3.0 %
a_1a_3	$398.1 \cdot 10^{-6}$	$74.5 \cdot 10^{-6}$				
$a_1a_3$log	−34.0 dB					
a_2a_3	$1.000 \cdot 10^{-3}$	$187 \cdot 10^{-6}$				
$a_2a_3$log	−30.0 dB					
G_1	−9.091 dBi	0.498 dBi				
d_{Aniso}	0.0 dB	0.289 dB	rectangular	−1.0	−0.29 dB m^{-1}	33.6 %
Z_0	376.99 V/A					
Z_L	50.0 V/A					
AF	20.148 dB m^{-1}	0.498 dB m^{-1}				

EDVE LOC 26 Orbit 7NM -10°

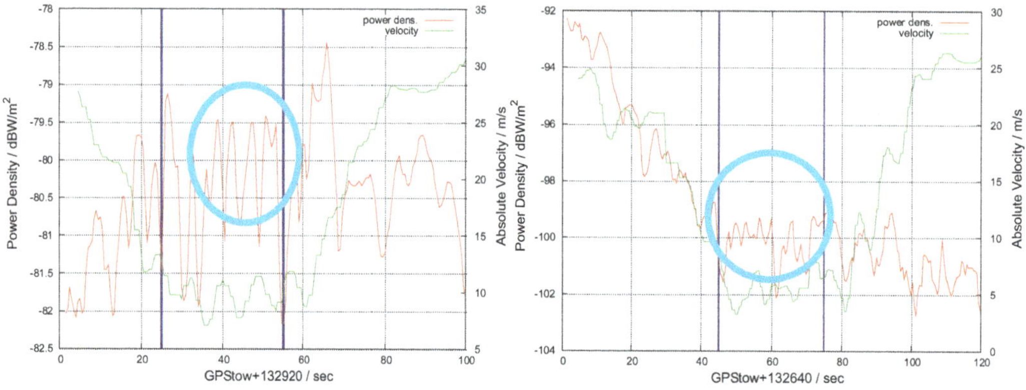

Fig. 7. Reference power densities on Approach 26 and Orbit 7NM.

Practical application of the reference antenna

In the first practical applications at different airports (Braunschweig and Bückeburg, Germany) we obtained measurement results of absolute electrical field strength levels from regular flight inspection employing the typical FI aircraft and from the measurement with the helicopter setup using the calibrated reference antenna. Some sample checks on different LOC frequencies were performed. This revealed a satisfactory agreement (max. deviation 2 dB) between the power densities gained for the localizer transmitters of instrument landing systems. Even despite the fact, that we have used two diametric airborne measurement setups on the same physical value, we found very satisfactory results. This ensures flight inspection and correspondent absolute electrical field strength measurements at VHF frequencies at a high level of confidence.

Acknowledgements. The authors thank Kai Baaske from PTB and Jörg Follop from FCS for their support during the measurement campaigns.

References

Balanis, C. A., Antenna Theory, 2 Edn., New York: John Wiley & Sons Inc., 1982.

Bredemeyer, J., Schrader, T., Kleine-Ostmann, T., and Garbe, H.: Quasi-stationary Signal-in-Space Measurements using Traceable Antennas, in: Proceedings of the 17th International Flight Inspection Symposium (IFIS) 2012, Braunschweig, Germany, 4–8 June, 2012.

Bredemeyer, J., Kleine-Ostmann, T., Schrader, T., Münter, K., and Ritter, J.: Airborne field strength monitoring, Adv. Radio Sci., 5, 49–55, doi:10.5194/ars-5-49-2007, 2007.

CISPR 16-1-5, Specification for radio disturbance and immunity measuring apparatus and methods – Part 1-5: Radio disturbance and immunity measuring apparatus – Antenna calibration test sites for 30 MHz to 1000 MHz, 2003.

CISPR 16-1-6, Specification for radio disturbance and immunity measuring apparatus and methods – Part 1-6: Radio disturbance and immunity measuring apparatus – EMC-Antenna calibration, Draft standard, not published yet.

Hemming, L. H.: Electromagnetic Anechoic Chambers, IEEE Press, Canada, 2002.

ICAO Annex 10, Volume I, Radio Navigation Aids, Sixth Edition, July 2006.

ICAO DOC 8071, Volume I, Testing of Ground-based Radio Navigation Systems, Fourth Edition – 2000.

JCGM 100:2008. Evaluation of measurement data — Guide to the expression of uncertainty in measurement (GUM).

Metrodata GmbH, 79576 Weil am Rhein, Germany, GUM Workbench Pro tool, 2010.

Newell, A. C.: Error Analysis Techniques for Planar Near-Field Measurements, IEEE Trans. Antennas Propagation, 36, 254–268, June 1988, 1988.

Wöger, W., Remarks on the E_n-Criterion Used in Measurement Comparisons, PTB-Mitteilungen, Braunschweig, 1, 24–27, 1999.

Fractional-N PLL based FMCW sweep generator for an 80 GHz radar system with 24.5 GHz bandwidth

T. Jaeschke[1], **C. Bredendiek**[1], **M. Vogt**[2], **and N. Pohl**[1]

[1]Ruhr-Universität Bochum, Institute of Integrated Systems, 44780 Bochum, Germany
[2]Ruhr-Universität Bochum, High Frequency Engineering Research Group, 44780 Bochum, Germany

Correspondence to: T. Jaeschke (timo.jaeschke@rub.de)

Abstract. A phase-locked loop (PLL) based frequency synthesizer capable of generating highly linear broadband frequency sweeps as signal source of a high resolution 80 GHz FMCW radar system is presented. The system achieves a wide output range of 24.5 GHz starting from 68 GHz up to 92.5 GHz. High frequencies allow the use of small antennas for small antenna beam angles. The wide bandwidth results in a radar system with a very high range resolution of below 1.5 cm. Furthermore, the presented synthesizer provides a very low phase noise performance of −80 dBc/Hz at 80 GHz carrier frequency and 10 kHz offset, which enables high precision distance measurements with low range errors. This is achieved by using two nested phase-looked loops with high order loop filters. The use of a fractional PLL divider and a high phase frequency discriminator (PFD) frequency assures an excellent ramp linearity.

1 Introduction

Frequency-modulated continuous-wave (FMCW) radar systems are widely used in a large field of applications. The most important industrial markets are e.g. automotive radars and high precision range measurement radars. This applications presume special requirements for the radar sensor.

Automotive radars need a high spatial resolution to divert the antenna beam to separate cars on different lanes in distances of about 150 m. Because of design specifications, the antennas have to be very small, and these requirements can only be fulfilled by using high frequencies in the region of 80 GHz or even higher.

Special range measurement radars for challenging applications with many disturbing objects also need a high beam directivity and a small antenna at the same time to be focused to the target. For these reasons it is also favorable to use high frequency ranges in this field of application.

In order to separate interfering reflections of disturbing objects from the desired signal of the radar target, a high range resolution is necessary. The radar range resolution ΔR defines the ability to distinguish two targets close to each other. It can be calculated as:

$$\Delta R = \frac{c_0 \cdot B_\mathrm{w}}{2(f_\mathrm{max} - f_\mathrm{min})} = \frac{c_0 \cdot B_\mathrm{w}}{2 \cdot \Delta f} , \quad (1)$$

where c_0 is the speed of light, B_w a factor to describe the influence of the window function used in signal processing, and Δf the bandwidth of the frequency ramp. With the commonly used Hanning window to suppress sidelobes and using the −6 dB width to define two separable targets $B_\mathrm{w} = 2$.

Table 1 shows the range resolution for several bandwidths. Standard industrial range measurement radar sensors typically have a bandwidth smaller than 2 GHz due to the complexity of broadband radar systems. State of the art research FMCW radar system like the COBRA94 (Fraunhofer FHR, Germany) allow a bandwidth up to 8 GHz (Essen et al., 2005, 2008). The current record in bandwidth of 10 GHz is set by Nicolson et al. (2008). The presented frequency synthesizer is capable of generating FMCW sweeps with more than 24.5 GHz bandwidth, so the range resolution is improved by more than a factor of 2. This allows a better separation of two near targets, where one of them is the wanted signal like shown in Fig. 1. Here, the advantage of ultra high resolution and high bandwidth FMCW radar systems is clearly visible. With 4 GHz bandwidth, it is impossible to separate the

Table 1. Range resolution for different ramp bandwidths using a Hanning window function in signal processing. Typically, the accuracy of single target distance measurements is decades better.

Bandwidth	2 GHz	5 GHz	10 GHz	25 GHz
Resolution	150 mm	60 mm	30 mm	12 mm

two targets. By using 24.5 GHz bandwidth, the signal of the wanted target can easily be separated from the disturbing target.

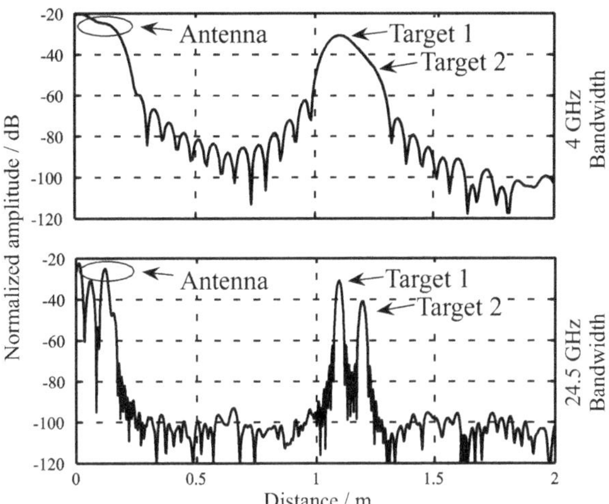

Fig. 1. Simulation of a radar scenario with 4 GHz and 24.5 GHz sweep bandwidth. Target 1 is positioned 1 m behind the antenna, followed by Target 2 another 0.1 m behind it.

2 Concept of the FMCW-Synthesizer

2.1 Overview

The simplified block diagram of the realized FMCW radar signal synthesizer is shown in Fig. 2. A SiGe monolithic microwave integrated circuit (MMIC) with a broadband millimeterwave voltage controlled oscillator (VCO) produces the output signal in a frequency range from 68 GHz to 92.5 GHz (VCO80G, Fig. 2) (Pohl et al., 2009; Pohl, 2010). The MMIC also includes a mixer, a second VCO at 24 GHz (VCO24G, Fig. 2) and two fixed frequency dividers (Pohl et al., 2011).

For stabilization of the VCOs two PLLs with commercially available off-the-shelf frequency synthesizer chips (HMC701LP6CE from Hittite Microwave) are used. The frequency of the output signal of the auxiliary VCO (VCO24G) is divided by 8 to get a low frequency signal, which is well suited for use with commercial frequency synthesizers. An external active loop filter is used to close the loop. A 100 MHz ultra low phase noise temperature controlled crystal oscillator (TCXO) provides the reference signal in order to allow low divider factors in the PLL. For best phase noise performance, an integer divider PLL synthesizer is used. The output frequency is fixed to 24 GHz.

The mmWave VCO (VCO80G) is stabilized in a different way. First, the output signal is divided by a factor of 4 to obtain a signal in a frequency range from 17 GHz to 23.125 GHz, then this signal is downconverted with the fixed 24 GHz output of the auxiliary PLL. Due to image frequencies this results in an output frequency in the range from 0.875 GHz to 7 GHz, which is well suited for commercial PLL synthesizer ICs. The loop is closed with an external loopfilter again. For highly configurable ramp generation the fractional mode of the commercial frequency synthesizer chip with a high reference frequency is used. Frequency ramps are generated by changing the fractional divider inside the synthesizer IC with the build in ramp generator.

The two PLLs are programmed using a microcontroller which can be interfaced with a serial to USB converter to a computer. This assures flexible ramp generation possibilities in relation to start and stop frequency, bandwidth and slope of the frequency ramp.

2.2 mmWave-Module

The RF-Module contains the high frequency parts of the system. It is based on a Rogers RT/duroid 5880 high frequency substrate mounted on a brass block for a good mechanical stability, and heat transfer. The MMIC is glued into a milled hole and connected to the substrate with bond wires as shown in Fig. 3. The inductive influence of the short and well-defined bond wires for the 68 GHz to 92.5 GHz output is compensated using an Monte-Carlo technique optimized on-chip matching network (Pohl et al., 2012). To reduce influences of the ground plane, to lower the supply voltage swing, and to assure robust signals, all high frequency transmission lines are connected using differential outputs (Rein and Moller, 1996). Bond wires on the left and right side connect the supply and the tuning voltage for the two VCOs.

Figure 4 shows the complete mmW-Module. A rat race coupler is used to transform the differential 80 GHz signal to a single ended signal. A pad for a wafer prober (110H-GSG-150 from PicoProbe) was designed to connect the 80 GHz output with external measurement equipment. The divide-by-4 outputs of the 80 GHz signal can also be connected to measurement equipment by using SMA-connectors to allow noise characterization with commercially available spectrum analyzers.

2.3 PLL-Module

A photo of the PLL-Module is shown in Fig. 5. It consists of a low noise power supply to prevent degradation of the good noise performance. In addition, both PLLs are placed

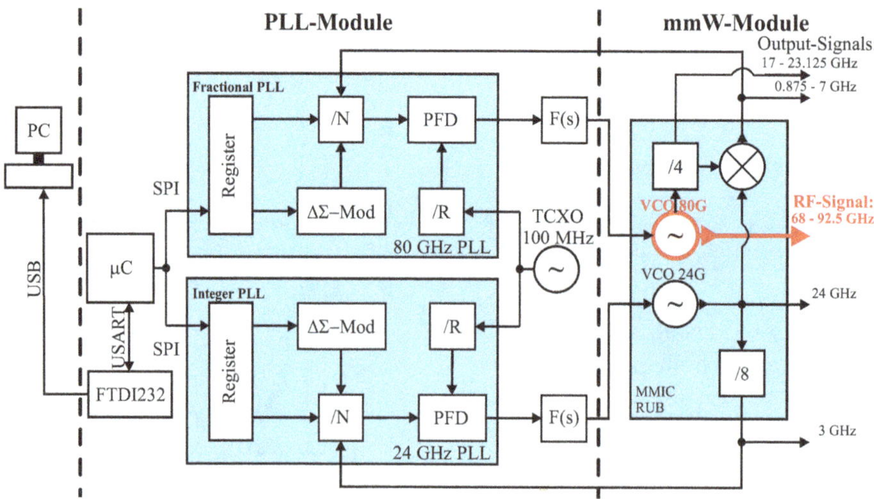

Fig. 2. Block diagram of the frequency synthesizer used to generate the wide bandwidth 68 GHz to 92.5 GHz output signal.

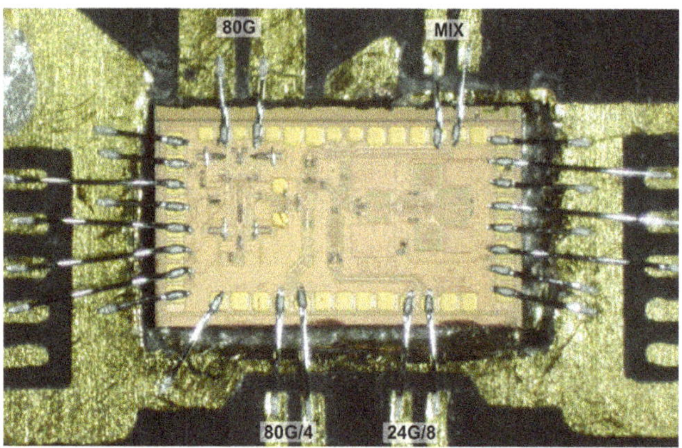

Fig. 3. Section of the MMIC with the two VCOs, the mixer and the two fixed dividers mounted on a Rogers RT/duroid 5880 high frequency laminate.

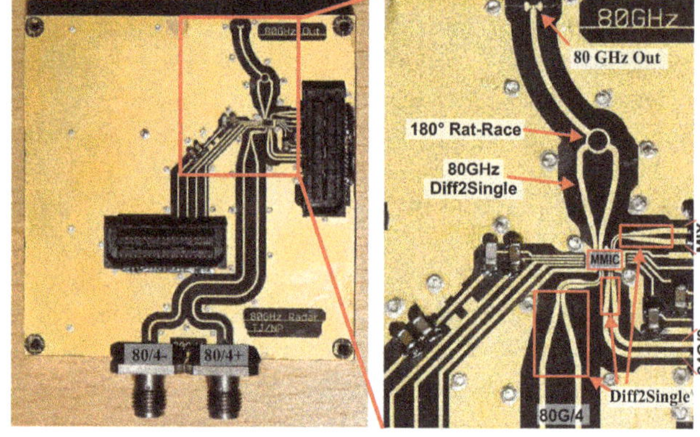

Fig. 4. Photo of the complete mmW-Module with the MMIC and two Samtec QSE high frequency connectors to connect the PLL-Module.

on this module. Each of them includes a PLL synthesizer IC and an optimized high order (PLL24: 4th order, PLL80: 5th order) active loop filter. The PLL ICs are programmed via a serial peripheral interface (SPI) using a microcontroller (ATXMEGA128 from Atmel) on the back side of the module. Connection to a computer can be achieved with the build in Mini-USB-Connector on the lower right side. On the upper left side the SMA-connector for a TCXO reference is visible.

Figure 6 shows the detailed schematic of the loop filter. A low noise operational amplifier (OpAmp) is used to prevent system noise degradation caused by the high tuning sensitivity of the VCO. The inverting Op Amp circuit is biased at half of the charge pump (CP) supply with R_{b1} and R_{b2} to achieve the best operating point. A first part of the filter is placed in front of the OpAmp to pre-smooth the hard current pulses of the charge pump. The loop bandwidth was chosen to achieve a minimal integrated phase noise. For the auxiliary VCO (VCO24G, 50 MHz PFD-frequency) a bandwidth of 530 kHz with a phase margin of 50° and for the mmWave VCO (VCO80G, 20 MHz PFD-frequency) a bandwidth of 270 kHz with a phase margin of 55° has been obtained.

3 Measurements

3.1 Phase noise

In order to test and to characterize the synthesizer, phase noise measurements have been performed in different temperature ranges.

Figure 7 shows the measured phase noise at a center frequency of 80 GHz and at a temperature of 20 °C. Further measurements show that the degradation over temperature

t]

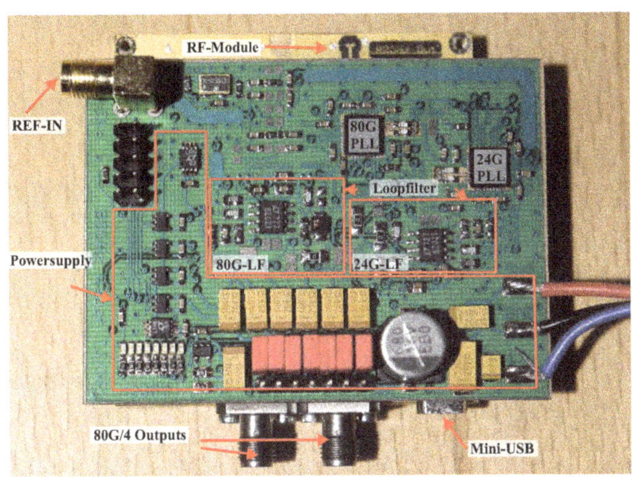

Fig. 5. Photo of the PLL-Module with low noise power supply, the commercial Hittite HMC701LP6CE PLL synthesizer ICs and the active high order loop filter for both phase-locked loops.

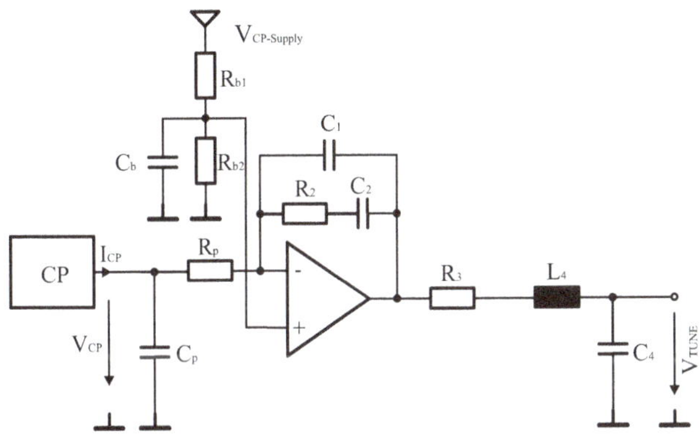

Fig. 6. Schematic of the 5th order active loop filter for optimal phase noise characteristics.

(−45 °C to 90 °C) is below 3 dB. Measurements have been done with a spectrum analyzer (8565E from Agilent) using the divide-by-4 outputs and adding 12 dB to compensate an influence of the fixed divider.

The inband phase noise of about −80 dBc/Hz is determined by the PLL noise floor. It almost perfectly fits the predicted phase noise values. The difference at offset frequencies lower than 2 kHz is due to measurement inaccuracy because of the carrier drift. At offset frequencies greater than the loop bandwidth, the phase noise approaches the free running VCO's phase noise.

3.2 Ramp generation

Almost the complete VCO bandwidth can be used in FMCW ramping mode. Figure 8 shows the tuning voltage in continuous sawtooth ramping mode. A FMCW bandwidth of

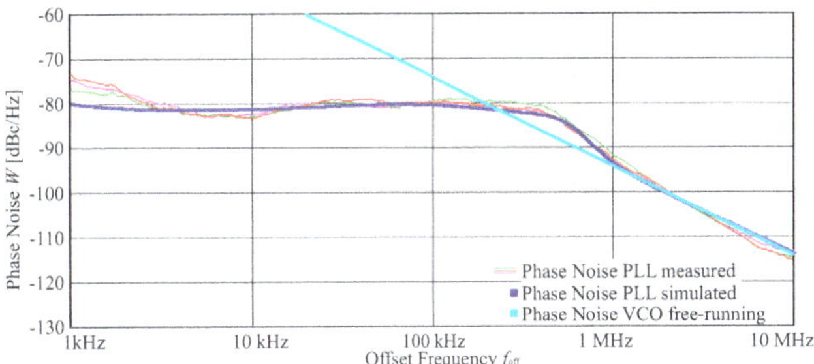

Fig. 7. Measured phase noise of the mmW-VCO output signal at 80 GHz and 20 °C against previously simulated and free-running VCO phase noise.

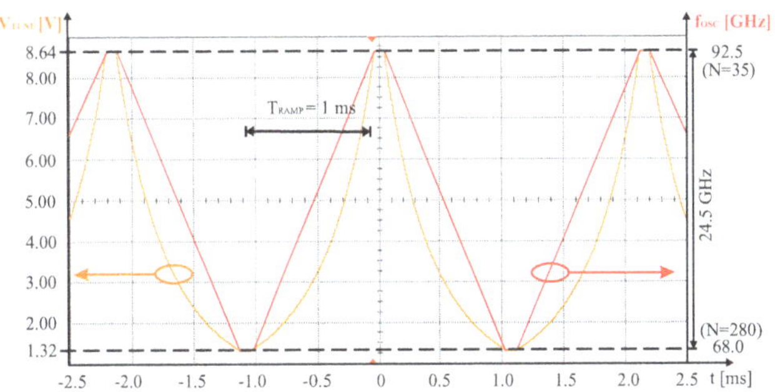

Fig. 8. Measured tuning voltage in continuous FMCW sawtooth ramping mode with a ramp time of 1 ms.

24.5 GHz with an output frequency from 68 GHz to 92.5 GHz is achieved with fractional frequency divider factors between $N = 35$ and $N = 280$. In Fig. 8, an example with 1 ms ramp-time is shown. The ramp duration can be programmed in wide ranges.

Figure 9 shows the measured spectrum of the divide-by-4 output while in ramping mode. Here, the wide bandwidth of 24.5 GHz ((23.125–17 GHz) = 6.125 GHz = 24.5 GHz/4 at the divide-by-4 output) is clearly visible. The output power of the mmW output is 10 dBm to 12 dBm and much smoother than the output power of the unbuffered divide-by-4 output (Pohl et al., 2009).

3.3 Radar performance

For demonstrating the FMCW synthesizer performance, a mixer and an IF-stage including analog digital conversion and signal processing were added to complete the FMCW radar (Pohl et al., 2012).

Figure 10 shows a first measurement of a radar scenario with 24.5 GHz FMCW bandwidth and a ramp duration of 4 ms. A W-band waveguide transition is used to connect the antenna to the FMCW system. The complete signal processing is done by using MathWorks MATLAB after digitization of the IF-signal with an 1 MSPS ADC and transferring the data to the computer over USB. Reflections caused by the

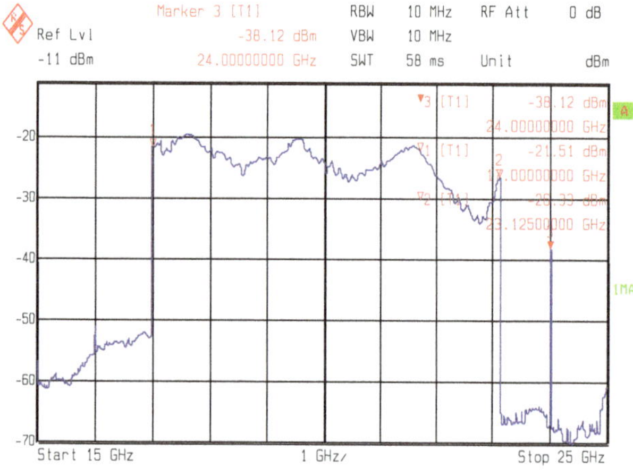

Fig. 9. Spectrum of the divide-by-4 measurement output in max-hold mode while ramping from 68 GHz to 92.5 GHz. The output power of the 80 GHz mmW generator is much smoother and shows a very small variation of ≈2 dB (10 dBm to 12 dBm).

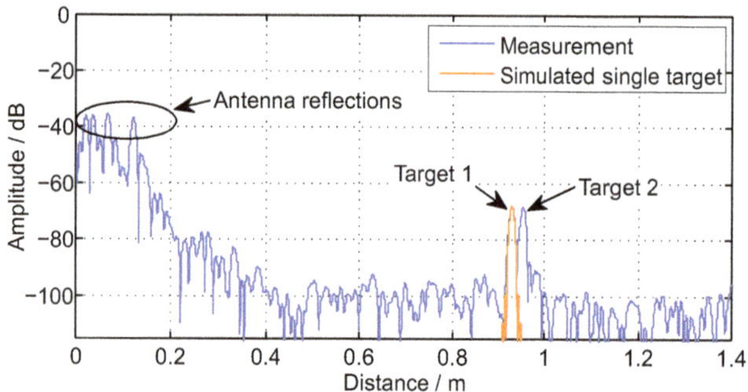

Fig. 10. IF spectrum of the measured radar scenario with two 3 mm diameter steel rods as targets.

antenna are visible at small distances. The targets 1 and 2 are steel rods with 3 mm diameter in about 2.58 cm distance from each other. These targets have a small radar cross section (RCS), but can easily be detected and separated by the sensitive high resolution FMCW radar.

4 Conclusions

A configurable PLL based frequency synthesizer for generation of highly linear broadband frequency ramps was presented, and its usability was confirmed by different measurements. Furthermore, the suitability for use as a FMCW signal source was approved by first measurements in a complete FMCW radar system with an outstanding maximum bandwidth of 24.5 GHz. The use of commercially available semiconductor components in combination with the SiGe MMIC allows a cost-effective realization of ultra high resolution FMCW radar systems, which is suitable for mass production. This allows new fields of application for future FMCW radar sensors.

Acknowledgements. This work has been supported by the Ministry of Economic Affairs and Energy of the State of North Rhine-Westphalia (Grant 315-43-02/2-005-WFBO-009) and KROHNE Messtechnik. The authors would also like to thank Infineon Technologies for fabricating the MMIC chips. Special thanks goes to Thomas Musch, Institute of Electronic Circuits, Ruhr-Universität Bochum (Germany) for his support and the good cooperation.

References

Essen, H., Konrad, O., Wahlen, A., and Sommer, R.: COBRA 94 – ultra broadband experimental radar for ISAR applications, in: Proc. Joint 30th Int. Conf. Infrared and Millimeter Waves and 13th Int. Conf. Terahertz Electronics IRMMW-THz 2005, 2, 355–356, doi:10.1109/ICIMW.2005.1572559, 2005.

Essen, H., Biegel, G., Sommer, R., Wahlen, A., Johannes, W., and Wilcke, J.: High Resolution Tower-Turntable ISAR with the Millimetre Wave Radar COBRA (35/94/220 GHz), in: Synthetic Aperture Radar (EUSAR), 2008 7th European Conference on Synthetic Aperture Radar – EUSAR 2008, 80–86, synthetic Aperture Radar (EUSAR), 2008 7th European Conference on, 2008.

Nicolson, S. T., Chevalier, P., Sautreuil, B., and Voinigescu, S. P.: Single-Chip W-band SiGe HBT Transceivers and Receivers for Doppler Radar and Millimeter-Wave Imaging, Journal of Solid-State Circuits, IEEE, 43, 2206–2217, doi:10.1109/JSSC.2008.2002934, 2008.

Pohl, N.: Systemkonzepte und SiGe-Bipolarschaltungen für ein 80-GHz-Radarsystem mit hoher Bandbreite, Ph.D. thesis, Ruhr-Universität Bochum, 2010.

Pohl, N., Rein, H.-M., Musch, T., Aufinger, K., and Hausner, J.: SiGe Bipolar VCO With Ultra-Wide Tuning Range at 80 GHz Center Frequency, Journal of Solid-State Circuits, IEEE, 44, 2655–2662, doi:10.1109/JSSC.2009.2026822, 2009.

Pohl, N., Klein, T., Aufinger, K., and Rein, H.-M.: A low-power 80 GHz FMCW radar transmitter with integrated 23 GHz downconverter VCO, in: Proc. IEEE Bipolar/BiCMOS Circuits and Technology Meeting (BCTM), 215–218, doi:10.1109/BCTM.2011.6082785, 2011.

Pohl, N., Jaeschke, T., and Aufinger, K.: An Ultra-Wideband 80 GHz FMCW Radar System Using a SiGe Bipolar Transceiver Chip Stabilized by a Fractional-N PLL Synthesizer, IEEE T. Microw. Theory, 60, 757–765, doi:10.1109/TMTT.2011.2180398, 2012.

Rein, H.-M. and Moller, M.: Design considerations for very-high-speed Si-bipolar IC's operating up to 50 Gb/s, Journal of Solid-State Circuits, IEEE, 31, 1076–1090, doi:10.1109/4.508255, 1996.

31

Comparison of the Extended Kalman Filter and the Unscented Kalman Filter for Magnetocardiography activation time imaging

H. Ahrens, F. Argin, and L. Klinkenbusch

Institut für Elektrotechnik und Informationstechnik, Christian-Albrechts-Universität zu Kiel, Germany

Correspondence to: H. Ahrens (ha@tf.uni-kiel.de)

Abstract. The non-invasive and radiation-free imaging of the electrical activity of the heart with Electrocardiography (ECG) or Magnetocardiography (MCG) can be helpful for physicians for instance in the localization of the origin of cardiac arrhythmia. In this paper we compare two Kalman Filter algorithms for the solution of a nonlinear state-space model and for the subsequent imaging of the activation/depolarization times of the heart muscle: the Extended Kalman Filter (EKF) and the Unscented Kalman Filter (UKF). The algorithms are compared for simulations of a (6×6) magnetometer array, a torso model with piecewise homogeneous conductivities, 946 current dipoles located in a small part of the heart (apex), and several noise levels. It is found that for all tested noise levels the convergence of the activation times is faster for the UKF.

1 Introduction

The localization of the origin of heart arrhythmia is an important part of a successful treatment. For instance, in the context of the Wolff-Parkinson-White syndrome (Nenonen et al., 1991) a pathological accessory pathway is located parallel to the atrioventricular node and can cause serious heart arrhythmia like tachycardia. A non-invasive radiation-free localization of the accessory pathway with ECG or MCG is helpful because it shortens the invasive and X-ray-based catheter mapping procedure. One of the main problems in imaging the electrical activity of the heart is the non-uniqueness of the inverse problem (Fokas et al., 2004) caused by the fact that the number of current dipoles in the heart to be estimated is typically much larger than the number of ECG/MCG sensors. Consequently, the system of linear equations to be solved is ill-posed in general. While in potential imaging the state vector of the state-space model is the transmembrane potential at all heart voxels (Schulze et al., 2009), in activation time imaging the state vector includes the activation times of all heart voxels. The idea behind activation time imaging (He et al., 2002) is to include physiological action potential information, e.g. the wavefront velocity and the upstroke velocity of the depolarization wavefront, to reduce the number of heart model parameters to be estimated without loosing too much accuracy in the calculated sensor signal. The physiological information is incorporated in a cellular automaton model (Weixue et al., 1993) approximating the depolarization wavefront with 3 states and neglecting repolarization. The corresponding nonlinear state-space model can be solved using Kalman Filter algorithms (Liu et al., 2011). It has been shown that activation time imaging on a 3-D myocardium is more stable with respect to measurement noise than potential imaging (Cheng et al., 2003) and can provide an averaged localization error of ~ 3 mm (Liu et al., 2011) which is small compared to ~ 1 cm for single dipole localization (Nenonen et al., 1991). In this paper the performances of two different Kalman Filters for the solution of the nonlinear state-space model of activation time imaging are compared for Magnetocardiography: The Extended Kalman Filter and the Unscented Kalman Filter. The convergence of activation times is compared for a quadratic plane (6×6) magnetometer array and several measurement noise levels. In Sect. 2 the MCG signal as a function of the activation times is derived, in Sect. 3 the state-space model is explained, and Sect. 4 summarizes the used Kalman Filter algorithms. Section 5 shows the used sensor array, torso model and sources, in Sect. 6.1 parameter tests for the Kalman Filters are described, while in Sect. 6.2 the EKF and UKF are compared for several measurement noise levels.

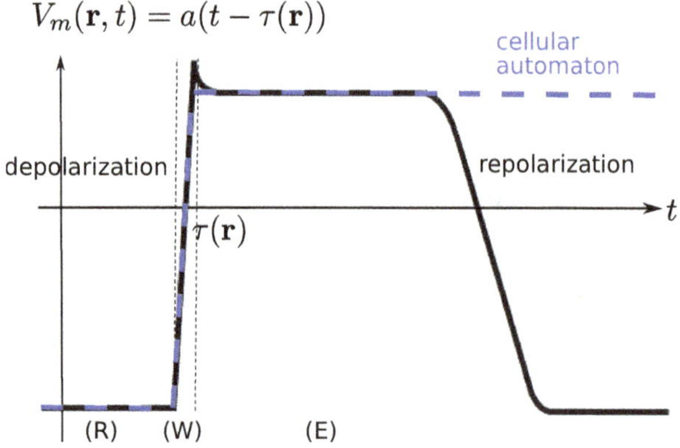

Fig. 1. Approximation of action potential in cellular automaton model: 3 states Resting (R), Wavefront (W) and Excited (E). $\tau(\boldsymbol{r})$ is the activation time at position $\boldsymbol{r}$ in myocardium.

2 From activation times on 3-D myocardium to sensor signal

Following the bidomain theory (Miller and Geselowitz, 1978) we first have to solve the electrical volume conduction problem:

$$\nabla \cdot (\boldsymbol{j}_P + \boldsymbol{j}_V) = 0 \tag{1}$$

$$\boldsymbol{j}_P = -\sigma_i \nabla V_m \tag{2}$$

with the transmembrane potential V_m of the heart muscle cells, primary current density $\boldsymbol{j}_P$, volume current density $\boldsymbol{j}_V$ and intracellular conductivity σ_i. The volume conduction problem (1) can be solved using the Finite Element Method (FEM) on a 3-D tetrahedral mesh. For the FEM simulation the SimBio neurofem program (SimBio Development Group, 2013) was used. The magnetic field has to be calculated from the total current density $\boldsymbol{j} = \boldsymbol{j}_P + \boldsymbol{j}_V$ using the Biot-Savart law. For n primary current dipoles

$$\mathbf{J}_p(t) = (\boldsymbol{j}_p(\boldsymbol{r}_1, t), \ldots, \boldsymbol{j}_p(\boldsymbol{r}_n, t))^T \tag{3}$$

(where the superscript T denotes the transpose) and q magnetometers at a given polarization the superposition principle yields the MCG signal vector caused by the primary currents at the time t:

$$\varphi_B(t) = \mathbf{L}\mathbf{J}_p(t) \tag{4}$$

$\mathbf{L}$ is the corresponding $(q \times 3n)$ leadfield matrix calculated from the volume conduction problem. The cellular automaton approximation for the transmembrane potential at position $\boldsymbol{r}$ in myocardium

$$V_m(\boldsymbol{r}, t) = a(t - \tau(\boldsymbol{r})) \tag{5}$$

is shown in Figs. 1 and 2. Now inserting Eq. (5) into (2), then Eq. (2) into (3) and Eq. (3) into (4), and define the activation time state vector as

$$\mathbf{x} = (\tau(\boldsymbol{r}_1), \ldots, \tau(\boldsymbol{r}_n))^T \equiv (x^1, \ldots, x^n)^T \tag{6}$$

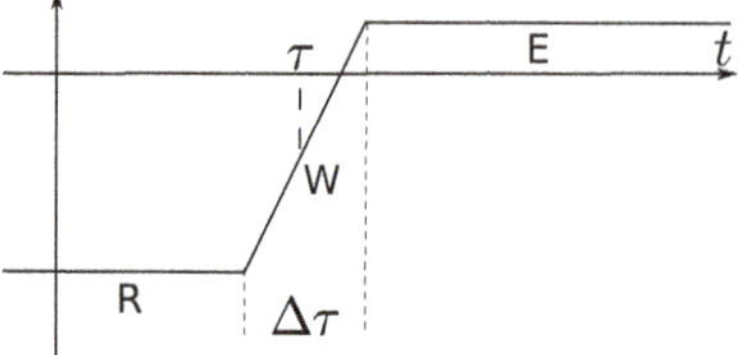

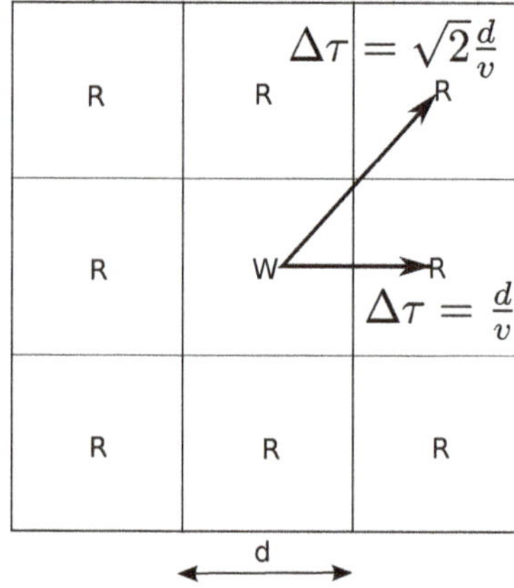

Fig. 2. Diffusion of activation in the cellular automaton model (Weixue et al., 1993). $\Delta\tau = 3$ ms is the difference of activation times between adjacent voxels.

as well as the corresponding data time series $t_1, \ldots, t_N$. Then the $(q \times N)$ MCG signal matrix $\phi_B = (\varphi_B(t_1), \ldots, \varphi_B(t_N))$ finally is described by

$$\phi_B = \mathbf{h}(\mathbf{x}) \tag{7}$$

where $\mathbf{h}$ $(a(t - \tau(\boldsymbol{r}))$ is representing a nonlinear function of $\tau(\boldsymbol{r})$.

3 State-space model

To tackle the discussed non-uniqueness of Eq. (7) the following state-space model is formulated:

$$\begin{aligned} \mathbf{x}_k &= \mathbf{f}(\mathbf{x}_{k-1}) + \mathbf{w}_{k-1} \\ \phi_{B,k} &= \mathbf{h}(\mathbf{x}_k) + \mathbf{v}_k \end{aligned} \tag{8}$$

where $\mathbf{x}_k$ is the activation time state vector of iteration step k, $\mathbf{f}$ is a process function that predicts the k-th state from the $(k-1)$-th state, $\phi_{B,k}$ is the MCG signal matrix of the k-th heart beat and $\mathbf{v}_k$ and $\mathbf{w}_{k-1}$ are the measurement noise and process noise, respectively. Both are assumed as Gaussian white noise with the following normal distributions:

$$p(\mathbf{w}) \sim N(0, \mathbf{Q}) \qquad p(\mathbf{v}) \sim N(0, \mathbf{R}) \tag{9}$$

where $\mathbf{R}$ and $\mathbf{Q}$ are the noise covariance matrices of the measurement and process, respectively. The process function $\mathbf{f}$ is determined by the following rules:

$$x_k^j = \begin{cases} x_{k-1}^j, & \text{if } \left| x_{k-1}^j - \frac{1}{Z}\sum_{n=1}^{Z} x_{k-1}^{i_n} \right| \leq \frac{\overline{d}}{v} \\ \frac{1}{Z}\sum_{n=1}^{Z} x_{k-1}^{i_n}, & \text{if } \left| x_{k-1}^j - \frac{1}{Z}\sum_{n=1}^{Z} x_{k-1}^{i_n} \right| > \frac{\overline{d}}{v} \end{cases} \tag{10}$$

where Z is the number of next neighbours ($i_n; n = 1, \ldots, Z$) within a distance $\overline{d}$, and v the isotropic activation wavefront velocity. Consequently, the activation time of a voxel tends to the average activation time of the adjacent voxels. In our calculation we chose for the distance to next neighbours $\overline{d} = \sqrt{3}d$ where d is the lattice constant of a cubic grid (Fig. 2).

4 Kalman Filters

Generally, the application of Kalman Filters is limited to linear processes, and different methods have been proposed to also treat nonlinear ones – as stated by Eqs. (9) and (7) – by means of this powerful technique. Two of these methods will be presented in the following.

4.1 Extended Kalman Filter (EKF)

The idea of the Extended Kalman Filter is the linearization of the nonlinear state-space model Eq. (9) by calculating the Jacobian matrices **H**, **F** of the functions **h**, **f** and approximating the first partial derivatives with difference quotients (Liu et al., 2011):

$$H_{ij} = \frac{h_i(x^1, \ldots, x^j + \rho, \ldots, x^n) - h_i(x^1, \ldots, x^j - \rho, \ldots, x^n)}{2\rho} \tag{11}$$

with the temporal resolution ρ and omitting the iteration step index k. The Extended Kalman Filter now minimizes the error covariance matrix defined by:

$$\mathbf{P}_k = < (\mathbf{x}_k - \mathbf{x}_{\text{true}})(\mathbf{x}_k - \mathbf{x}_{\text{true}})^T > \tag{12}$$

where $\mathbf{x}_{\text{true}}$ is the true activation sequence and $\mathbf{x}_k$ the estimated activation of step k. In the "predict" procedure of the algorithm a priori estimates of the state vector and the error covariance matrix are projected from the last step:

$$\mathbf{x}_k^- = \mathbf{f}(\mathbf{x}_{k-1}) \tag{13}$$

$$\mathbf{P}_k^- = \mathbf{F}_k \mathbf{P}_{k-1} \mathbf{F}_k^T + \mathbf{Q} \tag{14}$$

In the "correct" procedure, the Kalman gain $\mathbf{K}_k$ is calculated and the state vector and error covariance matrix are updated with the measured data $\phi_{B,k}$:

$$\mathbf{K}_k = \mathbf{P}_k^- \mathbf{H}_k^T (\mathbf{H}_k \mathbf{P}_k^- \mathbf{H}_k^T + \mathbf{R})^{-1} \tag{15}$$

$$\mathbf{x}_k = \mathbf{x}_k^- + \mathbf{K}_k(\phi_{B,k} - \mathbf{h}(\mathbf{x}_k^-)) \tag{16}$$

$$\mathbf{P}_k = (\mathbf{1} - \mathbf{K}_k \mathbf{H}_k)\mathbf{P}_k^- \tag{17}$$

$\mathbf{x}_k$ and $\mathbf{P}_k$ are then used for the next iteration step.

4.2 Unscented Kalman Filter (UKF)

The Unscented Kalman Filter (LaViola, 2003) generates a deterministic set of sampling points, stored in the $n \times (2n+1)$ sigma point matrix $\mathcal{X}_{k-1}$. The columns of $\mathcal{X}_{k-1}$ are calculated by:

$$\begin{aligned} (\mathcal{X}_{k-1})_1 &= \mathbf{x}_{k-1} \\ (\mathcal{X}_{k-1})_i &= \mathbf{x}_{k-1} + (\sqrt{(n+\lambda)\mathbf{P}_{k-1}})_i, \quad i = 2, \ldots, n \\ (\mathcal{X}_{k-1})_i &= \mathbf{x}_{k-1} - (\sqrt{(n+\lambda)\mathbf{P}_{k-1}})_{i-n}, i = n+1, \ldots, 2n+1 \end{aligned} \tag{18}$$

where $(\sqrt{(n+\lambda)\mathbf{P}_{k-1}})_i$ is the i-th column of the matrix square root and λ is defined by:

$$\lambda = \alpha^2(n+\kappa) - n \tag{19}$$

where α and κ are scaling parameters that determine the spread of the sigma points. The square root **A** of a matrix **B** satisfies $\mathbf{B} = \mathbf{A}\mathbf{A}^T$ and for the symmetric and positive definite matrix $\mathbf{B} = (n+\lambda)\mathbf{P}_{k-1}$ it can be calculated using a Cholesky decomposition (Rhudy et al., 2011). In the "predict" procedure the sigma points are propagated by the process function:

$$(\mathcal{X}_k)_i = \mathbf{f}((\mathcal{X}_{k-1})_i), \quad i = 1, \ldots, 2n+1 \tag{20}$$

Then the a priori state estimate is calculated by:

$$\mathbf{x}_k^- = \sum_{i=1}^{2n+1} W_i^{(m)} (\mathcal{X}_k)_i \tag{21}$$

where $W_i^{(m)}$ are weights defined by

$$W_1^{(m)} = \frac{\lambda}{(n+\lambda)} \tag{22}$$

$$W_i^{(m)} = \frac{1}{2(n+\lambda)}, \quad i = 2, \ldots, 2n+1 \tag{23}$$

and the a priori error covariance matrix is calculated by

$$\mathbf{P}_k^- = \sum_{i=1}^{2n+1} W_i^{(c)} [(\mathcal{X}_k)_i - \mathbf{x}_k^-][(\mathcal{X}_k)_i - \mathbf{x}_k^-]^T + \mathbf{Q} \tag{24}$$

with the process error covariance matrix **Q** and the following weights:

$$W_1^{(c)} = \frac{\lambda}{(n+\lambda)} + (1 - \alpha^2 + \beta) \tag{25}$$

$$W_i^{(c)} = \frac{1}{2(n+\lambda)}, \quad i = 2, \ldots, 2n+1 \tag{26}$$

β is another scaling parameter to adjust the speed of convergence. In the "correct" procedure first the sigma points are transformed by the measurement function:

$$(\mathcal{Z}_k)_i = \mathbf{h}((\mathcal{X}_k)_i), \quad i = 1, \ldots, 2n+1 \tag{27}$$

$$\mathbf{z}_k^- = \sum_{i=1}^{2n+1} W_i^{(m)} (\mathcal{Z}_k)_i \tag{28}$$

With the field vector $\mathbf{z}_k^-$ we can now compute the a posteriori state estimate:

$$\mathbf{x}_k = \mathbf{x}_k^- + \mathbf{K}_k(\phi_{B,k} - \mathbf{z}_k^-) \tag{29}$$

where the Kalman gain $\mathbf{K}_k$ of the UKF is defined by

$$\mathbf{K}_k = \mathbf{P}_{\mathbf{x}_k\mathbf{z}_k}\mathbf{P}_{\mathbf{z}_k\mathbf{z}_k}^{-1} \tag{30}$$

with

$$\mathbf{P}_{\mathbf{z}_k\mathbf{z}_k} = \sum_{i=1}^{2n+1} W_i^{(c)}[(\mathcal{Z}_k)_i - \mathbf{z}_k^-][(\mathcal{Z}_k)_i - \mathbf{z}_k^-]^T + \mathbf{R} \tag{31}$$

$$\mathbf{P}_{\mathbf{x}_k\mathbf{z}_k} = \sum_{i=1}^{2n+1} W_i^{(c)}[(\mathcal{X}_k)_i - \mathbf{x}_k^-][(\mathcal{Z}_k)_i - \mathbf{z}_k^-]^T \tag{32}$$

where $\mathbf{R}$ is the measurement noise covariance matrix. In the last step the error covariance matrix has to be updated:

$$\mathbf{P}_k = \mathbf{P}_k^- - \mathbf{K}_k\mathbf{P}_{\mathbf{z}_k\mathbf{z}_k}\mathbf{K}_k^T \tag{33}$$

The computation times of the EKF and the UKF are identical.

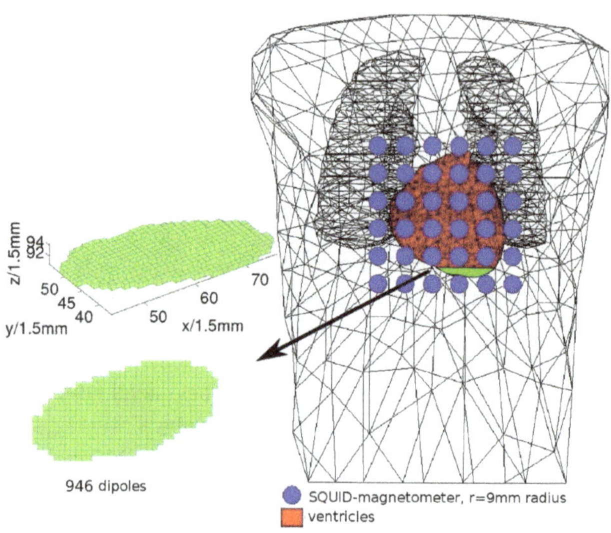

Fig. 3. Left: small heart model with 946 dipoles distibuted on a 3-D cubic grid with lattice constant $d = 1.5$ mm, Right: (6×6) magnetometers with radius $r = 9$ mm in a quadratic plane array 1 cm above the torso. The magnetic flux is calculated with 16 integration points per sensor. The data sampling rate is $\Delta t = t_{j+1} - t_j = 3$ ms.

5 MCG sensor array and dipole sources

Figure 3 shows the (6×6) array of circular magnetometers with radius $r = 9$ mm and the torso, lung and heart surfaces used in the simulations. For test purposes the activity is restricted to a small part of the apex of the heart with 946 dipoles distributed on a cubic grid with a lattice constant of $d = 1.5$ mm. The conductivities of the body tissues were set as in (Liu et al., 2011): torso (0.20 S m^{-1}), lungs (0.08 S m^{-1}) and cardiac tissue (average of conductivity parallel and transverse to muscle fibre directions: 0.6 S m^{-1}). The non-uniform tetrahedral FEM grid was separated in 48 780 (cardiac tissue), 8045 (lungs) and 238 305 (torso) tetrahedra. The activation wavefront velocity was assumed to be isotropic and we chose the average $v = 0.5$ m s^{-1} of the velocity parallel and transverse to the muscle fibre orientation (He et al., 2002) so in the cellular automaton model the difference between the activation times of adjacent voxels is $\Delta\tau = d/v = 3$ ms.

6 Results

6.1 Parameter selection, initial state and true state

For the optimization of the convergence of the Kalman Filters several parameters need to be optimized: the scaling parameters α, β, κ for the UKF and the temporal resolution ρ for the EKF. To measure the speed of convergence we define the Relative Error (RE) between the actual state vector $\mathbf{x}_k$ of step k and the true state vector $\mathbf{x}_{\text{true}}$:

$$\text{RE} = \frac{||\mathbf{x}_k - \mathbf{x}_{\text{true}}||}{||\mathbf{x}_{\text{true}}||} \tag{34}$$

with the Euclidian norm $||\cdot||$. Both the initial and the true state are calculated with the cellular automaton model. In praxis the location of the origin of the initial state can be calculated with a single dipole localization applied on the delta wave (Nenonen et al., 1991) of Wolff-Parkinson-White syndrome. The parameters are optimized for the case of measurement noise covariance $\mathbf{R} = 10^{-4} \cdot \mathbf{1}$, process noise covariance $\mathbf{Q} = \mathbf{0}$ and the initial value of the error covariance matrix is the unity matrix ($\mathbf{P} = \mathbf{1}$). Figure 4 shows the parameter test for the temporal resolution ρ of the EKF. The temporal resolution $\rho = 3$ms with the fastest convergence of the Relative Error is selected for all further EKF calculations. Figure 5 shows the test for the parameter α of the UKF. The other parameters of the UKF are fixed ($\beta = 0 = \kappa$). To enable a stable and fast convergence of the activation times the spread parameter $\alpha = 5$ was chosen for all further UKF calculations. β and κ have a smaller influence on RE than α.

6.2 Comparison of the EKF and the UKF for several measurement noise levels

Figure 6 depicts the convergence of RE as a function of the iteration step k for the EKF (diamonds) and the UKF (dots) for several measurement noise levels with the noise covariance matrix $\mathbf{R} = \sigma_n^2 \cdot \mathbf{1}$. The calculated standard deviations of the Gaussian white noise are $\sigma_n = 0.01, 0.03, 0.05, 0.1, 0.5$ while the minimum of the calculated signal is $\phi_{B,\min} = -3.085$ and the maximum of the calculated signal is $\phi_{B,\max} = 1.077$. For all calculated noise levels Fig. 6 shows that the UKF enables a faster convergence of the RE than the EKF, especially during the first 100 steps. Both for the

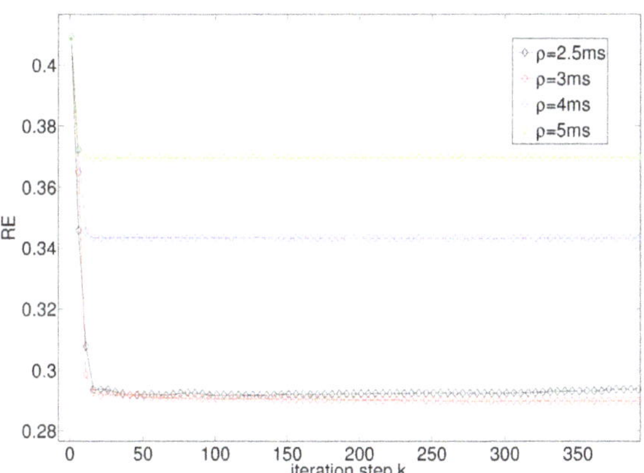

Fig. 4. Relative Error $\mathrm{RE} = \frac{||\mathbf{x}_k - \mathbf{x}_{\text{true}}||}{||\mathbf{x}_{\text{true}}||}$ as a function of the iteration step k for the EKF and several temporal resolutions ρ for the difference quotients of the Jacobians $\mathbf{H}, \mathbf{F}$ ($\mathbf{Q} = \mathbf{0}, \mathbf{R} = 10^{-4} \cdot \mathbf{1}$). Every 5th iteration step is shown.

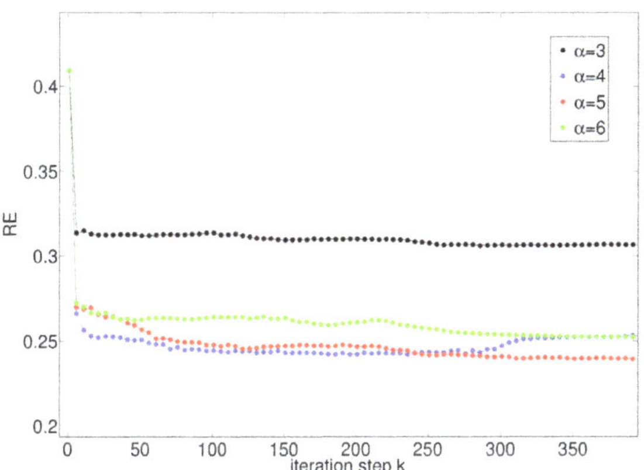

Fig. 5. Relative Error $\mathrm{RE} = \frac{||\mathbf{x}_k - \mathbf{x}_{\text{true}}||}{||\mathbf{x}_{\text{true}}||}$ as a function of the iteration step k for the UKF and several scaling parameters α for the spread of the sigma points ($\mathbf{Q} = \mathbf{0}, \mathbf{R} = 10^{-4} \cdot \mathbf{1}, \beta = 0 = \kappa$). Every 5th iteration step is shown.

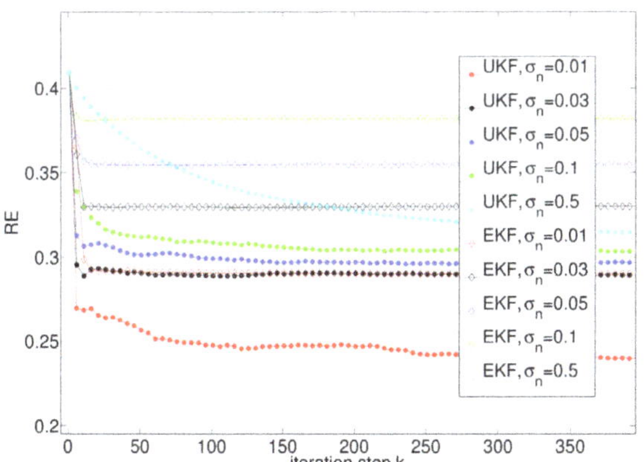

Fig. 6. Relative Error $\mathrm{RE} = \frac{||\mathbf{x}_k - \mathbf{x}_{\text{true}}||}{||\mathbf{x}_{\text{true}}||}$ as a function of the iteration step k for the EKF (diamonds) and UKF (dots) and several measurement noise levels with noise covariance matrix $\mathbf{R} = \sigma_n^2 \cdot \mathbf{1}$ ($\mathbf{Q} = \mathbf{0}$). Every 5th iteration step is shown.

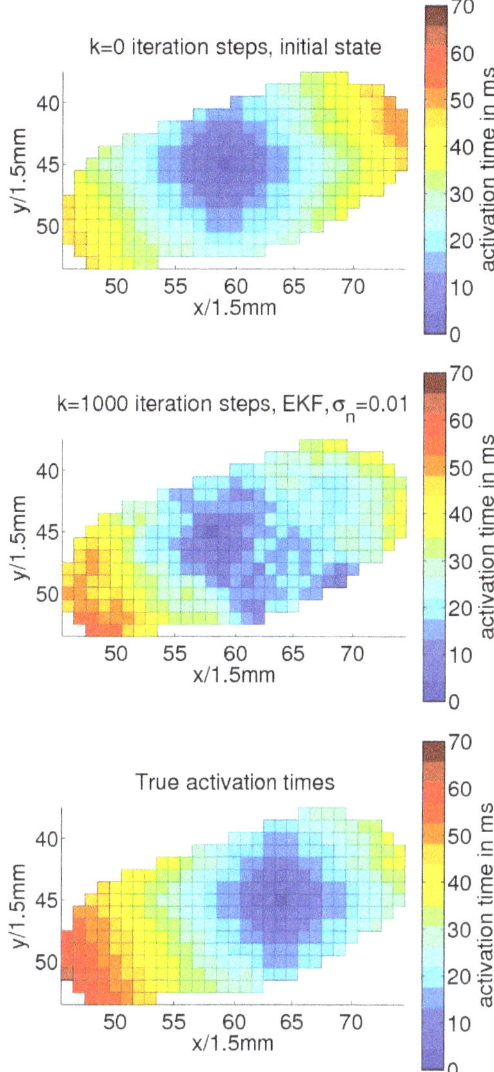

Fig. 7. Activation times for the EKF and measurement noise $\sigma_n = 0.01$ ($\mathbf{Q} = \mathbf{0}, \rho = 3\,\text{ms}$). Top: initial state, middle: 1000 iteration steps, bottom: true state.

UKF and the EKF the converged value of RE increases proportional to the noise σ_n. The convergence of the activation times for the EKF and the measurement noise $\sigma_n = 0.01$ is shown in Fig. 7. The convergence of the activation times for the UKF and the measurement noise $\sigma_n = 0.01$ is shown in Fig. 8.

7 Conclusions

In this paper the performance of cardiac activation time imaging with the help of a cellular automaton model including physiological information and Kalman Filter algorithms for the solution of a nonlinear state-space model have been investigated. The comparison of the EKF and the UKF in

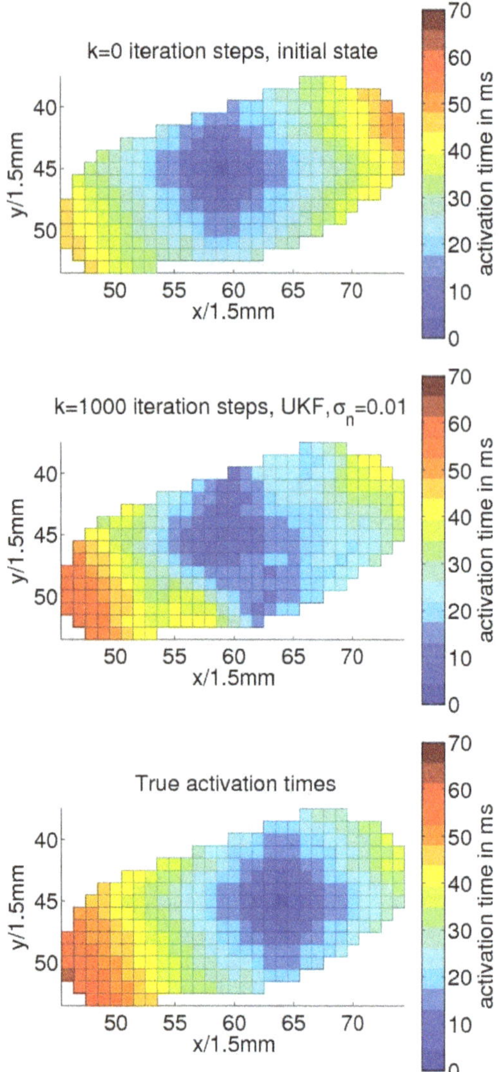

Fig. 8. Activation times for the UKF and measurement noise $\sigma_n = 0.01$ ($\mathbf{Q} = \mathbf{0}, \alpha = 5, \kappa = 0 = \beta$). Top: initial state, middle: 1000 iteration steps, bottom: true state.

Fig. 6 yields a faster convergence of the activation times of the UKF for all measurement noise levels. The signal and process functions **h** and **f** are nonlinear and even non-smooth functions of the activation times. The EKF is an algorithm of 1st order accuracy while the UKF captures the mean and covariance of the Gaussian distributed state vector to the 3rd order Taylor series (Wan and van der Merwe, 2000). In the next steps the heart model needs to be extended to the total ventricles and a muscle fibre model needs to be introduced to include an anisotropic activation wavefront velocity with respect to muscle fibre orientations. For an application to realistic data the heart model parameters, e.g. the process noise covariance $\mathbf{Q}$, need to be estimated from the data. This can be done e.g. with the Expectation-Maximization (EM) algorithm (Khan and Dutt, 2007).

Acknowledgements. This work was supported by the SFB 855 *Magnetoelectric Composites – Future Biomagnetic Interfaces* of the Deutsche Forschungsgemeinschaft.

References

Cheng, L. K., Bodley, J. M., and Pullan, A. J.: Comparison of Potential- and Activation-Based Formulations for the Inverse Problem of Electrocardiology, IEEE T. Bio-Med. Eng., 50, 11–22, 2003.

Fokas, A. S., Kurylev, Y., and Marinakis, V.: The unique determination of neuronal currents in the brain via magnetoencephalography, Institute of Physics Publishing, Inverse Problems, 20, 1067–1082, 2004.

He, B., Li, G., and Zhang, X.: Noninvasive three-dimensional activation time imaging of ventricular excitation by means of a heart-excitation model, Phys. Med. Biol., 47, 4063–4078, 2002.

Khan, M. E. and Dutt, D. N.: An expectation-maximization algorithm based Kalman smoother approach for event-related desynchronization (ERD) estimation from EEG, IEEE T. Bio-Med. Eng., 54, 1191–1198, 2007.

LaViola Jr., J. J.: A Comparison of Unscented and Extended Kalman Filtering for Estimating Quaternion Motion, P. Am. Contr. Conf., 3, 2435–2440, 2003.

Liu, C. and He, B.: Noninvasive Estimation of Global Activation Sequence Using the Extended Kalman Filter, IEEE T. Bio-Med. Eng., 58, 541–549, 2011.

Miller, W. T. and Geselowitz, D. B.: Simulation studies of the electrocardiogram. I. The normal heart, Circ. Res., 43, 301–315, 1978.

Nenonen, J., Purcell, C. J., Horacek, B. M., Stroink, G., and Katila, T.: Magnetocardiographic Functional Localization Using a Current Dipole in a Realistic Torso, IEEE T. Bio-Med. Eng., 38, 658–664, 1991.

Rhudy, M., Gu, Y., Gross, J., and Napolitano, M. R.: Evaluation of Matrix Square Root Operations for UKF within a UAV GPS/INS Sensor Fusion Application, Int. J. Navigation and Observation, 2011, Article ID 416828, doi:10.1155/2011/416828, 2011.

Schulze, W., Farina, D., Jiang, Y., and Dössel, O.: A Kalman filter with integrated Tikhonov-regularization to solve the inverse problem of electrocardiography, IFMBE Proceedings World Congress on Medical Physics and Biomedical Engineering, 25/2, 821–824, 2009.

SimBio Development Group: SimBio: A generic environment for bio-numerical simulations, online, https://www.mrt.uni-jena.de/simbio, last access: 15 January, 2013.

Wan, E. A. and van der Merwe, R.: The Unscented Kalman Filter for Nonlinear Estimation, Adaptive Systems for Signal Processing, Communications, and Control Symposium 2000, AS-SPCC, The IEEE 2000.

Weixue, L., Zhengyao, X., and Yingjie, F.: Microcomputer-based cardiac field simulation model, Med. Biol. Eng. Comput., 31, 384–387, 1993.

32

A simple evaluation procedure of the TAN calibration and the influence of non-ideal calibration elements on VNA S-parameter measurements

U. Stumper

Physikalisch-Technische Bundesanstalt, Bundesallee 100, 38116 Braunschweig, Germany

Abstract. For the 7-term general TAN (Through-Attenuator-Network) self-calibration method of a four-sampler vector network analyser (VNA), and for all derived calibration methods like TLN, TRL, TRM, TAR, or TMN, it is shown that a very simple evaluation procedure of the seven error terms is possible, even if the Through connection is replaced by a reflectionless network with known transmission. Expressions for the deviations of the measured S-parameters of two-port test objects (d.u.t.s) from the true values, which are caused by deviations of the modeled S-parameters of non-ideal calibration elements ("standards") from their true values, are also presented. Additionally, it is shown that a TAN calibration is also possible in case of unequal reflections of the Network.

1 Introduction

The 7-term general TAN self-calibration procedure of a 4-sampler VNA introduced by Eul and Schiek (1991), uses three two-ports as calibration elements, namely a reflectionless connection ("Through", T), a reflectionless two-port ("Attenuator", A) with unknown transmission coefficients, and a symmetrical two-port ("Network", N) with unknown S-parameters, but equal reflection coefficients at VNA test ports 1 and 2. From the uncorrected ("raw") reflection and transmission values determined for these three calibration elements, the seven error terms characterizing the VNA are calculated, and additionally, the five S-parameters of A and N are obtained.

The Attenuator A may be replaced by matched terminations (M) or by a matched precision line (L), and the Network (N) by two equally reflecting terminations (R), e.g. open circuits (O). Thus, from the TAN calibration other frequently used calibration procedures like TLN (Through-Line-Network), TRL (Through-Reflect-Line), TRM (Through-Reflect-Match), TOM (Through-Open-Match), TAR (Through-Attenuator-Reflect), or TMN (Through-Match-Network) are derived.

The Through can also be replaced by a network where all four S-parameters are known. If instead of the Through connection (T) a network *without reflection*, but with known transmission parameters (including total transmission) is considered, a very simple evaluation procedure of the seven error terms is possible. This will be shown in Sect. 2 of this paper.

The error terms will deviate from their true values (as given by the VNA hardware), if modeling of the calibration elements T, A, and N is non-ideal, i.e. if their S-parameters deviate from the "ideal" values as assumed by the error correction software:

1. With the Through (T), small reflection coefficients δT_{11},δT_{22}, and small deviations $\delta T_{12}, \delta T_{21}$ of the transmission coefficients from their assumed values T_{12}, T_{21} may occur, caused by transition resistance due to imperfect contacts, by cross-sectional discontinuities, or by small particles (e.g. lints) between the end planes of the connectors,
2. the Attenuator (A) may be of nonzero reflection δM_1, δM_2,
3. the Network (N) may be not perfectly symmetrical, e.g. may have unequal reflection coefficients $C + \delta C_1$, $C + \delta C_2$.

Consequently, with a non-ideal calibration, the measured S-parameters of a device under test (d.u.t.) will also deviate from their true values S_{ik}. In Sect. 3, these deviations δS_{ik}, assigned to the deviations from the true S-parameters associated with the calibration elements T, A, and N, are derived

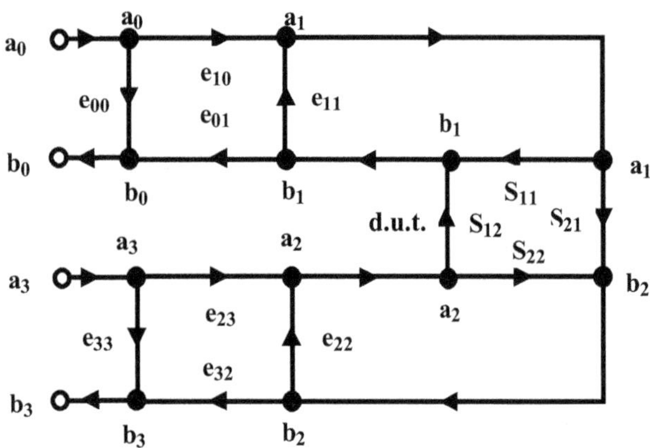

Fig. 1. Flow graph of 8-term error model. For a TAN calibration, the d.u.t. is replaced by the calibration elements: **T**hrough connection, **A**ttenuator and **N**etwork. Signals a_1, b_1 refer to VNA test port 1 and a_2, b_2 to test port 2, respectively.

for the general TAN calibration procedure and for all calibration procedures derived from TAN. The influence of noise, non-linearity, and cross-talk effects on the raw values is not considered here.

2 A simple calibration and measurement evaluation procedure

The characteristics of a VNA are most commonly described in terms of cascaded matrices of the error terms (e.g. in Gronau, 2001). However, as shown here, the use of scattering matrices in the "linear-in-T" form depicted by Rytting (2001) allows, for the general TAN procedure, a *direct* derivation of the error terms themselves. This is different from the method described by Eul and Schiek (1991), where the S-parameters of Attenuator and Network are determined in a first step and are then used as "known" standard values in a subsequent calculation of the error terms.

2.1 System equations

Using the "linear-in-T" form of the VNA system equations, the S-parameters S_{ik} of calibration elements or of a d.u.t. are linearly related to the S-parameters measured by the VNA (raw values m_{ik}) by 7 error terms a, b, c, d, e, f, and g as

$$a + S_{11}m_{11}bg - S_{11}cg + S_{21}m_{12}d = m_{11} \tag{1}$$

$$S_{12}m_{11}bg - S_{12}cg + S_{22}m_{12}d - m_{12}g = 0 \tag{2}$$

$$S_{11}m_{21}bg + S_{21}m_{22}d - S_{21}f = m_{21} \tag{3}$$

$$S_{12}m_{21}bg + eg + S_{22}m_{22}d - S_{22}f - m_{22}g = 0. \tag{4}$$

The a, ..., g are reduced from 8 error terms e_{ik} which are illustrated in Fig. 1. The reduction

$a = e_{00}, b = e_{11}e_{23}/e_{10}, c = (e_{00}e_{11} - e_{10}e_{01}) \cdot (e_{23}/e_{10}),$ $d = e_{22} \cdot e_{10}/e_{23}, e = e_{33}, f = (e_{22}e_{33} - e_{32}e_{23}) \cdot (e_{10}/e_{23}), g = e_{10}/e_{23}$

enables a simple calculation of a, ..., g from the raw values and furthermore allows the determination of the deviations of a, ..., g.

2.2 Calibration

Ideally, the S-matrices of the Through connection (T), the Attenuator (A), and the Network (N) are

$$\overline{\mathbf{T}} = \begin{pmatrix} 0 & T_{12} \\ T_{21} & 0 \end{pmatrix}, \quad \overline{\mathbf{A}} = \begin{pmatrix} 0 & A \\ B & 0 \end{pmatrix}, \text{ and } \overline{\mathbf{N}} = \begin{pmatrix} C & D \\ E & C \end{pmatrix}$$

whereas the T_{12}, T_{21} are known and A, B, C, D, and E are unknown. The calibration elements T, A, and N may be non-reciprocal.

2.2.1 Determination of a, b, c, d, e, and f using Through (T) and Attenuator (A)

Inserting the entries of $\overline{\mathbf{T}}$ and $\overline{\mathbf{A}}$ into the system Eqs. (1) to (4), for the Through (Index T) we obtain:

$$a + T_{21}m_{12\mathrm{T}}d = m_{11\mathrm{T}} \tag{5}$$

$$T_{12}m_{11\mathrm{T}}b - T_{12}c - m_{12\mathrm{T}} = 0 \tag{6}$$

$$T_{21}m_{22\mathrm{T}}d - T_{21}f = m_{21\mathrm{T}} \tag{7}$$

$$T_{12}m_{21\mathrm{T}}b + e - m_{22\mathrm{T}} = 0 \tag{8}$$

and for the Attenuator (Index A):

$$a + Bm_{12\mathrm{A}}d = m_{11\mathrm{A}} \tag{9}$$

$$Am_{11\mathrm{A}}b - Ac - m_{12\mathrm{A}} = 0 \tag{10}$$

$$Bm_{22\mathrm{A}}d - Bf = m_{21\mathrm{A}} \tag{11}$$

$$Am_{21\mathrm{A}}b + e - m_{22\mathrm{A}} = 0. \tag{12}$$

To calculate the first six error terms a, b, c, d, e, and f, only the equation sets for these two calibration elements are necessary! In a first step B is calculated from Eqs. (9) and (11). Using of Eqs. (5) and (7) we get:

$$B = [(m_{11\mathrm{A}} - m_{11\mathrm{T}}) + T_{21}d \cdot m_{12\mathrm{T}}] / (d \cdot m_{12\mathrm{A}}) \tag{13}$$

$$B = T_{21}m_{21\mathrm{A}} / [T_{21}d \cdot (m_{22\mathrm{A}} - m_{22\mathrm{T}}) + m_{21\mathrm{T}}]. \tag{14}$$

Equating these expressions yields a quadratic equation which can be solved for d:

$$T_{21}^2 d^2 \cdot (m_{22\mathrm{A}} - m_{22\mathrm{T}}) \cdot m_{12\mathrm{T}} + (m_{11A} - m_{11T}) \cdot m_{21T} + T_{21}d \cdot \begin{bmatrix} (m_{11\mathrm{A}} - m_{11\mathrm{T}}) \cdot (m_{22\mathrm{A}} - m_{22\mathrm{T}}) \\ + m_{12\mathrm{T}}m_{21\mathrm{T}} - m_{21\mathrm{A}}m_{12\mathrm{A}} \end{bmatrix} = 0. \tag{15}$$

Parameter A is calculated from Eqs. (10) and (12). Using Eqs. (6) and (8) we get:

$$A = T_{12}m_{12A}/\left[T_{12}b\cdot(m_{11A}-m_{11T})+m_{12T}\right] \tag{16}$$

$$A = \left[(m_{22A}-m_{22T})+T_{12}b\cdot m_{21T}\right]/(b\cdot m_{21A}). \tag{17}$$

Equating these expressions again yields a quadratic equation which can be solved for b:

$$T_{12}^2b^2\cdot(m_{11A}-m_{11T})\cdot m_{21T}+(m_{22A}-m_{22T})\cdot m_{12T} +T_{12}b\cdot\begin{bmatrix}(m_{11A}-m_{11T})\cdot(m_{22A}-m_{22T})\\ +m_{21T}m_{12T}-m_{12A}m_{21A}\end{bmatrix}=0. \tag{18}$$

With known d and b, the parameters a, c, e, and f are subsequently calculated using the Through Eqs. (5) to (8):

$$a=(m_{11T}-T_{21}m_{12T}d) \tag{19}$$

$$c=(T_{12}m_{11T}b-m_{12T})/T_{12} \tag{20}$$

$$e=(m_{22T}-T_{12}m_{21T}b) \tag{21}$$

$$f=(T_{21}m_{22T}d-m_{21T})/T_{21}. \tag{22}$$

2.2.2 Transmission symmetry of Attenuator (A)

In case of transmission symmetry of the Attenuator (A=B), we equate Eqs. (13) and (17) and obtain

$$(m_{11A}-m_{11T})\cdot m_{21A}b-(m_{22A}-m_{22T})\cdot m_{12A}d -(T_{12}m_{12A}m_{21T}-T_{21}m_{21A}m_{12T})\cdot bd=0. \tag{23}$$

Equating Eqs. (14) and (16) yields

$$T_{12}T_{21}\cdot(m_{11A}-m_{11T})\cdot m_{21A}b -T_{12}T_{21}\cdot(m_{22A}-m_{22T})\cdot m_{12A}d -(T_{12}m_{12A}m_{21T}-T_{21}m_{21A}m_{12T})=0. \tag{24}$$

The subtraction of Eq. (24) from Eq. (23) yields:

$$(1-T_{12}T_{21})\cdot\begin{bmatrix}(m_{11A}-m_{11T})\cdot m_{21A}b\\ -(m_{22A}-m_{22T})\cdot m_{12A}d\end{bmatrix} +(1-bd)\cdot(T_{12}m_{12A}m_{21T}-T_{21}m_{21A}m_{12T})=0. \tag{25}$$

Since both Eqs. (23) and (24) describe linear relations between b and d, only *one* of the quadratic Eqs. (15) or (18) has to be solved.

For a direct Through connection (T_{12}=T_{21}=1), Eq. (25) becomes

$$(1-bd)\cdot(m_{12T}m_{21A}-m_{21T}m_{12A})=0. \tag{26}$$

The error terms $|b|$ and $|d|$ are small compared to 1, viz $bd\neq1$:

$$m_{12T}m_{21A}-m_{21T}m_{12A}=0. \tag{27}$$

This relation between the raw values of Through and Attenuator is equivalent to the relation (33) of Engen (1979). Moreover, from Eq. (23) or (24) we obtain

$$b=\frac{(m_{22A}-m_{22T})\cdot m_{12A}}{(m_{11A}-m_{11T})\cdot m_{21A}}\cdot d. \tag{28}$$

2.2.3 Determination of g

To obtain the remaining error term g, we insert the entries of the Network matrix $\overline{\mathbf{N}}$ (Index N) into Eqs. (1) to (4):

$$a+Cm_{11N}bg-Ccg+Em_{12N}d=m_{11N} \tag{29}$$

$$Dm_{11N}bg-Dcg+Cm_{12N}d-m_{12N}g=0 \tag{30}$$

$$Cm_{21N}bg+Em_{22N}d-Ef=m_{21N} \tag{31}$$

$$Dm_{21N}bg+eg+Cm_{22N}d-Cf-m_{22N}g=0. \tag{32}$$

First, the Network parameters D and E are eliminated. For D, from Eqs. (30) and (32), respectively, we obtain:

$$D=m_{12N}\cdot(g-Cd)/(m_{11N}bg-cg) \tag{33}$$

$$D=\left[(m_{22N}-e)\cdot g-C\cdot(m_{22N}d-f)\right]/(m_{21N}bg). \tag{34}$$

Equating these expressions yields

$$C=\frac{\left[(m_{11N}b-c)\cdot(m_{22N}-e)-m_{12N}m_{21N}b\right]\cdot g}{\left[(m_{11N}b-c)\cdot(m_{22N}d-f)-m_{12N}m_{21N}bd\right]}. \tag{35}$$

For E, from Eqs. (29) and (31), respectively, we obtain:

$$E=\left[(m_{11N}-a)-Cg\cdot(m_{11N}b-c)\right]/(m_{12N}d) \tag{36}$$

$$E=(m_{21N}-Cm_{21N}bg)/(m_{22N}d-f). \tag{37}$$

Equating these expressions yields a second equation for C:

$$C=\frac{\left[(m_{11N}-a)\cdot(m_{22N}d-f)-m_{12N}m_{21N}d\right]}{\left[(m_{11N}b-c)\cdot(m_{22N}d-f)-m_{12N}m_{21N}bd\right]\cdot g}. \tag{38}$$

Eliminating C by equating Eqs. (35) and (38) gives

$$g^2=\frac{\left[(m_{11N}-a)\cdot(m_{22N}d-f)-m_{12N}m_{21N}d\right]}{\left[(m_{11N}b-c)\cdot(m_{22N}-e)-m_{12N}m_{21N}b\right]} \tag{39}$$

where a, b, c, d, e, and f are already known. The terms A, B, D, E, and C are not further used.

2.2.4 Transmission symmetry of Network (N)

In case of transmission symmetry of the Network (N) ($D=E$), we can formulate a second relation between the Through (T) and Network (N) raw values. Inserting Eq. (35) into Eqs. (33) or (34) yields

$$D=\frac{(de-f)\cdot m_{12N}}{\left[(m_{11N}b-c)\cdot(m_{22N}d-f)-m_{12N}m_{21N}bd\right]}, \tag{40}$$

and inserting Eq. (38) into Eqs. (36) or (37)

$$E=\frac{(ab-c)\cdot m_{21N}}{\left[(m_{11N}b-c)\cdot(m_{22N}d-f)-m_{12N}m_{21N}bd\right]}. \tag{41}$$

With D=E:

$$(de-f)\cdot m_{12N}=(ab-c)\cdot m_{21N}. \tag{42}$$

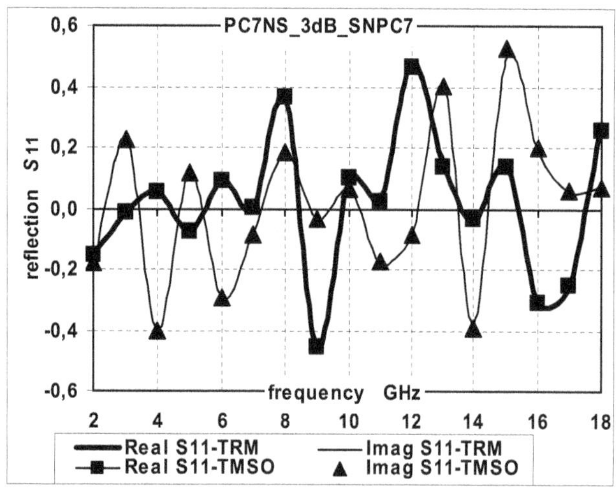

Fig. 2. Measured reflection coefficient S_{11} of a high-reflective 3 dB attenuator using TRM calibration and TMSO calibration.

This relation is even valid if the reflection coefficients of the Network (N) are not equal.

Inserting Eqs. (19) to (22) into Eq. (42) yields:

$$(T_{12}T_{21}bd-1)\cdot(T_{21}m_{21\mathrm{N}}m_{12\mathrm{T}}-T_{12}m_{12\mathrm{N}}m_{21\mathrm{T}})=0. \quad (43)$$

For a direct Through connection (T_{12}=T_{21}=1) we have

$$(bd-1)\cdot(m_{21\mathrm{N}}m_{12\mathrm{T}}-m_{12\mathrm{N}}m_{21\mathrm{T}})=0, \quad (44)$$

and, since $bd\neq 1$:

$$m_{21\mathrm{N}}m_{12\mathrm{T}}-m_{12\mathrm{N}}m_{21\mathrm{T}}=0 \quad (45)$$

has to be fulfilled, similar to the relation (27), but here between the raw values of Through and Network.

2.2.5 Calibration methods derived from TAN

For the Through-Line-Network (TLN) calibration method, A=B=L=$\exp(-\gamma\cdot l)$, where l and γ are length and propagation coefficient of the transmission line.

For the Through-Attenuator-Reflect (TAR) calibration method, we get D=E=0. In this case, $m_{12\mathrm{N}}$=$m_{21\mathrm{N}}$=0 is valid in Eq. (39).

For the Line-Reflect-Line (LRL) calibration method, T_{12}=T_{21}=L_1=$\exp(-\gamma\cdot l_1)$, A=B=L_2=$\exp(-\gamma\cdot l_2)$ and D=E=0 is fulfilled.

For the Through-Reflect-Line (TRL) calibration method, T_{12}=T_{21}=1, A=B=L=$\exp(-\gamma\cdot l)$ and D=E=0.

For the Through-Match-Network (TMN) calibration method, A=B=0. Hence, a direct determination of b and d from Eqs. (5) to (9) and Eq. (12) is possible.

For the Through-Reflect-Match (TRM) calibration method, A=B=0 and D=E=0.

2.3 Evaluation of measurements

With known error terms, the S-parameters S_{ik} of a d.u.t. are obtained using the equations

$$S_{11}=[(m_{11}-a)\cdot(m_{22}d-f)-m_{12}m_{21}d]\,/\,(M\cdot g) \quad (46)$$

$$S_{12}=m_{12}\cdot(de-f)\,/M \quad (47)$$

$$S_{21}=m_{21}\cdot(ab-c)\,/M \quad (48)$$

$$S_{22}=[(m_{11}b-c)\cdot(m_{22}-e)-m_{21}m_{12}b]\cdot g/M \quad (49)$$

$$M=(m_{11}b-c)\cdot(m_{22}d-f)-m_{12}m_{21}bd. \quad (50)$$

(cf. Rytting, 2001; Stumper, 2005c), where the m_{ik} are now the raw values of the d.u.t. We recognize that for transmission measurements, *only* the error terms a,..., f and the calibration steps using Through and Attenuator are needed, while g – and the calibration step using the Network – are only relevant for reflection measurements. The correct sign of g is found by measuring a d.u.t. where the value of its reflection coefficient is approximately known (e.g. a short or open circuit).

2.4 Verification

To verify the new procedure, measurements have been performed, either using a TRM calibration method (derived from TAN with A=B=D=E=0) where T_{12}=T_{21}=1, or the well-known TMSO (or SOLT) calibration method. With the latter method, a 10-term VNA model is considered, which uses – in addition to the Through (T) – well-defined matched terminations (M), short-circuits (S), and open-circuits (O) as calibration standards.

To compare the measurements carried out with both calibration methods, a 8510B-type VNA, a single calibration kit including a broadband match (M), and the same d.u.t.s. were used. The raw values for TRM calibration and subsequent measurements were calculated according to Schiek (1999) from the four sampler signals provided by the VNA for both signal directions. For the TMSO calibration and subsequent measurements, the internal VNA firmware was used.

The d.u.t.s. were a set of high-reflecting 7 mm coaxial two-port devices with PC-7 connectors including attenuators of (nominal) attenuation 3 dB to 60 dB which were sandwiched between the side arms of two T-junctions with the feeding arms shortened (cf. Stumper, 2005b). Real and imaginary parts of the S-parameters of these devices varied between approximately –0.5 and +0.5 in the frequency range 2–18 GHz. A second d.u.t. set included low-reflective PC-7 attenuator pads of (nominal) attenuation 10 dB, 20 dB, 40 dB, and 70 dB. Examples of results of the comparison are shown in Fig. 2 to Fig. 5. Considering the moderate reflection (max. 0.06 at 18 GHz) of the broadband Match, we observed a good coincidence of measurements carried out with the two calibrations.

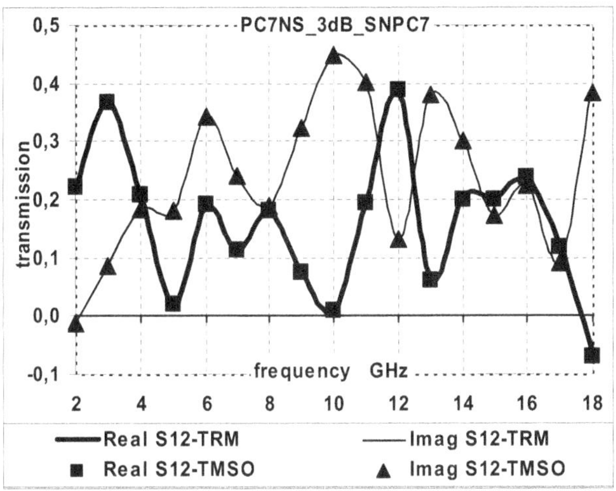

Fig. 3. Measured transmission coefficient S_{12} of a high-reflective 3 dB attenuator using TRM calibration and TMSO calibration.

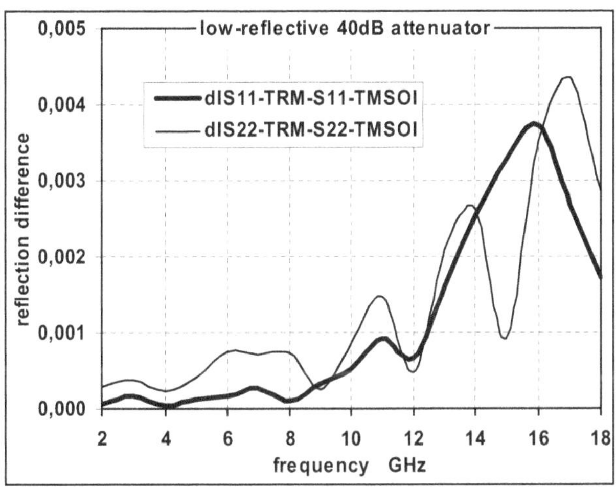

Fig. 4. Measured magnitude of vector difference of reflection coefficient S_{11} (and of S_{22} as well) of a low-reflective 40 dB attenuator using TRM calibration and TMSO calibration.

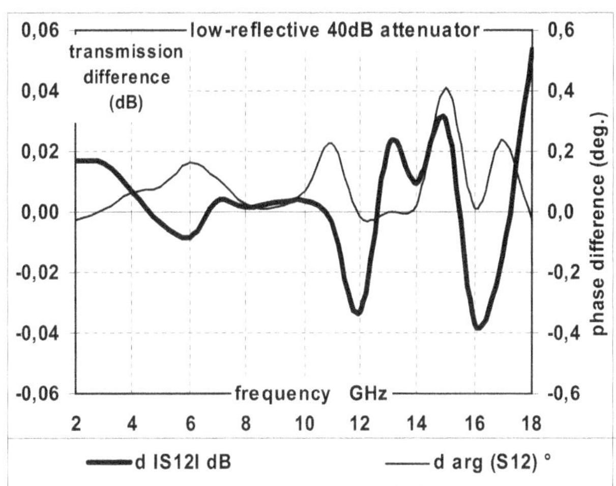

Fig. 5. Difference of transmission coefficient S_{12} in dB, and transmission phase difference $\Delta\arg(S_{12})$ in degree, of a low-reflective 40 dB attenuator, as measured using TRM calibration and TMSO calibration.

3 Derivation of sensitivity coefficients for the TAN calibration

3.1 Calculation of the δS_{ik}

In a first step, the deviations δS_{ik} of a d.u.t. as functions of the deviations δa, δb, δc, δd, δe, δf, and δg are calculated. The procedure has already been described by Stumper (2005b, c).

3.2 Calculation of the deviation of error terms

In a second step, the deviations δa, ..., δg are calculated as functions of the *deviations* from the true S-parameters which are associated with the non-ideal calibration elements. The *disturbed* S-matrices of the non-ideal Through (T), Attenuator (A), and Network (N) are

$$\overline{\mathbf{T}}_{\mathrm{d}} = \begin{pmatrix} \delta T_{11} & T_{12}+\delta T_{12} \\ T_{21}+\delta T_{21} & \delta T_{22} \end{pmatrix}, \quad \overline{\mathbf{A}}_{\mathrm{d}} = \begin{pmatrix} \delta M_1 & A \\ B & \delta M_2 \end{pmatrix},$$

$$\text{and } \overline{\mathbf{N}}_{\mathrm{d}} = \begin{pmatrix} C+\delta C_1 & D \\ E & C+\delta C_2 \end{pmatrix}.$$

At first, we calculate disturbed raw values $m_{ik}+\delta m_{ik}$ by inserting the entries of the disturbed $\overline{\mathbf{T}}_{\mathrm{d}}$, $\overline{\mathbf{A}}_{\mathrm{d}}$, or $\overline{\mathbf{N}}_{\mathrm{d}}$, respectively, into the Eqs. (51) to (55) (Rytting, 2001; Stumper, 2005c) for S_{11}, S_{12}, S_{21}, and S_{22}, respectively:

$$m_{11} = [(a - S_{11}cg)\cdot(g - S_{22}d) - S_{12}S_{21}cdg]/Q \tag{51}$$

$$m_{12} = S_{12}\cdot(ab - c)\cdot g/Q \tag{52}$$

$$m_{21} = S_{21}\cdot(de - f)\cdot g/Q \tag{53}$$

$$m_{22} = [(1 - S_{11}bg)\cdot(eg - S_{22}f) - S_{12}S_{21}bfg]/Q \tag{54}$$

$$Q = (1 - S_{11}bg)\cdot(g - S_{22}d) - S_{12}S_{21}bdg. \tag{55}$$

Next, the disturbed values $m_{ik}+\delta m_{ik}$ are inserted into the expressions (15), (18) to (22), and (39) for a, ..., g to obtain the deviations δa,..., δg.

As an example, the derivation of the deviation δg_{N} in case of a non-ideal Network (N) is shortly outlined here. The terms a, b, c, d, e, and f are neither depending on the Network parameters D, E, nor on the raw values $m_{ik\mathrm{N}}$. Inserting the entries of $\overline{\mathbf{N}}_{\mathrm{d}}$ in Eq. (51), we obtain

$$m_{11\mathrm{N}}+\delta m_{11\mathrm{N}} = \frac{[a-(C+\delta C_1)\cdot cg]\cdot[g-(C+\delta C_2)\cdot d]-DEcdg}{[1-(C+\delta C_1)\cdot bg]\cdot[g-(C+\delta C_2)\cdot d]-DEbdg}, \tag{56}$$

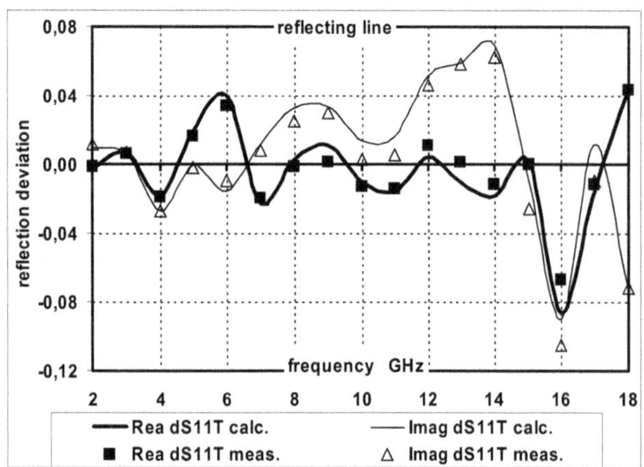

Fig. 6. Comparison of calculated and measured deviations δS_{11T} for TRM calibration and a high-reflective short line, using a non-ideal Through (copper foil at the inner conductors at the VNA test ports), instead of an ideal Through.

and finally we get

$$\delta m_{11N}=\frac{(ab-c)\cdot g\cdot\left[(g-Cd)^2\cdot\delta C_1+DEd^2\cdot\delta C_2\right]}{[(1-Cbg)\cdot(g-Cd)-DEbdg]^2}. \quad (57)$$

For δm_{12N}, δm_{21N}, and δm_{22N} we obtain similar expressions.

From Eq. (39) we then establish

$$(g+\delta g_N)^2=\frac{\left\{\begin{matrix}[(m_{11N}+\delta m_{11N})-a]\cdot[(m_{22N}+\delta m_{22N})\cdot d-f]\\-(m_{12N}+\delta m_{12N})\cdot(m_{21N}+\delta m_{21N})\cdot d\end{matrix}\right\}}{\left\{\begin{matrix}[(m_{22N}+\delta m_{22N})-e]\cdot[(m_{11N}+\delta m_{11N})\cdot b-c]\\-(m_{12N}+\delta m_{12N})\cdot(m_{21N}+\delta m_{21N})\cdot b\end{matrix}\right\}}. \quad (58)$$

Inserting the expressions obtained for δm_{11N}, δm_{12N}, δm_{21N}, and δm_{22N} into Eq. (58), we finally obtain:

$$\delta g_N=\frac{g}{2\cdot C}\cdot(\delta C_1-\delta C_2). \quad (59)$$

3.3 Resulting sensitivity coefficients

Inserting the results of the second step (Sect. 3.2) into Eqs. (32) to (35) as given in Stumper (2005c) (for $r_{ik}=S_{ik}$ there), we obtain sensitivity coefficients for the four S-parameters of a d.u.t. which are separately given for the deviations associated with the non-ideal Through (Index T), Attenuator (A), and Network (N). The expressions are *not dependent on the error terms* (i.e. not dependent on the VNA hardware):

$$\delta S_{11T}^{TAN}=\left\{\begin{matrix}\frac{(AB-S_{12}S_{21})}{(T_{12}T_{21}-AB)}\cdot\delta T_{11}\\-\frac{S_{11}^2}{(T_{12}T_{21}-AB)}\cdot\delta T_{22}\\-\frac{S_{11}\cdot(AB+C^2-DE)}{2\cdot C\cdot(T_{12}T_{21}-AB)}\cdot(\delta T_{11}-\delta T_{22})\\-\frac{S_{11}}{2}\cdot\left(\frac{\delta T_{12}}{T_{12}}+\frac{\delta T_{21}}{T_{21}}\right)\end{matrix}\right\} \quad (60)$$

$$\frac{\delta S_{12T}^{TAN}}{S_{12}}=\left\{\begin{matrix}-\frac{S_{22}}{(T_{12}T_{21}-AB)}\cdot\delta T_{11}-\frac{\delta T_{12}}{T_{12}}\\-\frac{S_{11}}{(T_{12}T_{21}-AB)}\cdot\delta T_{22}\end{matrix}\right\} \quad (61)$$

$$\delta S_{11A}^{TAN}=\left\{\begin{matrix}-\frac{(T_{12}T_{21}-S_{12}S_{21})}{(T_{12}T_{21}-AB)}\cdot\delta M_1\\+\frac{S_{11}^2}{(T_{12}T_{21}-AB)}\cdot\delta M_2\\+\frac{S_{11}\cdot(T_{12}T_{21}+C^2-DE)}{2\cdot C\cdot(T_{12}T_{21}-AB)}\cdot(\delta M_1-\delta M_2)\end{matrix}\right\} \quad (62)$$

$$\frac{\delta S_{12A}^{TAN}}{S_{12}}=\frac{\delta S_{21A}^{TAN}}{S_{21}}=\frac{S_{22}}{(T_{12}T_{21}-AB)}\cdot\delta M_1+\frac{S_{11}}{(T_{12}T_{21}-AB)}\cdot\delta M_2. \quad (63)$$

To obtain the expressions for δS_{21} and δS_{22}, index 1 has to be replaced by 2 and vice versa in these equations.

Nonequal reflection of the Network (N) influences only the error term g and, consequently, influences only the reflection coefficients of the d.u.t.:

$$\delta S_{11N}^{TAN}=\frac{S_{11}}{2\cdot C}\cdot(\delta C_2-\delta C_1) \quad (64)$$

$$\delta S_{22N}^{TAN}=\frac{S_{22}}{2\cdot C}\cdot(\delta C_1-\delta C_2). \quad (65)$$

There is no influence on transmission, viz $\delta S_{12N}^{TAN}=\delta S_{21N}^{TAN}=0$.

These expressions for the non-ideal Network remain the same for *all calibration methods* derived from TAN.

For the Line-Reflect-Line (LRL) calibration method we obtain the same sensitivity coefficients as given by Stumper (2005c) for small e_{11}. For the Through-Reflect-Line (TRL) calibration method we obtain the same sensitivity coefficients as given by Stumper (2005a).

For the Through-Reflect-Match (TRM) calibration method where $A=B=D=E=0$, and a Through (T) of total transmission ($T_{12}=T_{21}=1$), we obtain the simple expressions:

$$\delta S_{11T}^{TRM}=\left\{\begin{matrix}-S_{12}S_{21}\cdot\delta T_{11}-S_{11}^2\cdot\delta T_{22}\\-\frac{S_{11}\cdot C}{2}\cdot(\delta T_{11}-\delta T_{22})\\-\frac{S_{11}}{2}\cdot(\delta T_{12}+\delta T_{21})\end{matrix}\right\} \quad (66)$$

$$\frac{\delta S_{12T}^{TRM}}{S_{12}}=-S_{22}\cdot\delta T_{11}-\delta T_{12}-S_{11}\cdot\delta T_{22} \quad (67)$$

$$\delta S_{11A}^{TRM}=\left\{\begin{matrix}-(1-S_{12}S_{21})\cdot\delta M_1+S_{11}^2\cdot\delta M_2\\+\frac{S_{11}\cdot(1+C^2)}{2\cdot C}\cdot(\delta M_1-\delta M_2)\end{matrix}\right\} \quad (68)$$

$$\frac{\delta S_{12A}^{TRM}}{S_{12}}=\frac{\delta S_{21A}^{TRM}}{S_{21}}=S_{22}\cdot\delta M_1+S_{11}\cdot\delta M_2. \quad (69)$$

To obtain the expressions for δS_{21} and δS_{22}, index 1 has to be replaced by 2 and vice versa in these equations.

3.4 Experimental results

For the TRL calibration method, Eqs. (60) to (65) have already been verified in the frequency range 2–18 GHz (see Stumper, 2005a, b, c) showing good agreement between theory and experiment.

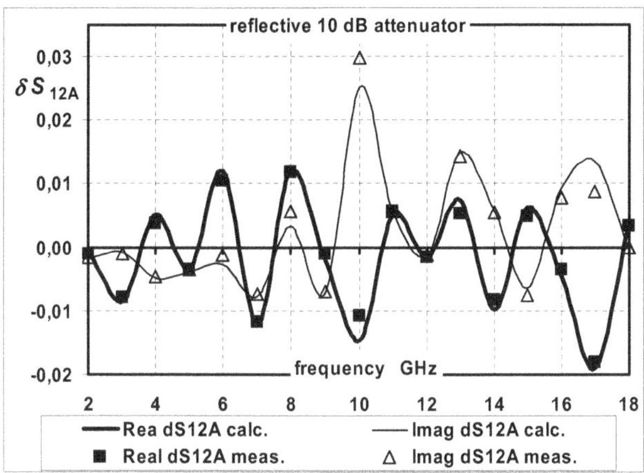

Fig. 7. Comparison of calculated and measured deviations δS_{12A} for TRM calibration and a high-reflective 10 dB attenuator, using a mismatch of reflection 0.2 at the VNA test port 2, instead of an ideal Match.

For the TRM calibration method, Eqs. (64) to (69) have also been experimentally verified by use of the same set of coaxial (PC-7) high- and low-reflective two-ports as described by Stumper (2003a, 2005a, b, c). Only one of the elements Through, Reflect, and Match at a time was considered non-ideal. Examples are shown in Fig. 6 and Fig. 7. The agreement of theory and experiment is also good, with exception of the Match in the case of high-reflective d.u.t.s where some differences between calculated and measured deviations were observed.

3.5 Comparison of TRL, TRM, and TMSO methods

It is interesting to compare the influence of non-ideal calibration elements used for different calibration methods including the TMSO (or SOLT) calibration method. Some results are given here:

For TRM and for TMSO (Stumper, 2003a), a non-ideal Match has nearly the same influence on the deviations of the S-parameters for both high- and low-reflective d.u.t.s.

The influence of a non-ideal Through is different for TRL, TRM, and TMSO, and depends on the reflection of the d.u.t. In Fig. 8, the large influence of a non-ideal Through (with an 0.1 mm thick hair in the measuring plane (Stumper, 2003b)) on S_{11} for a high-reflective attenuator is shown for TRL and TRM, whereas the influence is very low for TMSO.

For a low-reflective d.u.t., the influence on S_{11} (and S_{22}) depends on its attenuation, and the effect for TRL is rather strong, especially for large attenuation (small $S_{12}S_{21}$),

$$\delta S_{11\mathrm{T}}^{\mathrm{TRL}} \approx \frac{\left(L^2 - S_{12}S_{21}\right)}{\left(1 - L^2\right)} \cdot \delta T_{11}, \tag{70}$$

whereas the effect is small for TRM and for TMSO:

$$\delta S_{11\mathrm{T}}^{\mathrm{TRM}} \approx -\delta S_{11\mathrm{T}}^{\mathrm{TMSO}} \approx -S_{12}S_{21} \cdot \delta T_{11}. \tag{71}$$

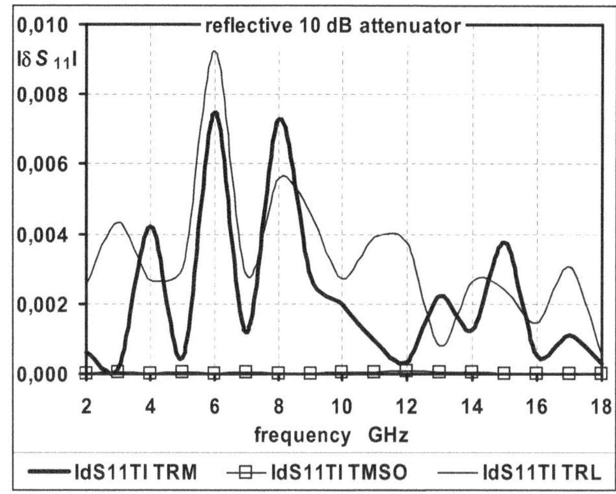

Fig. 8. Comparison of calculated deviations $|\delta S_{11\mathrm{T}}|$ for TRM, TMSO, and TRL calibrations and a high-reflective 10 dB attenuator, using a non-ideal Through with a 0.1 mm thick hair in the measuring plane, instead of an ideal Through.

The influence on S_{12} (and S_{21}) is nearly equal for TRL, TRM, and TMSO (there with opposite sign):

$$\delta S_{12\mathrm{T}}^{\mathrm{TRL}}/S_{12} \approx \delta S_{12\mathrm{T}}^{\mathrm{TRM}}\Big/S_{12} \approx -\delta S_{12\mathrm{T}}^{\mathrm{TMSO}}/S_{12} \approx -\delta T_{12}. \tag{72}$$

4 Determination of *g* for unequal Network reflections

It can be shown that a determination of the error term g is also possible if the reflection coefficients of the Network (N) are not equal, viz $C_1 \neq C_2$. In this case, an additional connection of the reversed Network to the VNA test ports is necessary. However, such a reversed connection is only possible for equal connector sex at both test ports. For example, this is the case for calibration elements and d.u.t.s in microstrip technique which can be fixed on a test fixture.

For the Network (N) connected directly to the test ports (Index q) and connected reversed (Index r), the S-matrices are

$$\overline{\mathbf{N}}_{\mathrm{q}} = \begin{pmatrix} C_1 & D \\ E & C_2 \end{pmatrix} \text{ and } \overline{\mathbf{N}}_{\mathrm{r}} = \begin{pmatrix} C_2 & E \\ D & C_1 \end{pmatrix}, \text{ respectively.}$$

We insert the entries of these two matrices into Eqs. (1) to (4) and obtain for the direct connection:

$$a + C_1 m_{11\mathrm{Nq}} bg - C_1 cg + E m_{12\mathrm{Nq}} d = m_{11\mathrm{Nq}} \tag{73}$$

$$D m_{11\mathrm{Nq}} bg - Dcg + C_2 m_{12\mathrm{Nq}} d - m_{12\mathrm{Nq}} g = 0 \tag{74}$$

$$C_1 m_{21\mathrm{Nq}} bg + E m_{22\mathrm{Nq}} d - Ef = m_{21\mathrm{Nq}} \tag{75}$$

$$D m_{21\mathrm{Nq}} bg + eg + C_2 m_{22\mathrm{Nq}} d - C_2 f - m_{22\mathrm{Nq}} g = 0, \tag{76}$$

and for the reverse connection:

$$a + C_2 m_{11\mathrm{Nr}} bg - C_2 cg + D m_{12\mathrm{Nr}} d = m_{11\mathrm{Nr}} \tag{77}$$

$$Em_{11\mathrm{Nr}}bg - Ecg + C_1 m_{12\mathrm{Nr}}d - m_{12\mathrm{Nr}}g = 0 \tag{78}$$

$$C_2 m_{21\mathrm{Nr}}bg + Dm_{22\mathrm{Nr}}d - Df = m_{21\mathrm{Nr}} \tag{79}$$

$$Em_{21\mathrm{Nr}}bg + eg + C_1 m_{22\mathrm{Nr}}d - C_1 f - m_{22\mathrm{Nr}}g = 0. \tag{80}$$

We eliminate D and E from these equations and obtain

$$C_1 = H_\mathrm{r} \cdot g / F_\mathrm{r} \tag{81}$$

$$C_2 = G_\mathrm{r} / (I_\mathrm{r} \cdot g) \tag{82}$$

$$C_1 = G_\mathrm{q} / \left(I_\mathrm{q} \cdot g\right) \tag{83}$$

$$C_2 = H_\mathrm{q} \cdot g / F_\mathrm{q} \tag{84}$$

where

$$F_i = (m_{22\mathrm{Ni}}d - f) \cdot (m_{11\mathrm{Ni}}b - c) - m_{12\mathrm{Ni}}m_{21\mathrm{Ni}}bd \tag{85}$$

$$G_i = (m_{11\mathrm{Ni}} - a) \cdot (m_{22\mathrm{Ni}}d - f) - m_{12\mathrm{Ni}}m_{21\mathrm{Ni}}d \tag{86}$$

$$H_i = (m_{22\mathrm{Ni}} - e) \cdot (m_{11\mathrm{Ni}}b - c) - m_{12\mathrm{Ni}}m_{21\mathrm{Ni}}b \tag{87}$$

$$I_i = (m_{11\mathrm{Ni}}b - c) \cdot (m_{22\mathrm{Ni}}d - f) - m_{12\mathrm{Ni}}m_{21\mathrm{Ni}}bd. \tag{88}$$

The index i is either q or r. By equating Eqs. (81) and (83) or Eqs. (82) and (84), we finally obtain:

$$g^2 = g_{\mathrm{qr}}^2 = F_\mathrm{q} G_\mathrm{r} / \left(H_\mathrm{q} I_\mathrm{r}\right) = F_\mathrm{r} G_\mathrm{q} / \left(H_\mathrm{r} I_\mathrm{q}\right). \tag{89}$$

The case D=E = 0 is trivial.

5 Conclusions

It has been shown that for the general TAN calibration and derived calibration methods, the error terms and also the deviations of S-parameters of d.u.t.s. due to non-ideal calibration elements can easily be calculated by using the linear-in-T form of the VNA system equations. This is also possible if the Through (T) is a reflectionless network with known transmission parameters. Simple expressions have been obtained for these deviations (sensitivity coefficients). They are suitable for establishing the type-B uncertainty budget for S-parameter measurements, as, in most cases, the deviations from the true S-parameters associated with Through (T), Attenuator (A), and Network (N) can be estimated. Moreover, it is now possible to compare the influence of non-ideal calibration elements for the different calibration methods based on TAN and the 10-term TMSO calibration method. A TAN calibration is also possible in case of unequal reflections of the Network (N).

Acknowledgements. The author would like to thank W. Peinelt, D. Schubert und T. Schrader for carrying out the precise verification measurements.

References

Engen, G. F. and Hoer, C. A.: Thru-Reflect-Line: An Improved Technique for Calibrating the Dual Six-port Automatic Network Analyzer, IEEE Trans. Microwave Theory Tech., 27, 987–993, 1979.

Eul, H.-J. and Schiek, B.: A Generalized Theory and New Calibration Procedures for Network Analyzer Self-Calibration, IEEE Trans. Microwave Theory Tech., 39, 724–731, 1991.

Gronau, G.: Höchstfrequenztechnik. Berlin, Springer, Ch. 8, 389–398, 2001.

Rytting, D.: Network Analyzer Error Models and Calibration Methods, ARFTG/NIST Short Course on RF Measurements for a Wireless World, San Diego, CA, Nov. 29–30, 2001.

Schiek, B.: Grundlagen der Hochfrequenz-Messtechnik. Berlin, Springer, Ch. 4.4, p. 159, 1999.

Stumper, U.: Influence of TMSO Calibration Standards Uncertainties on VNA S-Parameter Measurements, IEEE Trans. Instrum. Meas., 52, 311-315, 2003a.

Stumper, U.: Uncertainty of VNA S-Parameter Measurement due to Non-Ideal TMSO or LMSO Calibration Standards, Advances in Radio Science (Kleinheubacher Berichte), [Online], 1, 1–8, available: http://www.copernicus.org/URSI/ars/1/1.pdf, 2003b.

Stumper, U.: Uncertainty of VNA S-Parameter Measurement due to Non-Ideal TRL Calibration Items, IEEE Trans. Instrum. Meas., 55, 676–679, 2005a.

Stumper, U.: Fehlerfortpflanzung bei der LRL- oder TRL-Kalibrierung eines vektoriellen Vierstellen-Netzwerkanalysators, in: Neue Kalibrierverfahren im Nieder- und Hochfrequenzbereich – Vorträge des 206. PTB-Seminars, edited by: Bachmair, H., PTB-Bericht E-89, Braunschweig, May 2005b.

Stumper, U.: Influence of Non-ideal LRL or TRL Calibration Elements on VNA S-Parameter Measurements, Advances in Radio Science (Kleinheubacher Berichte), [Online], 3, 51–58, available: http://www.copernicus.org/URSI/ars/3/ars-3-51.pdf, 2005c.

Obstacle-based self-calibration techniques for the determination of the permittivity of liquids

I. Rolfes

Institut für Hochfrequenztechnik und Funksysteme, Leibniz Universität Hannover, Appelstraße 9A, 30167 Hannover, Germany

Abstract. In this contribution, different obstacle-based self-calibration techniques for the measurement of the dielectric properties of liquids are investigated at microwave frequencies. The liquid under test is contained inside a waveguide, which is connected to the ports of a vector network analyzer. The permittivity of the liquid is characterized on the basis of the measured scattering parameters.

In order to extract the material parameters precisely and to eliminate systematic errors of the setup, calibration measurements have to be performed. For this purpose, different self-calibration methods based on the displacement of an obstacle are considered. The presented methods differ in that way, that either transmission and reflection measurements or purely reflection measurements are performed. All these methods have in common that the material parameters are already calculable within a so-called self-calibration procedure. Thus, a full two-port calibration of the whole setup is not necessary. Furthermore, the methods can be realized effectively in a practical setup having the advantage that a rearrangement of the setup is not needed for the material parameter measurements and that the liquid under investigation can pass continuously through the measurement cell. This might be of interest for the application in an industrial process, enabling the continuous flow of the material while the parameter characterization can take place at the same time.

1 Introduction

At microwave frequencies, the dielectric properties of liquids can be measured by means of guided-wave travelling-wave methods Clarke et al. (2003). The material parameters are determined on the basis of the measured complex scattering parameters using a vector network analyzer (VNA). For this purpose, the dielectric sample is filled into a measurement cell, which can be a coaxial line or a rectangular or circular waveguide.

The measurements with the VNA are influenced by systematic errors due to transmission losses, reflections at the measurement ports and other non-idealities of the setup. These systematic errors can be represented by transmission matrices describing the error two-ports $\mathbf{G}^{-1}$ and $\mathbf{H}$, as known from the error description of two-port vector network analyzers according to the 7-term-model and as shown in Fig. 1. In order to eliminate these systematic errors and to obtain the error-corrected material parameters of the specimen, multiple calibration measurements have to be performed.

A well-known method for the material characterization is the TL-method (through, line) (Engen and Hoer, 1979; Eul and Schiek, 1991), which is based on the measurement of the scattering parameters of two lines with a difference Δl in length as shown in Fig. 2.

With the measured wave parameters m_1 to m_4 from Fig. 1 two measurement transmission matrices $\mathbf{M_T}$ and $\mathbf{M_L}$ can be constructed. The transmission matrix $\mathbf{M}$ can be written in its general form as:

$$\mathbf{M} = \begin{bmatrix} m_1' & m_1'' \\ m_2' & m_2'' \end{bmatrix} \begin{bmatrix} m_3' & m_3'' \\ m_4' & m_4'' \end{bmatrix}^{-1} = \mathbf{G}^{-1}\mathbf{T_D}\mathbf{H} \tag{1}$$

where the one-dashed and two-dashed terms represent the measured wave parameters in dependance of the feeding direction of the generator signal, either from port 1 or port 2. $\mathbf{T_D}$ is the transmission matrix of the device connected to the ports of the analyzer. Thus, the propagation constant γ of the line standard can be calculated by:

$$\begin{aligned} trace\{\mathbf{M_T}\mathbf{M_L}^{-1}\} &= trace\{(\mathbf{G}^{-1}\mathbf{T_T}\mathbf{H})(\mathbf{G}^{-1}\mathbf{T_L}\mathbf{H})^{-1}\} \\ &= trace\{\mathbf{T_T}\mathbf{T_L}^{-1}\} \\ &= trace\left\{\begin{bmatrix} 1 & 0 \\ 0 & 1 \end{bmatrix}\begin{bmatrix} e^{-\gamma\Delta l} & 0 \\ 0 & e^{\gamma\Delta l} \end{bmatrix}^{-1}\right\} \\ &= e^{\gamma\Delta l} + e^{-\gamma\Delta l} \ , \end{aligned} \tag{2}$$

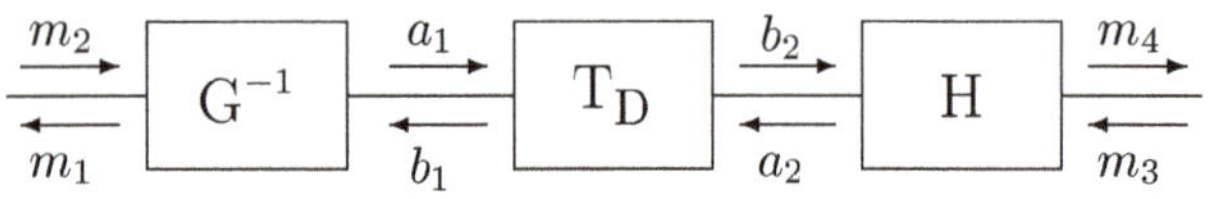

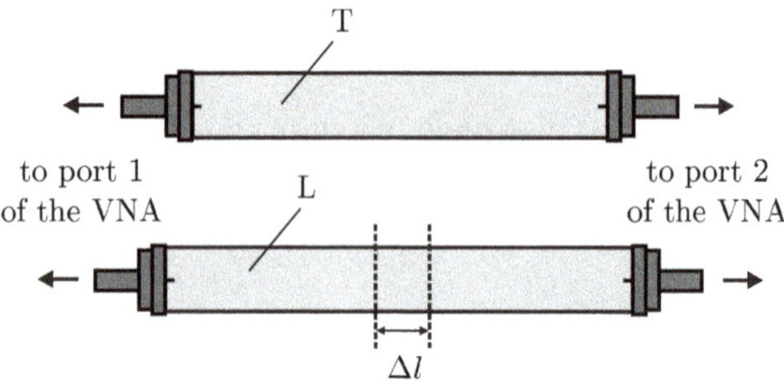

Fig. 1. Error model of the measurement system.

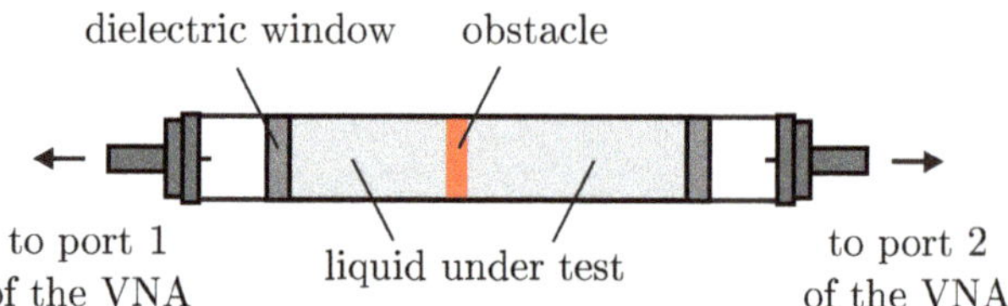

Fig. 2. Setup of the TL-method, based on two lines of different lengths.

where the operator *trace* stands for the calculation of the trace of a matrix. The complex permittivity $\epsilon_r = \epsilon_r' - j\epsilon_r''$ can thus be calculated for a TEM waveguide on the basis of the propagation constant γ:

$$\epsilon_r = \left(\frac{\gamma \cdot c}{2\pi f}\right)^2 , \tag{3}$$

where f is the frequency and c the speed of light in vacuum. One disadvantage of this method is that two lines of different lengths are needed which have to be connected to the ports of the analyzer. Thus, a rearrangement of the measurement setup becomes necessary, enlarging the effort of the measurements.

2 Obstacle-based methods

Based on the displacement of an obstacle within a waveguide being filled with the liquid under investigation and on the measurement of the scattering parameters for different positions of the obstacle, the permittivity of the liquid under test can be determined. In the following different variants of such obstacle-based methods will be presented.

2.1 LNN-method

In Fig. 3 a measurement setup for the line-network-method (LNN) (Heuermann and Schiek, 1997; Rolfes and Schiek, 2006) is shown, which enables a material characterization within the same setup without the need for a rearrangement. Instead of the two lines as necessary for the thru-line method, a so-called obstacle is used, which can be realized as a dielectric plate with unknown parameters. For the measurements,

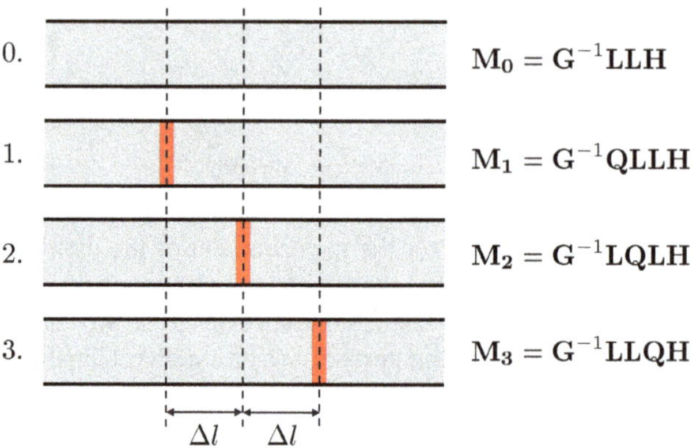

Fig. 3. Measurement system for the material characterization based on the line-network-method.

Fig. 4. Structures for a full two-port calibration according to the line-network-method.

the waveguide is filled with the specimen. The liquid under test can either be kept in a static position without any movement or in a dynamic state, passing through the waveguide. The latter variant is of interest for the application in an industrial process, enabling the continuous flow of the material, while the parameter characterization can take place at the same time. In order to keep the specimen in a defined position two dielectric windows can be used as shown in Fig. 3.

For a full two-port calibration of the whole system, the scattering parameters of the setup have to be measured for four different positions of the obstacle. First, the measurement of the waveguide without obstacle has to be performed (pos. 0 in Fig. 4). Next, the obstacle is placed at three consecutive, equidistant positions (1, 2, 3) within the waveguide, as shown in Fig. 4. The displacement of the obstacle can be accomplished with the help of a step motor. For the material characterization it is already sufficient to perform the network analyzer measurements with the three latter obstacle positions (1, 2, 3). This has the advantage, that the obstacle can be left completely within the waveguide, reducing the complexity of the measurement setup. Thus the test cell as shown in Fig. 3 can be used for the permittivity measurements, with the obstacle being left in the setup. For the determination of the material parameters, the measured values for the three different positions are arranged in measurement transmission matrices $\mathbf{M}_1$, $\mathbf{M}_2$ and $\mathbf{M}_3$, according to Eq. (1).

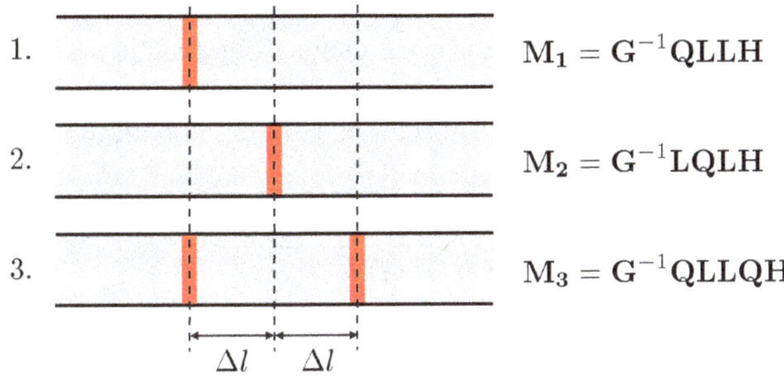

Fig. 5. Measurement structures of the multiple-N-method.

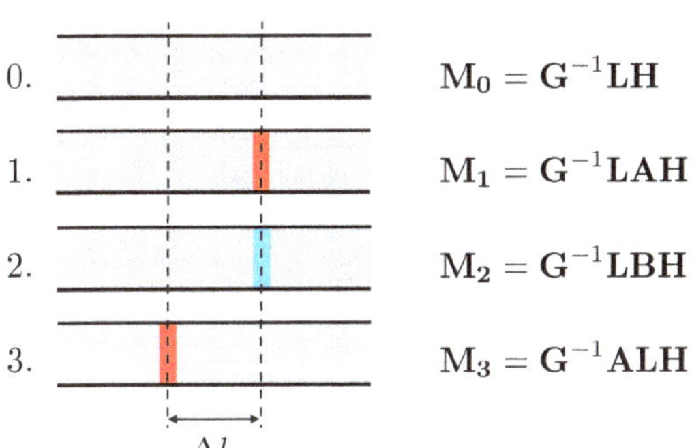

Fig. 6. Measurement structures of the 2-state-multiple-N-method.

As already described in Fig. 4, these measurement matrices are equal to the product of the transmission matrices $\mathbf{G}^{-1}$ and $\mathbf{H}$, representing the systematic errors of the setup including the VNA, and matrices $\mathbf{L}$ and $\mathbf{Q}$, representing the line elements between the obstacle positions and the obstacle, respectively. The obstacle needs not to be symmetrical. Furthermore, the obstacle parameters as well as the line propagation terms can be unknown. Calculating the traces of the following matrix combinations

$$\delta_1 = trace\{\mathbf{M_1 M_2^{-1}}\} \tag{4}$$

$$\delta_2 = trace\{\mathbf{M_1 M_3^{-1}}\} \tag{5}$$

and with

$$\delta = \frac{\delta_2 - 2}{\delta_1 - 2} \tag{6}$$

the propagation constant of the material within the waveguide can be calculated,

$$e^{-\gamma \cdot \Delta l} = \pm\frac{\sqrt{\delta}}{2} \pm \sqrt{\frac{\delta}{4} - 1}\ , \tag{7}$$

where Δl is the distance between two obstacle positions. As can be seen, Eq. 7 has four possible solutions. In order to choose the appropriate solution an a priori knowledge of the approximate value of $\gamma \cdot \Delta l$ is necessary. Thus, the permittivity can be determined, if the relationship between the propagation constant and the waveguide dimensions are known, also for a non-TEM waveguide.

2.2 Multiple-N method and 2-state-Multiple-N-method

Further obstacle-based methods with transmissive obstacles can be applied for the calibrated and error-corrected determination of the material parameters of a specimen within a waveguide. Possible measurement structures for the multiple-N and the 2-state-multiple-N-method (Rolfes and Schiek, 2002b) are depicted in Figs. 5 and 6. As well as for the line-network-method, obstacle networks have to be placed at different positions within the waveguide. For the multiple-N-method a second obstacle of the same type is needed whereas for the 2-state-multiple-N-method a second obstacle with different characteristics is needed. This can be realized by choosing obstacle materials with different permittivities. However, for these two methods it is necessary to perform measurements with all structures as shown in Figs. 5 and 6, having the drawback for the multiple-N-method that a further obstacle has to be inserted into the setup, and for the 2-state-multiple-N-method that two different obstacles characterized by matrices $\mathbf{A}$ and $\mathbf{B}$ have to be inserted into the waveguide. This makes the realization and the measurements more complex. Based on the determination of the propagation constant, the permittivity can be calculated, similar to the previously described line-network-method. For the realization of the test cell it is mandatory that higher order modes are neither able to propagate nor to be excited. This can be achieved by realizing obstacles which are completely filling the waveguide and have plane surfaces perpendicular to the guide. Furthermore, these obstacles have to be homogenous and their permittivity has to be chosen suitably, i.e. it does not have to be too high.

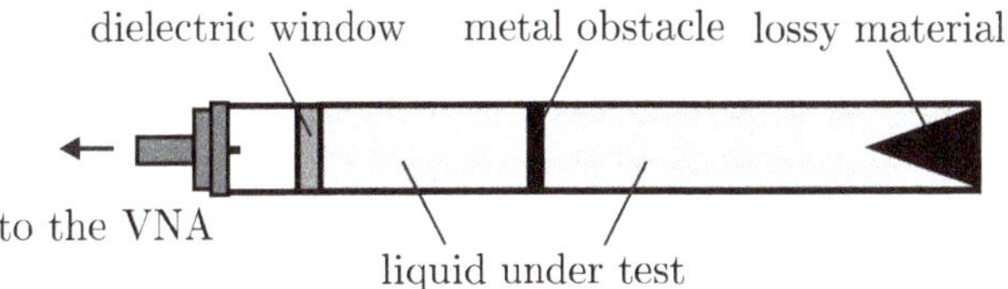

Fig. 7. One-port test cell for the reflective method.

2.3 Reflective method

For the reflective method, a one-port setup as shown in Fig. 7 is already sufficient. The setup consists of a waveguide, e.g. a rectangular waveguide, which is connected to one port of the VNA. A reflective obstacle, which can be realized as a metal plate, has to be placed at four consecutive positions within the waveguide, being filled with the specimen, as can

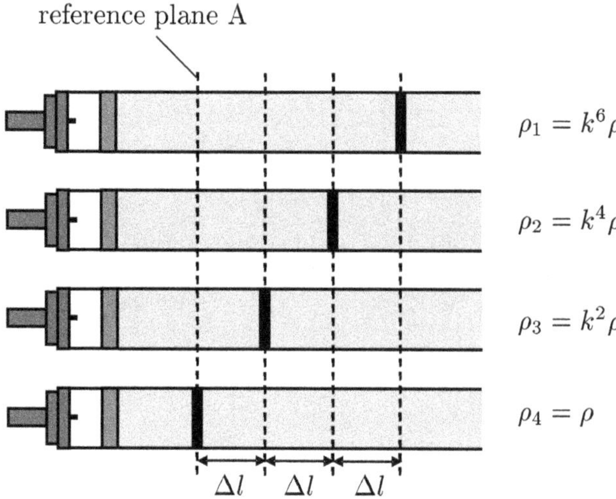

Fig. 8. Obstacle positions for the reflective methods.

be seen in Fig. 8. The waveguide is terminated on the other side by a lossy material, in order to inhibit resonances.

For the reflective method the calibration configurations are described with the help of the reflection coefficient ρ and the line parameter k, which can be unknown. Based on the setup in Figs. 7 and 8, the reflection coefficients ρ_i at the reference plane A are defined as follows:

$$\rho_1 = k^6\rho, \tag{8}$$
$$\rho_2 = k^4\rho, \tag{9}$$
$$\rho_3 = k^2\rho, \tag{10}$$
$$\rho_4 = \rho \tag{11}$$

With respect to the systematic errors according to the error model in Fig. 1 and with $\tilde{\mathbf{G}}=\mathbf{G}^{-1}$ the following equation results for the measurement of the i-th reflection coefficient ρ_i, $i=1,\ldots,4$:

$$\begin{bmatrix} m_{1,i} \\ m_{2,i} \end{bmatrix} = \tilde{\mathbf{G}} \begin{bmatrix} b_{1,i} \\ a_{1,i} \end{bmatrix} = \tilde{\mathbf{G}} \begin{bmatrix} \rho_i a_{1,i} \\ a_{1,i} \end{bmatrix} \tag{12}$$

In dependance of the measured reflection coefficients ν_i the following bilinear transformation follows:

$$\nu_i = \frac{m_{1,i}}{m_{2,i}} = \frac{\tilde{G}_{11}\rho_i a_{1,i} + \tilde{G}_{12} a_{1,i}}{\tilde{G}_{21}\rho_i a_{1,i} + \tilde{G}_{22} a_{1,i}} = \frac{\tilde{G}_{11}\rho_i + \tilde{G}_{12}}{\tilde{G}_{21}\rho_i + \tilde{G}_{22}} \tag{13}$$

also known as Möbius-transformation, being generally defined as,

$$x_j = \frac{C_1 y_j + C_2}{C_3 y_j + C_4} \tag{14}$$

where the two variables x_j and y_j correspond to the measurement value ν_i and the unknown calibration standard parameter ρ_i and the constants $C_1, \ldots, C_4$ represent the error two-port parameters of Eq. (13). On the basis of the measurement of four reflection coefficients, four equations of the type of Eq. (13) result, so that the unknown error parameter $\tilde{G}_{ik}$ can be eliminated. This can be performed with the help of the cross ratio

$$\frac{(y_1 - y_2)(y_3 - y_4)}{(y_1 - y_4)(y_3 - y_2)} = \frac{(x_1 - x_2)(x_3 - x_4)}{(x_1 - x_4)(x_3 - x_2)} \tag{15}$$

which generally holds for a bilinear transformation as given in Eq. (14). An equation can thus be constructed, which only depends on the unknown line parameter $k=e^{-\gamma\Delta l}$.

$$\begin{aligned} \nu &= \frac{(\nu_1 - \nu_2)(\nu_4 - \nu_3)}{(\nu_1 - \nu_3)(\nu_4 - \nu_2)} \\ &= \frac{(\rho_1 - \rho_2)(\rho_4 - \rho_3)}{(\rho_1 - \rho_3)(\rho_4 - \rho_2)} \\ &= \frac{(k^6\rho - k^4\rho)(\rho - k^2\rho)}{(k^6\rho - k^2\rho)(\rho - k^4\rho)} \\ &= \frac{k^2}{(1+k^2)^2} \end{aligned} \tag{16}$$

The unknown reflection coefficient ρ as well as the error two-port parameters are both eliminated and the following equation results for the determination of the propagation constant and thus the permittivity.

$$k = \pm\frac{1}{2\sqrt{\nu}} \pm \sqrt{\frac{1}{4\nu} - 1} \tag{17}$$

Like the line-network-method this method has a very compact solution and is easy to realize in a practical setup.

3 Simulation results

For the verification of the different obstacle based methods and for the determination of the material parameters of dielectric specimens, various simulations were performed. By means of an electromagnetic field simulator, CST Microwave Studio, the electromagnetic properties of the measurement setup, based on a realization in a rectangular waveguide, were investigated. An obstacle network filling completely the waveguide as well as a dielectric obstacle with holes, enabling the continuous flow of a probe liquid within the test fixture were considered. Based on the simulated scattering parameters describing the propagation in the measurement cell, the further evaluation of the material parameters was investigated. Some results for the determination of the permittivity of a specimen within a frequency range of 8.2 GHz to 12.5 GHz are shown exemplarily in Fig. 9. A rectangular X-band waveguide 200mm in length, a dielectric obstacle with ϵ_r=3.2 and 5 mm in length and liquids with different permittivities were considered. The permittivity of the obstacle must differ from the permittivity of the sample material. The application of the multiple-N- and the 2-state-multiple-N-method as well as the reflective method, led to similar results. Within the simulations, no errors occur.

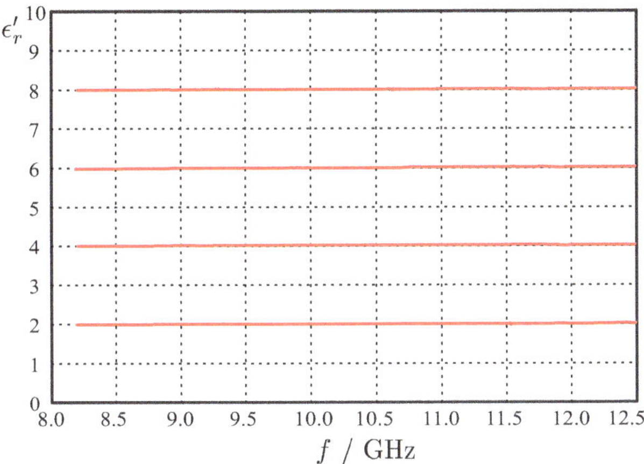

Fig. 9. Simulation results for the determination of the permittivity according to the line-network-method.

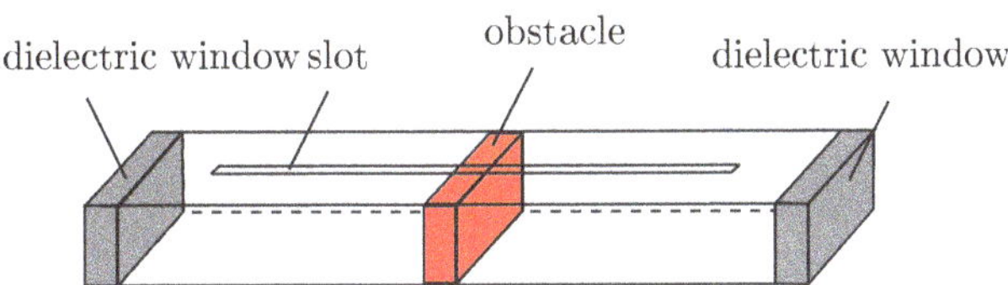

Fig. 10. Schematic setup of the measurement test cell for the line-network-method.

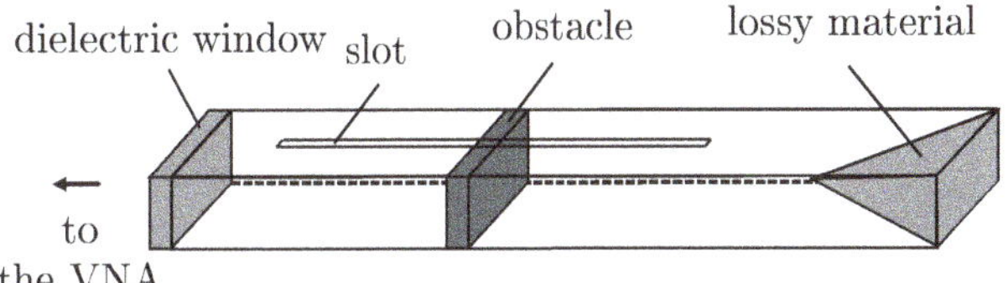

Fig. 11. Schematic setup of the measurement test cell for the reflective measurements.

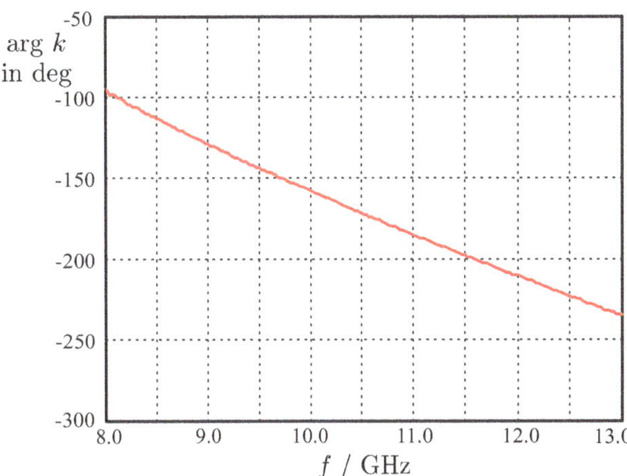

Fig. 12. Measured phase of air as a DUT based on the line-network-method.

4 Measurement results

Measurements were performed on the basis of a vector network analyzer and a rectangular waveguide measurement cell in X-band, as shown schematically in Fig. 10 for the line-network-method and in Fig. 11 for the reflective method.

The obstacle network is inserted into the waveguide. By a slot on top of the waveguide the obstacle can be displaced and oriented precisely. Furthermore, the slot can be used for the filling of the test cell with the specimen. On both ends of the measurement cell for the transmission measurements and on one end for the reflective measurements, respectively, a waveguide filled with a dielectric plate serves as a dielectric window. This allows to keep the liquid under test in a defined position within the test cell, filling it completely. This makes the measurements robust against level changes of the liquid as for example due to the displacement of the obstacle. In order to verify the functionality of the measurement system, measurements with air as a DUT, i.e. without any filling of the waveguide test cell, were performed first. For the line-network-method, an obstacle made from the commercial dielectric material PEI was used with a thickness of 4 mm, filling the waveguide nearly completely. Three separate measurement positions were considered, with the obstacle being displaced consecutively by a distance of 5 mm. In Fig. 12 the angle of the calculated propagation constant is shown. Based on Eq. (7) and an approximate knowledge of the term $\gamma \cdot \Delta l$, the appropriate solution was chosen. With the PEI obstacle it turned out that one solution was valid for the whole frequency range from 8 to 13 GHz. In Fig. 13 the determined permittivity is depicted. Differing from the TEM propagation in Eq. 3, a correction term $\epsilon_{r,corr}$ has to be added in order to account for the propagation in the rectangular waveguide

$$\epsilon_r' = \left(\frac{\gamma \cdot c}{2\pi f}\right)^2 + \left(\frac{c}{2a \cdot f}\right)^2 \tag{18}$$

where a=22.86 mm is the broadside width of the used rectangular X-band waveguide. The measurements show very good results, the measured dielectric constant of air is very close to 1, as can be observed in Fig. 13.

Similar measurements were performed on the basis of the reflective methods, leading to similar results.

Furthermore, measurements for the determination of the dielectric constant of different DUTs were performed also with varying obstacle position distances and also varying dielectric parameters for the obstacle. By Rolfes and Schiek (2006), measurement results for the determination of the permittivity of rape oil based on the line-network-method have already been presented Fig. 14. Here, better measurement

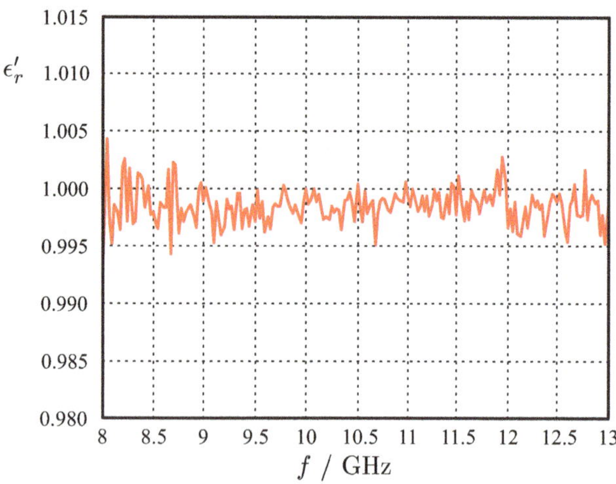

Fig. 13. Measured permittivity of air based on the line-network-method.

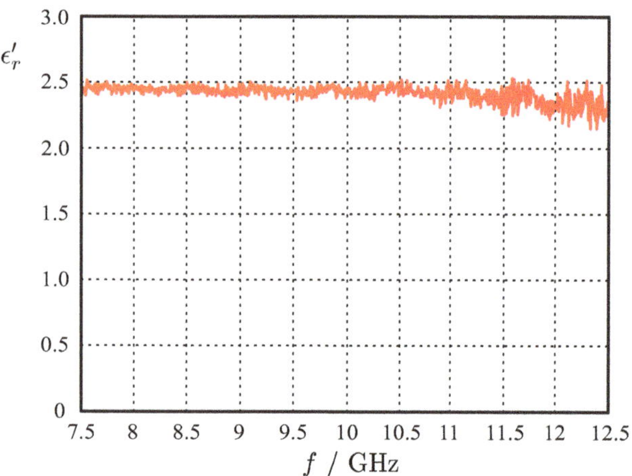

Fig. 14. Measured permittivity of rape oil as a DUT based on the LNN-method.

results for the reflective method with rape oil as a DUT are shown in Fig. 15. The obstacle was placed at four consecutive positions in a distance of 4 mm. In the frequency range of 8 GHz up to 13.5 GHz, the measured values are close to 2.4 and show a nearly constant characteristic over frequency. One of the main advantages of both line-network- and reflective methods is that a measurement of the test cell without obstacle is not needed. This helps to reduce the effort of the measurements. The DUT can be left within the test cell as well as the obstacle network for all measurements.

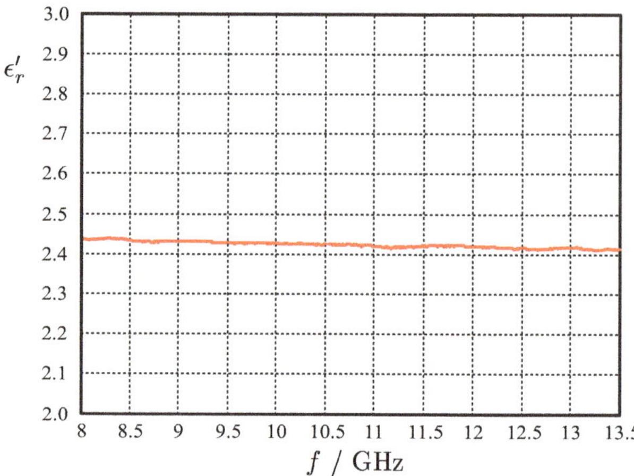

Fig. 15. Measured permittivity of rape oil as a DUT based on the reflective method.

5 Conclusions

For the determination of the permittivity of liquids based on the measurement of the scattering parameters using a vector network analyzer, different methods like the line-network-method and the reflective method are presented. Both methods are based on the displacement of an obstacle network within a measurement cell filled with the dielectric specimen. Both methods have in common that the measurements can be realized by means of a compact measurement setup. An extensive rearrangement of waveguide components is not necessary. The liquid under investigation can pass continuously through the measurement cell. This might be of interest for the application in an industrial process, enabling the continuous flow of the material while the parameter characterization can take place at the same time. Measurement results verify the robust functionality of the methods.

References

Clarke, B., Gregory, A., Cannell, D., Patrick, M., Wylie, S., Youngs, I., and Hill, G.: A Guide to the characterisation of dielectric materials at RF and microwave frequencies, National Physical Laboratory, Institute of Measurement and Control, London, ISBN 0 904457 38 9, 2003.

Eul, H.-J. and Schiek, B.: A Generalized Theory and New Calibration Procedures for Network Analyzer Self-Calibration, IEEE Trans. Microwave Theory Tech., 39, 724–731, 1991.

Engen, G. F. and Hoer, C. A.: Thru-Reflect-Line: An improved technique for calibrating the dual six port automatic network analyzer, IEEE Trans. Microwave Theory Tech., 27, 987–993, 1979.

Heuermann, H. and Schiek, B.: Line Network Network (LNN): An Alternative In-Fixture Calibration Procedure, IEEE Trans. Microwave Theory Tech., 45, 408–413, 1997.

Rolfes, I. and Schiek, B.: Calibration Methods for Free Space Dielectric Microwave Measurements with a 4-Channel-Network-Analzer, Proceedings of the 32nd European Microwave Conference, Mailand, Italien, 1077–1080, 2002a.

Rolfes, I. and Schiek, B.: The Multiple-N-Method for the Calibration of Vector Network Analyzer, Proceedings of the Conference on Precision Electromagnetic Measurement, Ottawa, Kanada, 134–135, 2002b.

Rolfes, I. and Schiek, B.: A Novel Method for the Determination of the Dielectric Properties of Liquids at Microwave Frequencies, Proceedings of the 36th European Microwave Conference, Manchester, UK, 399–402, 2006.

A new shielding effectiveness measurement method based on a skin-effect transmission line coupler

T. Kleine-Ostmann, K. Münter, and T. Schrader
Physikalisch-Technische Bundesanstalt, Braunschweig, Germany

Abstract. We propose a new convenient material shielding effectiveness measurement method based on a skin-effect transmission line coupler. The method is somewhat similar to the arrangement with two coupled TEM cells known from literature. The transmission line coupler consists of a pair of identical transmission line 2-port devices. Each device contains a coaxial waveguide, with a circular inner conductor and an outer conductor having a square cross section. One side of the outer conductor is left completely open as a slot. The slot is surrounded by a large metal housing to contact the two halves. As a measure for the shielding effectiveness the coupling between the two devices is measured in terms of scattering parameters after the test material is brought between the two halves. The devices can be used in a range from low frequencies to a few GHz.

1 Introduction

A variety of unconventional shielding materials finds vast applications in very diverse areas of our technical environment. Modern electronic equipment usually contains sensitive circuits and/or clock oscillators, and proper shielding of these components is often required to comply with electromagnetic compatibility regulations. Especially for high-volume production of consumer devices the conventional metal shielded enclosure is often prohibitively expensive, therefore alternatives had to be found, e.g. metal coated plastic materials. Another example of a rather unconventional shielding material is conductive textile used to make protective overalls for workers in possibly hazardous electromagnetic fields (Guy et al., 1987; Hoeft and Tokarsky, 2000). Also in traditional applications like shielded room construction new modular concepts rise the question how good the shielding effectiveness of jointed metal panels (Fig. 1) is, which are screwed together. In any case, the shielding effectiveness of such materials can only be calculated for simple shielding geometries (Klinkenbusch, 2005). In most cases it must be measured, according to a suitable method.

Depending on the purpose, some methods based on cables (ASTM D4935, 1999; Kinningham and Yenni, 1988), coupled TEM cells (Nishikata et al., 2006) or antennas (IEEE Std 299, 1997; MIL Std 285, 1956; EN 50147-1, 1997; Wilson et al., 1986) are described in literature, all with specific advantages and disadvantages. In many cases the results depend on the individual test arrangement, and it is difficult to separate these influences and assign an uncertainty to the results. Obtaining material specific attenuation properties that are geometry independent remains difficult. The method and equipment described in the present paper is based on a transmission line coupler consisting of two separate halves between which a planar sample can be pressed (Fig. 2). It is similar to an arrangement with two coupled TEM cells, but easier to handle. However, it has also certain limitations. The measured material sample must be thin enough (thickness at least in the order of the skin depth) and must have sufficiently high conductivity from low frequencies to a few GHz. However, advantages are: calibration and traceability with a reference material, a small required sample size of only a few cm^2, very little crosstalk and rapid measurements with a vector network analyzer (VNA), offering fast and simple traceable calibration.

In this paper we present our new measurement method and describe its numerical modelling based on the method of moments. In order to verify our calculations we compare shielding effectiveness measurements on steel foils of known thickness and conductivity to our numerical results. We find a good agreement of measurements and calculations, especially for low frequencies. Finally we present first applications of our new method by verifying the shielding effectiveness of assembled shielding panels and comparing the atten-

Fig. 1. Panels for modular construction of shielded rooms (left: separate zinc panels, right: copper panels screwed together). The measurement positions of the couplers at the position of a screw (dotted line) and between two screws (dashed line) are indicated.

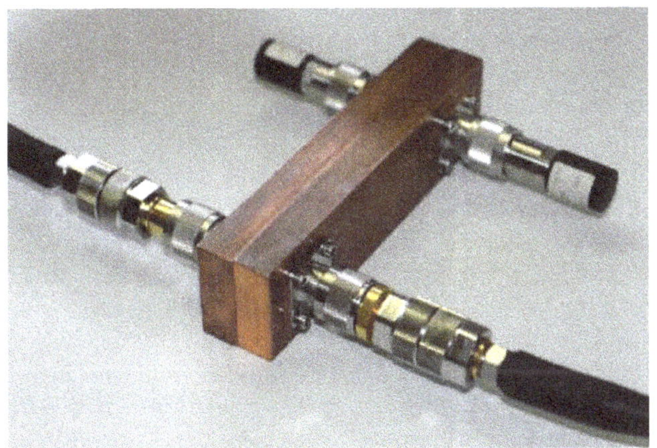

Fig. 2. Skin-effect transmission line coupler.

uation of different possible reflector materials for a new open area test site which is currently under construction.

2 Measurement method and setup

For the measurements a pair of identical transmission line 2-port devices are made from copper and fitted with N-connectors (Fig. 2). Each device contains a coaxial wave-guide, with a circular inner conductor (length l=105 mm, diameter d=3 mm) and an outer conductor having a square cross section (6.5 mm x 6.5 mm, length l=110 mm). One side of the outer conductor is entirely left open as a slot, while the housing consists of a large metal area to contact the devices. If the slot is closed with a solid metal plate, the device is completed to become a precision 50 Ω coaxial line with low insertion loss and a low input reflection coefficient. The measured input and output return loss shows that by the mechanical design, the coupler devices are well matched up to 3 GHz, ensuring a low voltage standing wave ratio (VSWR) on both coaxial lines. If both devices are attached to opposite sides of the separating metal plate, the crosstalk between the two coaxial compartments of the 4-port is strongly reduced depending on the shielding properties of the material. If the metal plate is removed and the devices are directly connected, there is a maximum coupling between the electric and magnetic fields in the two slots. In this configuration the 4-port behaves like a conventional directional coupler – but, of course, here directivity and bandwidth are not optimized for that purpose. Fig. 3 shows the 4-port representation of the transmission line coupler indicating the waves a_1, ..., a_4 in the forward direction and b_1, ..., b_4 in the backward direction. In the ideal case it is loss-less and reciprocal. Therefore it can be described by scattering parameters S_{ik}=S_{ki} with

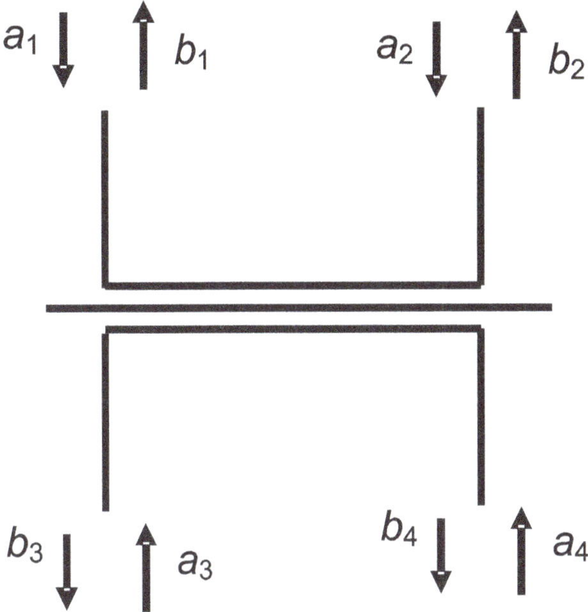

Fig. 3. Four-port representation of an ideal transmission line coupler.

$$\begin{pmatrix} b_1 \\ b_2 \\ b_3 \\ b_4 \end{pmatrix} = \begin{pmatrix} S_{11} \approx 0 & S_{12} \approx 1 & S_{13} & S_{14} \\ S_{21} \approx 1 & S_{22} \approx 0 & S_{23} & S_{24} \\ S_{31} & S_{32} & S_{33} \approx 0 & S_{34} \approx 1 \\ S_{41} & S_{42} & S_{43} \approx 1 & S_{44} \approx 0 \end{pmatrix} \cdot \begin{pmatrix} a_1 \\ a_2 \\ a_3 \\ a_4 \end{pmatrix} . \quad (1)$$

Of especial importance are the coupling strengths S_{31} and S_{41} since they can be used to characterize the shielding effectiveness of the tested material. In the case of directly connected couplers without metal plate between them, $|S_{41}|$ is in the range of −50 dB whereas $|S_{31}|$ varies between −25 dB and −50 dB, showing oscillatory behavior as a result of the resonance length of the device.

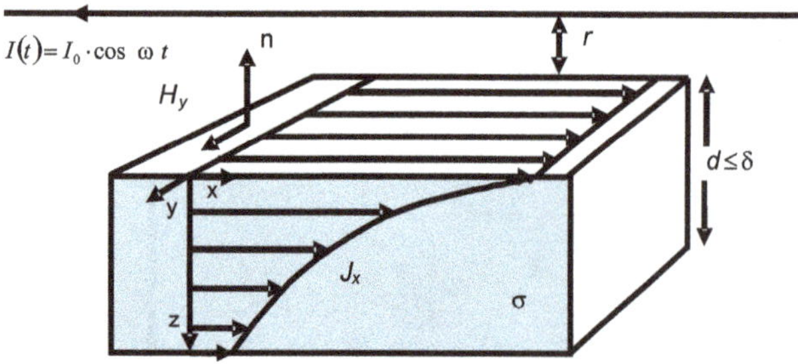

Fig. 4. Energy transport through metallic shielding material (surface normal n and distance r to the circular conductor) with specific conductivity and with thickness d smaller than the skin-depth δ. The magnetic field H_y of the circular conductor with alternating current $I(t)$ of amplitude I_0 and angular frequency ω causes a current density J_x in the shielding material.

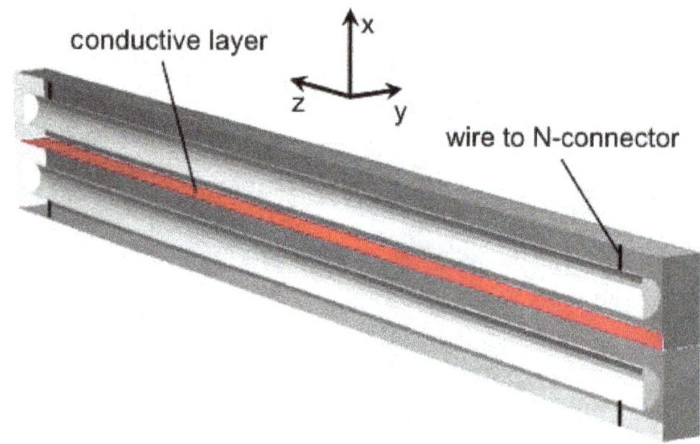

Fig. 5. Computer model of the skin-effect transmission line coupler.

Shielding effectiveness measurements can only be made in comparison to a reference material. The first reason why it is not desirable to use the direct connection as 0 dB-reference for measurements of shielding material is that the devices are no longer 50 Ω lines. The second reason follows from the fact that the coupling mechanism is different, if very thin, but highly conductive material is inserted between the two devices. As with the bulk metal plate described above, both sides are then again 50 Ω lines. The excitation signal on one side causes a current distribution which is nearly the same as if the coaxial lines were completely separated. In this case the coupling into the second compartment is no longer caused by electromagnetic fields in air, instead it may be best described by the skin effect, which makes a small fraction of the current distribution on the excited input line side "visible" at the other surface of the thin material. This situation is depicted in Fig. 4. The propagation of external fields into a conductive material under these conditions is discussed by Jackson (2002). The current density in the conductive material decays with the distance from the surface where the excitation takes place. The coaxial line at the output side is excited by a low-impedance, travelling wave current distribution on one of its surfaces – a situation which is difficult to model with conventional transmission line theory, because distributed generators would be required.

To test these assumptions, measurements as well as calculations based on a numerical simulation model were performed for the coupling devices described above, separated by foils of different thickness made from stainless steel. Materials of 10 μm, 20 μm, 25 μm, 30 μm and 40 μm thickness were available for this purpose.

The two ports of the VNA were connected to one port of each device, with the other port of the transmission lines each terminated with a precision 50 Ω load resistor. Two different arrangements were then measured, one with the port connectors in close proximity (referred as the forward direction, ports 1 and 3 as in Fig. 3), and the other with one line in reverse direction (ports 1 and 4 as in Fig. 3), with the VNA ports at opposite ends.

3 Numerical simulation model

For the numerical evaluation of the transmission properties of the empty coupler and the coupler with foil the method of moments implemented in the program Concept II (Singer and Brüns, 2006) seemed appropriate. This numerical program code offers a special feature, which seemed best suited to calculate the transmission through highly reflective and very thin layers. It accurately calculates the penetration of waves through highly reflective materials with high shielding attenuation – provided that the material is thin and highly conductive.

Under these circumstances the energy flow inside the material is perpendicular to the surface, and its propagation is well described by a transmission-line approach, where the attenuation along the line is derived from the skin-effect formula.

The problem was to define a geometrical model which was accurate enough to ensure an unidirectional energy flow and a proper smooth current distribution along the excited transmission line. Figure 5 shows the simulated geometry. The model consists of two different dielectric bodies of r=1 with ideally conducting metallic surfaces and embedded conductors which represent the two coupler halves. The connections from the N-connector to the inner conductor are simply modelled by 1 mm thick wires of 5 mm length which are connecting the conductors to the outer surface of the slot waveguide in which they are embedded. All wires are loaded with a 50 Ω impedance. A signal of constant power is fed in port 1. Scattering parameters are derived from the currents observed on the wires.

The two halves are separated by a conductive layer that represents the sample under investigation for which thickness and conductivity can be specified. The implementation

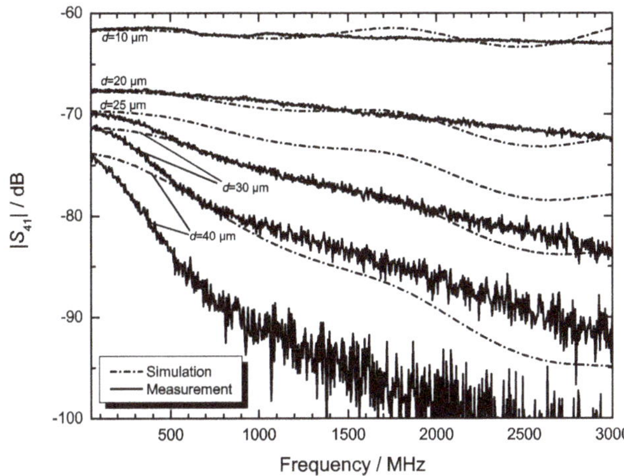

Fig. 6. Comparison between shielding measurements and numerical simulations for S_{41}.

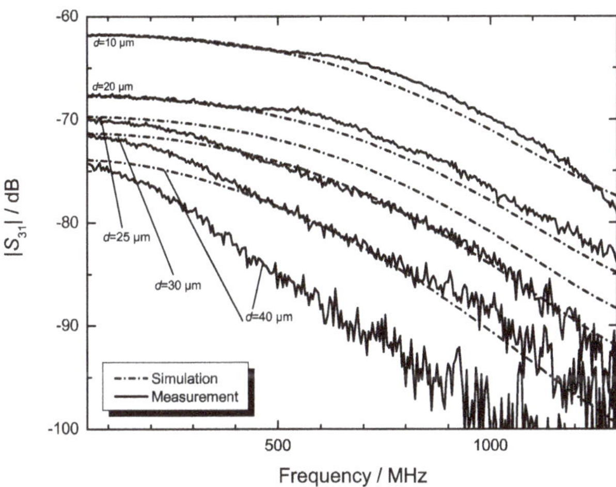

Fig. 7. Comparison between shielding measurements and numerical simulations for S_{31}.

benefits from the symmetry of the devices with regard to the cross section shown in Fig. 5 which reduces the amount of variables and computation time by a factor of two. The simulations were performed with approximately 13 500 complex variables on a PC with 2 GByte RAM using swap files on the hard disk. A single simulation run involved 60 frequency steps of 50 MHz up to 3 GHz and took approximately three days.

4 Comparison of measurements and simulations

We performed measurements with stainless steel foils of known thickness and conductivity to be able to relate material properties to their measured coupling strength and to

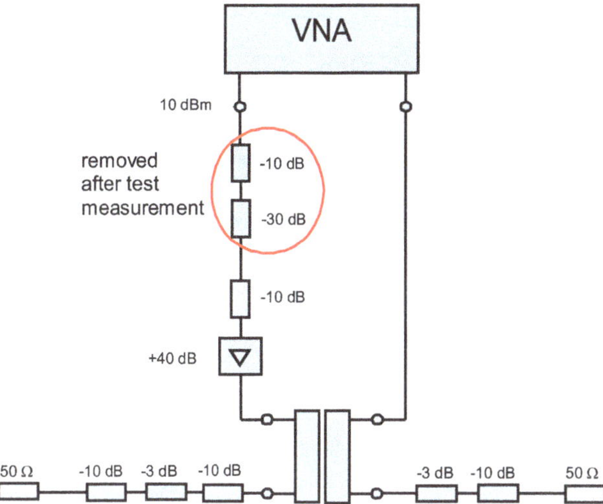

Fig. 8. Measurement setup with increased dynamics.

verify our numerical simulations. The measurement dynamics are in the order of 90 dB. Figure 6 shows a comparison of measurements and simulations for the transmission in reverse direction (S_{41}) whereas Fig. 7 shows the corresponding comparison for the forward direction (S_{31}). Before the measurement the VNA was calibrated. The 0 dB-line corresponds to a through connection between the VNA ports.

In both cases we see a good agreement between measurements and simulations for the lower frequencies, especially for 10 μm and 20 μm thick foils. As the frequency increases, the discrepancy between simulations and measurements becomes more pronounced, especially in the case of S_{31}. We attribute this to the fact that the modelled geometry lacks a sufficient level of detail to represent standing waves in the output section correctly. Furthermore, the measured coupling ranges below the simulated results. Obviously losses are not sufficiently accounted for as the coupler housing is assumed to be ideally conducting. Therefore, the reverse (S_{41}) direction should be preferred for the measurement of the shielding effectiveness. For thin foils (10 μm and 20 μm thickness), the coupling can be calculated with good agreement between theory and experiment up to 3 GHz, whereas for thicker foils accurate calculations are restricted to a few hundred MHz, depending on the thickness. The results show that the measured coupling strength can be reliably related to the thickness of the foil.

5 Applications

First, we use our new method to verify the shielding effectiveness of the panels shown in Fig. 1. The panels under investigation were stringed either with copper or with zinc foil. While the 200 μm thick foil has almost perfect shielding properties in both cases (beyond the limits set by the mea-

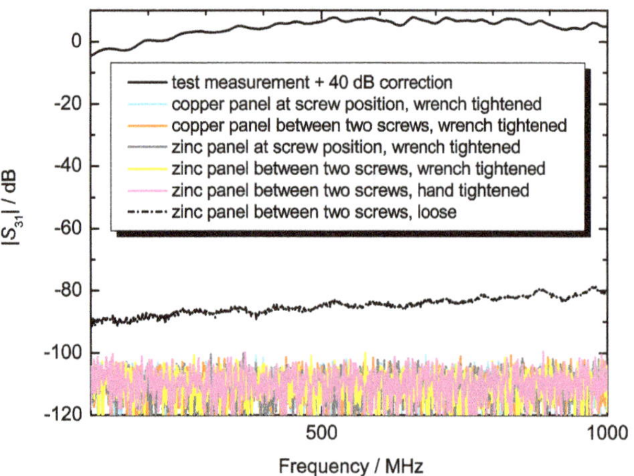

Fig. 9. Attenuation of different panel configurations.

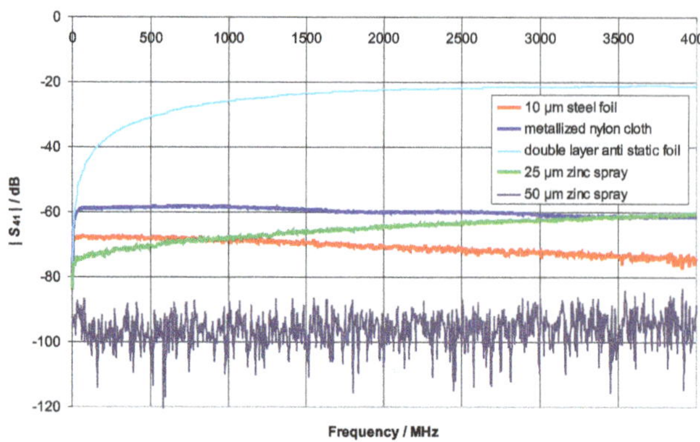

Fig. 10. Attenuation of different possible reflector materials for open area test sites.

surement dynamics), the connection between panels seems critical. The interface between two adjacent panels is simply established by screwing panels together. This way blank copper or zinc surfaces on the panel sides are pressed together.

In order to achieve higher measurement dynamics compared to a simple VNA measurement, we use the setup shown in Fig. 8. Precision attenuators of 50 dB followed by a 40 dB amplifier (Amplifier Research 10W1000) are connected to port 1 of the VNA. After calibration of the VNA and a test measurement of the empty coupler two of the attenuators are removed, increasing the measurement dynamics by 30 dB. Next, the couplers were placed orthogonally across the interconnection of two panels (see Fig. 1), forcing the currents to flow across the contact area. Figure 9 shows the results for different measurement positions (at the position of a screw and between two screws) and for different attachment strengths (loose, hand tightened and wrench tightened), both for copper and zinc panels. Only in the case of a loose screw which led to a visible crack between panels, a decoupling of approximately 80 dB was measured. In all other cases the decoupling was beyond the measurement dynamics of at least 100 dB.

As a second application we compare the shielding effectiveness of different surface coating materials. This measurement was done to evaluate the performance of possible reflector materials for a new open area test site which is currently under construction at the Physikalisch-Technische Bundesanstalt. Since we were not sure that a thin metal coating is sufficiently reflecting we tested its transparency for electromagnetic waves assuming that an intransparent metal layer would reflect most of the incoming radiation. Figure 10 shows the coupling factor S_{41} for metallized steel cloth, a double layer of anti-static foil, and two different zinc metal layers sprayed on paper using a electro-thermal process compared to the previously measured 10 μm zinc foil. The 50 μm zinc layer causes a complete decoupling, already. Although our method does not measure its reflectivity, good reflection properties of such a layer can be assumed as the transmission parameter $|S_{21}|$ is close to 0 dB, representing a low attenuation of the signal along the line. A comparable zinc layer will be used for the new open area test site in Braunschweig.

6 Conclusions

We demonstrated a new method to evaluate the shielding effectiveness of planar materials, based on a transmission line coupler. Couplers are broadband devices with well-matched transmission lines at both VNA ports, reducing uncertainty in measured transmission S-parameters. Compared to other methods, it is much easier to handle since it avoids antennas or resonant cavities and allows fast network analyzer measurements. Furthermore, with other methods the reflection coefficients at the VNA ports change significantly, when the material sample is inserted. The measurements can be used to specify a reference material to which other shielding materials can be compared.

Measured and calculated transmission coefficients for the transmission line couplers and different metallic shielding materials are in reasonable agreement already (especially at the low-frequency end), which indicates that the numerical model is adequate and suited for the problem discussed here. However, a further refinement of the simulated geometry will help to achieve a better match between measurements and theory. Further work that aims to relate material properties directly to the observed coupling is under way.

Acknowledgements. The authors thank H. Brüns from the Institute of Electromagnetic Theory at Technical University Hamburg-Harburg for his advice regarding the electromagnetic simulation of the device and A. Daneschnejad as well as Thomas Baron from the Physikalisch-Technische Bundesanstalt for the precise fabrication of the transmission line coupler.

References

ASTM D4935: Standard Test Method for Measuring the Electromagnetic Shielding Effectiveness of Planar Materials, 1999.

EN 50147-1: Anechoic chambers, Shield attenuation measurement, 1997.

Guy, A. W., Chung-Kwang Chou, McDougall, J. A., and Sorensen, C.: Measurement of Shielding Effectiveness of Microwave-Protective Suits, IEEE Trans. Microwave Theory Techn., 35, 984–994, 1987.

Hoeft, L. O. and Tokarsky, E. W.: Measured Electromagnetic Shielding Characteristics of Fabric Made from Metal Clad Aramid Yarn and Wire, EMC 2000 Symp. Rec., 2, 883–886, 2000.

IEEE Std 299: IEEE Standard Method for Measuring the Effectiveness of Electromagnetic Shielding Enclosures, 1997.

Jackson, J. D.: Klassische Elektrodynamik, De Gruyter, Berlin, 2002.

Kinningham, B. A. and Yenni, D. M.: Test methods for electromagnetic shielding materials, IEEE Intern. Symp. Electromagn. Comp., 223–230, 1988.

Klinkenbusch, L.: On the Shielding Effectiveness of Enclosures, IEEE Trans. Electromagn. Comp., 47, 589–601, 2005.

MIL Std 285: Attenuation Measurements for Enclosures, Electromagnetic Shielding, for Electronic Test Purposes, Method of, 1956.

Nishikata, A., Saito, R., and Yamanaka, Y.: Low Frequency Equivalent Circuit of Dual TEM Cell for Shielding Material Measurement, IEICE Trans. Electron., E89-C, 44–50, 2006.

Singer, H. and Brüns, H.: Concept II, www.tet.tu-harburg.de, 2006.

Wilson, P. F., Adams, J. W., and Ma, M. T.: Measurement of the electromagnetic shielding capabilities of materials, Proc. IEEE, 74, 112–115, 1986.

Considerations on the frequency resource of professional wireless microphone systems

S. Dortmund[1], M. Fehr[2], and I. Rolfes[1]

[1]Institute for Radiofrequency and Microwave Engineering, Leibniz Universität Hannover, Germany
[2]Sennheiser Electronic GmbH, Wedemark, Germany

Abstract. This Paper presents the results of spectral observations in the UHF TV Bands IV and V from 470 MHz up to 862 MHz with focus on the TV-Channels 61 to 63 and 67 to 69. Concerning the discussions on WRC (2007) this frequency range is in great demand of several applications and is usually treated as a "white space" in the TV-Bands. According to typical scenarios, two different spectral loads will be presented considering the requirements of professional wireless microphone receivers with respect to in-band intermodulation.

1 Introduction

In the course of the digitalisation of the UHF TV-Bands a reallocation of the UHF frequency spectrum will take place. This is based on the presumption that Digital Video Broadcasting (DVB-T) will occupy less spectrum than the analogue counterpiece. On the other hand the increasing demand on digital broadcasting for hand-held devices (DVB-H) may require the vacant frequency resource, especially in urban areas. Furthermore several providers are interested in installing new mobile multimedia-based services in UHF in so called "white spaces". Applications like Professional Wireless Microphone Systems (PWMS) which actually use these "white spaces" are not often mentioned as secondary users of the UHF frequency spectrum. Additionally unlicensed devices with cognitive skills may become access to the UHF frequency spectrum. On WRC (2007) the possibility to hold an auction on the upper TV channels was discussed. According to BNetzA (2005) PWMS are allowed to operate on these channels without individual permission until 2015, named channel 61 to 63 (f=790 MHz to f=814 MHz) and channel 67 to 69 (f=838 MHz to f=862 MHz) which were former used by military applications. PWMS is a strongly growing application with high demands on reliability, latency and sound quality which in practice can only be faced with an assured access to the frequency resource. For the protection of PWMS a default signal strength of (68 dBμV/m) is specified in CEPT (1997). This value is also used for several other governmental issues (e.g., ERC Report 88, 2000). PWMS manufacturers suppose that this level is about more than (10 dB) too high for a secure PWMS connection in on stage environments. According to ETSI (2006) a new suggestion for the default protected signal strength in PWMS for future frequency allocation in UHF is given, adapted from the demands on PWMS in Sect. 2 of this paper. In Sect. 3 the spectral loads of two different applicable cases are presented. Futhermore one of them is evaluated from an objective point of view by the aid of frequency availability and from a subjective point of view with respect to in-band intermodulation in PWMS receivers as well. Finally, in Sect. 4 the results are concluded.

2 PWMS – an application with high demands

PWMS is a common application in the cultural industry and includes wireless microphones for professional usage, In-Ear Monitoring Systems (IEM), Electronic News Gathering (ENG) and wireless audio systems. These applications can be found in studios, theatres, musicals, politics, sports, broadcasting and on stage. Thus PWMS is a core application in the production of multimedia content and is used for recording and archiving unique events. This leads to very high demands on the reliability and sound quality of PWMS. Furthermore sound engineers and musicians have high demands on latency. Therefore PWMS, unlike wireless microphones, headphones or loudspeakers for consumer applications, are not able to operate on Industrial, Scientific, and Medical Bands (ISM). In contrast to daily usage, today's mass events, for example in sports, generate a temporary much higher demand on a useful frequency resource.

2.1 Requirements of PWMS

According to ETSI (2006) a high audio quality can be achieved with an Audio Signal-to-Noise Ratio SNR_{AF} of

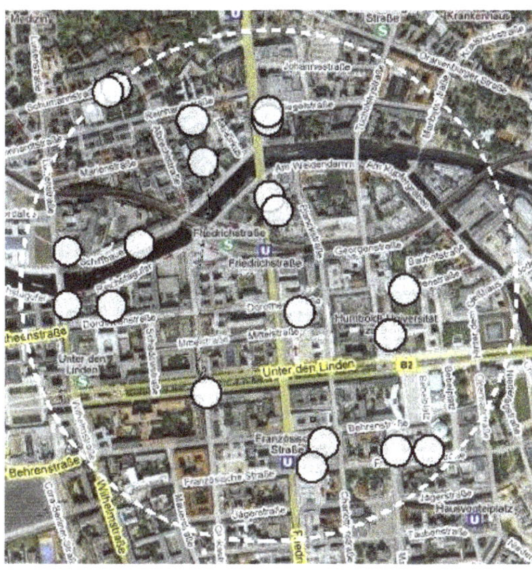

Fig. 1. Measurement location in the center of Berlin within a diameter of 1 km (marked by white, dotted circles), with known PWMS users (marked by black circle).

(80 dB (A)) at the NF-Path of a PWMS receiver. The associated Radio Frequency Signal-to-Noise Ratio SNR_{RF} can be calculated from

$$\mathrm{SNR}_{RF} = \frac{\mathrm{SNR}_{AF}}{K_{\mathrm{compand}}} - G_{\mathrm{demod}} \tag{1}$$

where K_{compand} represents a coefficient for the degree of companding (typically K_{compand}=2) and G_{demod} represents the demodulation gain of the used modulation scheme. Actually PWMS works with frequency modulation (FM) with G_{demod}=20 dB. This leads to a minimum $\mathrm{SNR}_{RF,min}$ of typically 20 dB. The thermal noise level $N_{thermal}$ for a PWMS channel with 200 kHz Bandwidth is about −120 dBm. According to ETSI (2006) other technical on-stage-applications, for example light installations, add a cumulative man-made-noise N_{ManMade} of 10 dB. With a typical noise figure F of a PWMS receiver the minimum receiving level P_{sens} can be calculated by

$$P_{\mathrm{sens}} = N_{\mathrm{thermal}} + N_{\mathrm{ManMade}} + F\ . \tag{2}$$

The output power of PWMS is limited to 17 dBm. With a typical stationary antenna gain G_{static}=5 dBi and a mobile antenna gain G_{mobile}=0 dBi of a hand-held or of a body-mounted device, the maximum path loss L for a gaussian channel is given with L_{gauss}=87 dB including the absorption of the human body D_{human}=30 dB defined in ETSI (2006). For a reliable PWMS link, fading break-ins up to 30 dB have to be considered and lead to L_{rayleigh}=57 dB. According to Meinke et al. (2005) the range r for f=800 MHz is calculated with

$$L = 32.44\mathrm{dB} + 20log\frac{r}{1000\mathrm{m}} + 20log\frac{f}{\mathrm{MHz}}\ . \tag{3}$$

Hence, the maximum range for a secure and reliable PWMS link considering the antennas in use, human body absorption and fading effects is about r=21.1 m without and $r_{\mathrm{diversity}}$=47.3 m with an additional diversity gain $G_{\mathrm{diversity}}$=7 dB. For the most applications, this range is barely adequate. Thus, we suppose P_{sens}=−80 dBm for the PWMS protection level. With $\mathrm{SNR}_{RF,\mathrm{min}}$=20 dB the resulting maximum interference level is P_{if}=−100 dBm. The corresponding field strengths are 51.5 dB$\mu V/m$ and 31.5 dB$\mu V/m$ with an antenna factor k=24.5 for isotropic antennas operating at f=800 MHz. The calculated value for the protected field strength is lower than the one given in CEPT (1997) for PWMS [1] with 68 dB$\mu V/m$.

3 Spectral load

For the spectral loads the concrete PWMS application is not necessary. However, it is important to distinguish two different applicable cases, on the one hand the daily use of PWMS in an urban area with a high concentration of PWMS users with a moderate frequency load, on the other hand a mass event with a high frequency load (see Sect. 2). For recording the daily use of PWMS, an area with a diameter of 1 km within the center of Berlin was chosen (see Fig. 1). In this area at least 20 PWMS users were known (marked with black circles). The measurement took place from 17th to 18th July 2007. For a mass event the biggest international music fair ("Musikmesse") was chosen. This Measurement took place from 28th to 31th March 2007 in the exhibition hall 4, Frankfurt/Main.

The spectral load was recorded in both cases using as receiving antenna a Sennheiser type A 5000 CP , which is an applicable, directional antenna installed by many PWMS users, connected to a Rohde & Schwarz FSP 3 spectrum analyzer. The data was measured with Δf=20 kHz, RMS detection and a reference level P_{ref}=−20 dBm. The analyzer was software controlled and the measured data stream was mapped with a measurement PC. The whole analysis was performed by use of MATLAB. The surface plots in Fig. 2 and Fig. 3 show the recorded signal levels $P_{\mathrm{Musikmesse}}$ and P_{Berlin} (color coded) over frequency (abscissa) and measurement cycle (ordinate) and will be discussed in detail in the following subsections.

3.1 PWMS in daily use

In Fig. 2, an overview of the spectral load in Berlin is displayed. The measurement equipment was placed in a car. The car stopped at numerous stop stations, as close as possible to the known PWMS users, and continued in place for several measurement cycles. The recording was continued during driving from one stop station to another. The light

[1] PWMS are part of Service Auxiliary Broadcasting (SAB) / Service Auxiliary Program making (SAP) in CEPT (1997)

blue horizontal bars between measurement cycle 350 and 400 result from a temporary increase of the noise floor because of problems with the measurement equipment. In spite of the at least 20 known PWMS users the spectral load is very weak. This leads to the presumption, that not all known devices were in use during the measurement periode. Probably this is caused by the fact that the measurement period coincided with the holiday season in Berlin. Thus a higher spectral load than the one displayed in Fig. 2 might be more representative for PWMS in daily use. Moreover, because of the overall measured weak levels, we suppose that not all PWMS in indoor use can be recorded by observing the outdoor area using the chosen measurement technique. The whole measurement took place in parallel to a measurement campaign of BNetzA which observed the same frequency range at the same stop stations. The results are nearly equal. By simply cumulating the amount of measured levels higher −100 dBm (excluding the light blue horizontal bars between measurement cycle 350 and 400) multiplicated with Δf=20 kHz the used frequency can be calculated. In maximum 5.04 MHz are in use simultanously. With a typical bandwidth of 200 Hz this value corresponds to 25 PWMS.

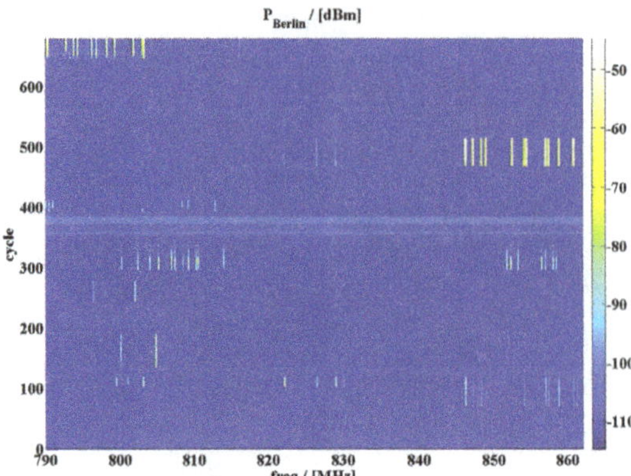

Fig. 2. Spectral load from 17th to 18th July 2007 between f=790 MHz and f=862 MHz in the center of Berlin.

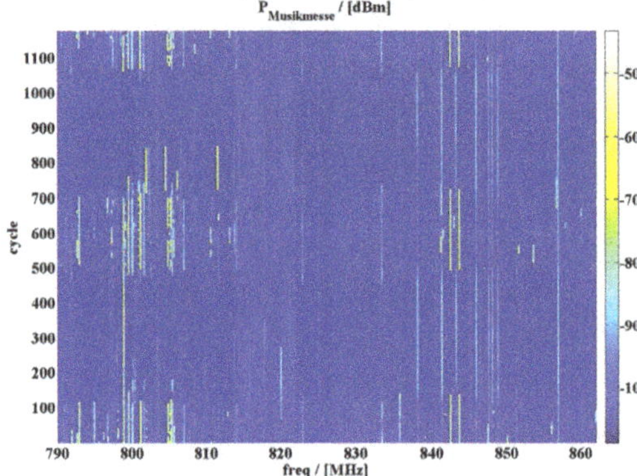

Fig. 3. Spectral load from 28th to 31th March 2007 between f=790 MHz and f=862 MHz in the exhibition hall4, "Musikmesse".

3.2 PWMS during mass event

In contrast to the low daily spectral load Fig. 3 gives an overview of the spectral load of a mass event. The measurement equipment was placed in an office at the face side of the exhibition hall with an antenna height of 5 m. The hall is 145 m in width and 133 m in length. With a 3 dB-angle of 60° and a 0dB-angle of nearly 180°, the antenna covers the whole exhibition hall. The lowest expected level was calculated to −70 dBm excluding attenuation or fading. Via a simple test measurement with a hand-held microphone, the level calculation could be confirmed. The spectral load in Fig. 3 has a very strong variance in time especially in the lower TV-channels (below f=814 MHz) in conjunction with the opening hours of the fair. The measurement started at the 28th March at about 01:00 p.m. After door closing at 07:00 p.m. only some PWMS did not switch off, because of corporate functions in the evening hours. The same behaviour can be seen after door opening at about 8am between cycle 500 to 900, analogue to the second measurement day as well as above cycle 1125, analogue to the third measurement day on 30th March till about 02:00 p.m. At the upper TV-channels (above f=838 MHz) also continuous PWMS signals can be identified, presumptively belonging to not battery powered IEM. On TV-channel 64 (f=814 MHz to f=822 MHz) a deadbeat DVBT signal can be recognized, because of an everlasting increase of the noise floor. Additional PWMS is apparent between f=814 MHz and f=838 MHz out of the operable frequency range according BNetzA (2005), including TV-channel 64. Additionaly, weak PWMS signals below the supposed levels belonging to PWMS placed inside the exhibition hall can be detected, probably transmitted from another exhibition hall or pavillion. By simply cumulating the amount of measured levels higher −100 dBm multiplicated with Δf=20 kHz the used frequency range can be calculated. In maximum 11.32 MHz are in use simultanously. With a typical bandwidth of 200 kHz this value corresponds to 56 PWMS. On the average f=5.58 MHz are in use, corresponding to 28 PWMS.

3.2.1 Frequency availability

In the following the spectral load of the "Musikmesse" will be evaluated more objectively. Therefore the frequency availability is calculated by

$$AV = \tilde{P}_{\text{meas}} + \mu\sigma_{P_{\text{meas}}} \quad (4)$$

where $\tilde{P}_{\text{meas}}$ represents the median of the measured spectral load and $\mu\sigma_{P_{\text{meas}}}$ the weighted variance in time. According to Konstantinos et al. (2005) a frequency availability of 99% corresponding with μ=2.33 is determined as a suitable avail-

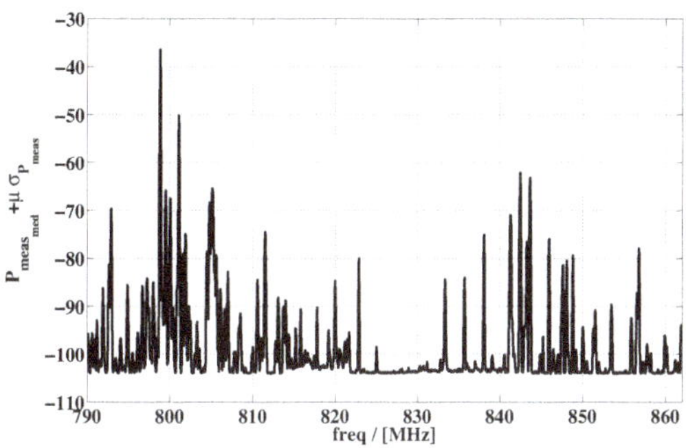

Fig. 4. 99% availability at "Musikmesse".

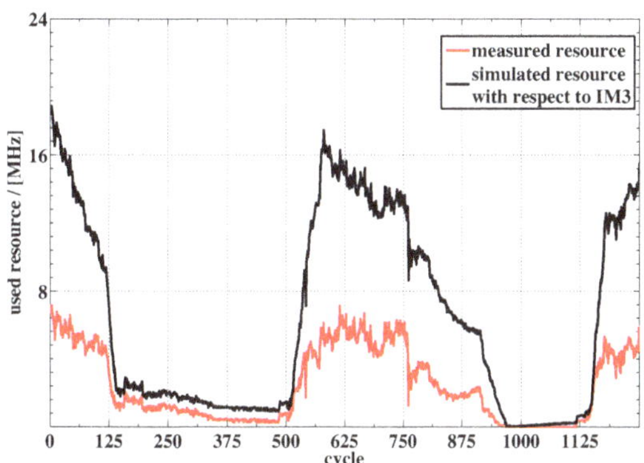

Fig. 5. Subjective spectral load with respect to IM3, "Musikmesse"

ability for PWMS.[2] The availability AV (see Fig. 4) is now compared with $P_{if}=-100$ dBm defined in section 2.1. This leads to

AV_{61-63}=11.88 MHz/24 MHz for the frequency range from f=790 MHz to f=814 MHz,

AV_{67-69}=18.06 MHz/24 MHz for the frequency range from f=838 MHz to f=862 MHz and

AV_{total} =50.78 MHz/72 MHz for the frequency range from f=790 MHz to f=862 MHz.

3.2.2 Spectral load with respect to in-band intermodulation

For the usage in different scenarios a PWMS receiver has to be designed with a tradeoff between frequency selectivity, to avoid in-band intermodulation, and frequency agility for the option to cover different operating frequencies. In Germany, a typical front end bandwidth is 24 MHz or 3 TV-channels. The spectral load of a mass event like the "Musikmesse" causes in-band intermodulation inside the receiver front end, resulting from the huge amount of PWMS in use. This leads to a subjective higher spectral load inside the PWMS receiver than the objective one presented in Sect. 3.2.1. The amount of the increase of spectral load due to in-band intermodulation cannot be generalized. Therefore a nonlinearity of a typical PWMS with an input intercept point IP_3=0 dBm was burdened with the measured spectral load at several measurement cycles in a system simulation, considering only intermodulation of the third order (IM3). The resulting spectral load was again compared with $P_{if}=-100$ dBm. The simulation results are displayed in Fig. 5. It is obvious, that the peak-periode of the spectral load with respect to IM3 at cycle 0 to 125, 500 to 750 and above 1125 exceeds twice the amount of the objective spectral load. In the cycles between 125 and 500, IM3 raises the spectral load insignificantly. The maximum load was detected for cycle 1 with a spectral load of 7.2 MHz excluding IM3 and 18 MHz including IM3.

[2]In the following this value will be used, although PWMS users and manufacturers suppose that a higher frequency availability might be necessary for a reliable PWMS link.

4 Conclusions

This paper shows that PWMS in daily use only requires a small part of the available spectrum and that not all PWMS in use can be made visible easily. In contrast the results based on the objective frequency availability as well as on the subjective spectral load including IM3 show that a generously determined headroom of the allocated frequency resource for PWMS is strongly required for mass events and the recording of unique multimedia content.

Acknowledgements. We thank K.-H. Schwaiger (IRT) for post-processing suggestions and M. Fiks (FHH) for the measurement support.

References

BNetzA: Allgemeinzuteilung von Frequenzen für drahtlose Mikrofone für professionelle Nutzung in den Frequenzbereichen 790–814 und 838–862 MHz, Bonn, Germany, 2005 (http://www.bundesnetzagentur.de/media/archive/4469.pdf).

CEPT: The Chester 1997 multilateral coordination agreement related to technical criteria, coordination principles and procedures for the introduction of Terrestrial Digial Video Broadcasting (DVB-T), Chester, England, 1997 (http://www.ero.dk/132D67A4-8815-48CB-B482-903844887DE3?frames=0).

ERC: Compatibility and sharing analysis between DVB-T and radio microphones in bands IV and V, Naples, Italy, 2000 (http://www.ero.dk/doc98/official/Word/REP088.DOC).

ETSI: Electromagnetic compatibility and Radio Spectrum Matters (ERM), Technical characteristics for Professional Wireless Microphone Systems (PWMS), System Reference Document, Sophia-Antipolis, France, 2006.

Konstantinos, M., Athanassios, A. V.,Petros, V. P., Elias, K. K., Nocolaos, R. E., and Constantinou, P.: A new methology for frequency coordination of wireless microphone systems over occupied TV-spectrum, IEEE Conference on wireless and mobile computing, networking and communication, Montreal, Canada, 2005.

Meinke H. and Gundlach, F. W.: Taschenbuch der Hochfrequenztechnik, Springer Verlag, Berlin, Germany, 1968.

WRC: World Radiocommunication Conference, Provisional Final Acts, International Communication Union, Geneva, Swiss, 2007.

Permissions

All chapters in this book were first published in ARS, by Copernicus Publications; hereby published with permission under the Creative Commons Attribution License or equivalent. Every chapter published in this book has been scrutinized by our experts. Their significance has been extensively debated. The topics covered herein carry significant findings which will fuel the growth of the discipline. They may even be implemented as practical applications or may be referred to as a beginning point for another development.

The contributors of this book come from diverse backgrounds, making this book a truly international effort. This book will bring forth new frontiers with its revolutionizing research information and detailed analysis of the nascent developments around the world.

We would like to thank all the contributing authors for lending their expertise to make the book truly unique. They have played a crucial role in the development of this book. Without their invaluable contributions this book wouldn't have been possible. They have made vital efforts to compile up to date information on the varied aspects of this subject to make this book a valuable addition to the collection of many professionals and students.

This book was conceptualized with the vision of imparting up-to-date information and advanced data in this field. To ensure the same, a matchless editorial board was set up. Every individual on the board went through rigorous rounds of assessment to prove their worth. After which they invested a large part of their time researching and compiling the most relevant data for our readers.

The editorial board has been involved in producing this book since its inception. They have spent rigorous hours researching and exploring the diverse topics which have resulted in the successful publishing of this book. They have passed on their knowledge of decades through this book. To expedite this challenging task, the publisher supported the team at every step. A small team of assistant editors was also appointed to further simplify the editing procedure and attain best results for the readers.

Apart from the editorial board, the designing team has also invested a significant amount of their time in understanding the subject and creating the most relevant covers. They scrutinized every image to scout for the most suitable representation of the subject and create an appropriate cover for the book.

The publishing team has been an ardent support to the editorial, designing and production team. Their endless efforts to recruit the best for this project, has resulted in the accomplishment of this book. They are a veteran in the field of academics and their pool of knowledge is as vast as their experience in printing. Their expertise and guidance has proved useful at every step. Their uncompromising quality standards have made this book an exceptional effort. Their encouragement from time to time has been an inspiration for everyone.

The publisher and the editorial board hope that this book will prove to be a valuable piece of knowledge for researchers, students, practitioners and scholars across the globe.

List of Contributors

D. Gerhardt
E-Plus Mobilfunk GmbH & Co. KG, E-Plus-Platz, D-40468 Düsseldorf, Germany

S. Korte
Institute of Electrical Engineering and Measurement Science, University of Hannover, Germany

H. Garbe
Institute of Electrical Engineering and Measurement Science, University of Hannover, Germany

J. Bredemeyer
FCS Flight Calibration Services GmbH, Braunschweig, Germany

T. Kleine-Ostmann
Physikalisch-Technische Bundesanstalt, Braunschweig, Germany

T. Schrader
Physikalisch-Technische Bundesanstalt, Braunschweig, Germany

K. Münter
Physikalisch-Technische Bundesanstalt, Braunschweig, Germany

J. Ritter
EADS Deutschland GmbH, Military Aircraft, Bremen, Germany

R. Krzikalla
Institute of Measurement Technology and Electromagnetic Compatibility, Hamburg, Germany

J. L. ter Haseborg
Institute of Measurement Technology and Electromagnetic Compatibility, Hamburg, Germany

H. Rabe
Leibniz Universität Hannover, Institut für Hochfrequenztechnik und Funksysteme, Appelstr. 9a, 30167 Hannover, Germany

E. Denicke
Leibniz Universität Hannover, Institut für Hochfrequenztechnik und Funksysteme, Appelstr. 9a, 30167 Hannover, Germany

G. Armbrech
Leibniz Universität Hannover, Institut für Hochfrequenztechnik und Funksysteme, Appelstr. 9a, 30167 Hannover, Germany

T. Musch
Ruhr-Universität Bochum, Lehrstuhl für Elektronische Mess- und Schaltungstechnik, Universitätsstr. 150, 44801 Bochum, Germany

I. Rolfes
Leibniz Universität Hannover, Institut für Hochfrequenztechnik und Funksysteme, Appelstr. 9a, 30167 Hannover, Germany

A. El Ouardi
Chair of Electromagnetic Theory, University of Wuppertal, 42097 Wuppertal, Germany

J. Streckert
Chair of Electromagnetic Theory, University of Wuppertal, 42097 Wuppertal, Germany

A. Lerchl
School of Engineering and Science, Jacobs University Bremen gGmbH, 28759 Bremen, Germany

K. Schwarzpaul
School of Engineering and Science, Jacobs University Bremen gGmbH, 28759 Bremen, Germany

V. Hansen
Chair of Electromagnetic Theory, University of Wuppertal, 42097 Wuppertal, Germany

K. Haake
Hamburg University of Technology, Institute for Metrology and EMC, Germany

J. L. ter Haseborg
Hamburg University of Technology, Institute for Metrology and EMC, Germany

B.Will
Ruhr-University Bochum, Institute of High Frequency Engineering, Universitätsstr. 150, 44801 Bochum, Germany

M. Gerding
Ruhr-University Bochum, Institute of High Frequency Engineering, Universitätsstr. 150, 44801 Bochum, Germany

S. Schultz
Ruhr-University Bochum, Institute of High Frequency Engineering, Universitätsstr. 150, 44801 Bochum, Germany

B. Schiek
Ruhr-University Bochum, Institute of High Frequency Engineering, Universitätsstr. 150, 44801 Bochum, Germany

C. Orlob
Leibniz Universität Hannover, Institut für Hochfrequenztechnik und Funksysteme, Appelstr. 9A, 30167 Hannover, Germany

D. Kornek
Leibniz Universität Hannover, Institut für Hochfrequenztechnik und Funksysteme, Appelstr. 9A, 30167 Hannover, Germany

S. Preihs
Leibniz Universität Hannover, Institut für Hochfrequenztechnik und Funksysteme, Appelstr. 9A, 30167 Hannover, Germany

I. Rolfes
Leibniz Universität Hannover, Institut für Hochfrequenztechnik und Funksysteme, Appelstr. 9A, 30167 Hannover, Germany

B.Will
Ruhr-Universität Bochum, Institute of High Frequency Engineering, Universitätsstr. 150, 44801 Bochum, Germany

M. Gerding
Ruhr-Universität Bochum, Institute of High Frequency Engineering, Universitätsstr. 150, 44801 Bochum, Germany

M. M. Leibfritz
Institute of Radio Frequency Technology (IHF), Universität Stuttgart, Germany

M. D. Blech
Institute of Radio Frequency Technology (IHF), Universität Stuttgart, Germany

F. M. Landstorfer
Institute of Radio Frequency Technology (IHF), Universität Stuttgart, Germany

T. F. Eibert
Institute of Radio Frequency Technology (IHF), Universität Stuttgart, Germany

B. Schetelig
Faculty of Electrical Engineering, Helmut-Schmidt-University/University of the Federal Armed Forces Hamburg, Germany

S. Parr
Faculty of Electrical Engineering, Helmut-Schmidt-University/University of the Federal Armed Forces Hamburg, Germany

S. Potthast
Bundeswehr Research Institute for Protective Technologies and NBC Protection (WIS) Munster, Germany

S. Dickmann
Faculty of Electrical Engineering, Helmut-Schmidt-University/University of the Federal Armed Forces Hamburg, Germany

M. Rohland
Physikalisch-Technische Bundesanstalt, 38116 Braunschweig, Germany
Institut für Mikrotechnik, Langer Kamp 8, 38106 Braunschweig, Germany

U. Arz
Physikalisch-Technische Bundesanstalt, 38116 Braunschweig, Germany

S. Büttgenbach
Institut für Mikrotechnik, Langer Kamp 8, 38106 Braunschweig, Germany

T. Schrader
Physikalisch-Technische Bundesanstalt, Braunschweig, Germany

K. Kuhlmann
Physikalisch-Technische Bundesanstalt, Braunschweig, Germany

R. Dickhoff
Physikalisch-Technische Bundesanstalt, Braunschweig, Germany

J. Dittmer
Physikalisch-Technische Bundesanstalt, Braunschweig, Germany

M. Hiebel
Rohde & Schwarz, Munich, Germany

C. Baer
Ruhr-Universität Bochum, Institute for Electronic Circuits, 44780 Bochum, Germany

T. Musch
Ruhr-Universität Bochum, Institute for Electronic Circuits, 44780 Bochum, Germany

M. Gerding
Ruhr-Universität Bochum, Institute for Electronic Circuits, 44780 Bochum, Germany

M. Vogt
Ruhr-Universität Bochum, High Frequency Engineering Research Group, 44780 Bochum, Germany

V. C. Motrescu
Institute of General Electrical Engineering, University of Rostock, Germany

U. van Rienen
Institute of General Electrical Engineering, University of Rostock, Germany

L. O. Fichte
Professur für Theoretische Elektrotechnik und Numerische Feldberechnung, Helmut-Schmidt-Universität Universität der Bundeswehr Hamburg, PO. Box 700 822, D-22 008 Hamburg, Germany

S. Lange
Professur für Theoretische Elektrotechnik und Numerische Feldberechnung, Helmut-Schmidt-Universität Universität der Bundeswehr Hamburg, PO. Box 700 822, D-22 008 Hamburg, Germany

M. Clemens
Professur für Theoretische Elektrotechnik und Numerische Feldberechnung, Helmut-Schmidt-Universität Universität der Bundeswehr Hamburg, PO. Box 700 822, D-22 008 Hamburg, Germany

T. Zelder
Institut für Hochfrequenztechnik und Funksysteme, Universität Hannover, Appelstraße 9A, 30167 Hannover, Germany

I. Rolfes
Institut für Hochfrequenztechnik und Funksysteme, Universität Hannover, Appelstraße 9A, 30167 Hannover, Germany

H. Eul
Institut für Hochfrequenztechnik und Funksysteme, Universität Hannover, Appelstraße 9A, 30167 Hannover, Germany

T. Zelder
Institut für Hochfrequenztechnik und Funksysteme, Leibniz Universität Hannover, Appelstr. 9A, 30167 Hannover, Germany

B. Geck
Institut für Hochfrequenztechnik und Funksysteme, Leibniz Universität Hannover, Appelstr. 9A, 30167 Hannover, Germany

I. Rolfes
Institut für Hochfrequenztechnik und Funksysteme, Leibniz Universität Hannover, Appelstr. 9A, 30167 Hannover, Germany

H. Eul
Institut für Hochfrequenztechnik und Funksysteme, Leibniz Universität Hannover, Appelstr. 9A, 30167 Hannover, Germany

B.Will
Ruhr-Universität Bochum, Institut für Hochfrequenztechnik, Universitätsstrasse 150, 44801 Bochum, Germany

I. Rolfes
Leibniz-Universität Hannover, Institut für Hochfrequenztechnik und Funksysteme, Appelstrasse 9A, 30167 Hannover, Germany

B. Schiek
Ruhr-Universität Bochum, Institut für Hochfrequenztechnik, Universitätsstrasse 150, 44801 Bochum, Germany

M. Vogt
Forschungsgruppe Hochfrequenztechnik, Ruhr-Universität Bochum, 44780 Bochum, Germany

M. Gerding
Lehrstuhl für Elektronische Schaltungstechnik, Ruhr-Universität Bochum, 44780 Bochum, Germany

T. Musch
Lehrstuhl für Elektronische Schaltungstechnik, Ruhr-Universität Bochum, 44780 Bochum, Germany

C. Jastrow
Physikalisch-Technische Bundesanstalt, Braunschweig and Berlin, Germany

T. Kleine-Ostmann
Physikalisch-Technische Bundesanstalt, Braunschweig and Berlin, Germany

T. Schrader
Physikalisch-Technische Bundesanstalt, Braunschweig and Berlin, Germany

J. Dittmer
Institute for Microtechnology, Technische Universität Braunschweig, Germany

R. Judaschke
Physikalisch-Technische Bundesanstalt, Braunschweig, Germany

S. Büttgenbach
Institute for Microtechnology, Technische Universität Braunschweig, Germany

I. Rolfes
Institut für Hochfrequenztechnik und Funksysteme, Leibniz Universität Hannover, Appelstraße 9A, 30167 Hannover, Germany

B.Will
Institut für Hochfrequenztechnik, Ruhr-Universität Bochum, Universitätsstraße 150, 44801 Bochum, Germany

B. Schiek
Institut für Hochfrequenztechnik, Ruhr-Universität Bochum, Universitätsstraße 150, 44801 Bochum, Germany

D. Piester
Physikalisch-Technische Bundesanstalt (PTB), Bundesallee 100, 38116 Braunschweig, Germany

M. Rost
Physikalisch-Technische Bundesanstalt (PTB), Bundesallee 100, 38116 Braunschweig, Germany

M. Fujieda
National Institute of Information and Communications Technology (NICT), Tokyo, Japan

T. Feldmann
Physikalisch-Technische Bundesanstalt (PTB), Bundesallee 100, 38116 Braunschweig, Germany

A. Bauch
Physikalisch-Technische Bundesanstalt (PTB), Bundesallee 100, 38116 Braunschweig, Germany

M. Vuchkovikj
Graduate School of Computational Engineering, TU Darmstadt, Darmstadt, Germany
Computational Electromagnetics Laboratory, TU Darmstadt, Darmstadt, Germany

I. Munteanu
Graduate School of Computational Engineering, TU Darmstadt, Darmstadt, Germany
Computational Electromagnetics Laboratory, TU Darmstadt, Darmstadt, Germany

T. Weiland
Graduate School of Computational Engineering, TU Darmstadt, Darmstadt, Germany
Computational Electromagnetics Laboratory, TU Darmstadt, Darmstadt, Germany

G. Armbrecht
Inst. für Hochfrequenztechnik und Funksysteme, Leibniz Universität Hannover, Appelstraße 9A, 30167 Hannover, Germany

E. Denicke
Inst. für Hochfrequenztechnik und Funksysteme, Leibniz Universität Hannover, Appelstraße 9A, 30167 Hannover, Germany

I. Rolfes
Inst. für Hochfrequenztechnik und Funksysteme, Leibniz Universität Hannover, Appelstraße 9A, 30167 Hannover, Germany

N. Pohl
Lehrstuhl für Integrierte Systeme, Ruhr-Universität Bochum, Universitätsstraße 150, 44801 Bochum, Germany

T. Musch
Institut für Hochfrequenztechnik, Ruhr-Universität Bochum, Universitätsstraße 150, 44801 Bochum, Germany

B. Schiek
Institut für Hochfrequenztechnik, Ruhr-Universität Bochum, Universitätsstraße 150, 44801 Bochum, Germany

I. Gaspard
Hochschule Darmstadt, FB Elektrotechnik und Informationstechnik, Darmstadt, Germany

T. Schrader
Physikalisch-Technische Bundesanstalt (PTB), Bundesallee 100, 38116 Braunschweig, Germany

T. Kleine-Ostmann
Physikalisch-Technische Bundesanstalt (PTB), Bundesallee 100, 38116 Braunschweig, Germany

J. Bredemeyer
Flight Calibration Services FCS GmbH, Hermann-Blenk-Straße 32 A, 38108 Braunschweig, Germany

T. Jaeschke
Ruhr-Universität Bochum, Institute of Integrated Systems, 44780 Bochum, Germany

C. Bredendiek
Ruhr-Universität Bochum, Institute of Integrated Systems, 44780 Bochum, Germany

M. Vogt
Ruhr-Universität Bochum, High Frequency Engineering Research Group, 44780 Bochum, Germany

N. Pohl
Ruhr-Universität Bochum, Institute of Integrated Systems, 44780 Bochum, Germany

H. Ahrens
Institut für Elektrotechnik und Informationstechnik, Christian-Albrechts-Universität zu Kiel, Germany

F. Argin
Institut für Elektrotechnik und Informationstechnik, Christian-Albrechts-Universität zu Kiel, Germany

L. Klinkenbusch
Institut für Elektrotechnik und Informationstechnik, Christian-Albrechts-Universität zu Kiel, Germany

U. Stumper
Physikalisch-Technische Bundesanstalt, Bundesallee 100, 38116 Braunschweig, Germany

I. Rolfes
Institut für Hochfrequenztechnik und Funksysteme, Leibniz Universität Hannover, Appelstraße 9A, 30167 Hannover, Germany

T. Kleine-Ostmann
Physikalisch-Technische Bundesanstalt, Braunschweig, Germany

K. Münter
Physikalisch-Technische Bundesanstalt, Braunschweig, Germany

T. Schrader
Physikalisch-Technische Bundesanstalt, Braunschweig, Germany

S. Dortmund
Institute for Radiofrequency and Microwave Engineering, Leibniz Universität Hannover, Germany

M. Fehr
Sennheiser Electronic GmbH, Wedemark, Germany

I. Rolfes
Institute for Radiofrequency and Microwave Engineering, Leibniz Universität Hannover, Germany

www.ingramcontent.com/pod-product-compliance
Lightning Source LLC
Chambersburg PA
CBHW080651200326
41458CB00013B/4815
* 9 7 8 1 6 8 2 8 5 0 6 8 8 *